Communications and Control Engineering

Series Editors
E.D. Sontag • M. Thoma • A. Isidori • J.H. van Schuppen

Published titles include:

Stability and Stabilization of Infinite Dimensional Systems with Applications
Zheng-Hua Luo, Bao-Zhu Guo and Omer Morgul

Nonsmooth Mechanics (Second edition)
Bernard Brogliato

Nonlinear Control Systems II
Alberto Isidori

L_2-Gain and Passivity Techniques in Nonlinear Control
Arjan van der Schaft

Control of Linear Systems with Regulation and Input Constraints
Ali Saberi, Anton A. Stoorvogel and Peddapullaiah Sannuti

Robust and H_∞ Control
Ben M. Chen

Computer Controlled Systems
Efim N. Rosenwasser and Bernhard P. Lampe

Control of Complex and Uncertain Systems
Stanislav V. Emelyanov and Sergey K. Korovin

Robust Control Design Using H_∞ Methods
Ian R. Petersen, Valery A. Ugrinovski and Andrey V. Savkin

Model Reduction for Control System Design
Goro Obinata and Brian D.O. Anderson

Control Theory for Linear Systems
Harry L. Trentelman, Anton Stoorvogel and Malo Hautus

Functional Adaptive Control
Simon G. Fabri and Visakan Kadirkamanathan

Positive 1D and 2D Systems
Tadeusz Kaczorek

Identification and Control Using Volterra Models
Francis J. Doyle III, Ronald K. Pearson and Babatunde A. Ogunnaike

Non-linear Control for Underactuated Mechanical Systems
Isabelle Fantoni and Rogelio Lozano

Robust Control (Second edition)
Jürgen Ackermann

Flow Control by Feedback
Ole Morten Aamo and Miroslav Krstić

Learning and Generalization (Second edition)
Mathukumalli Vidyasagar

Constrained Control and Estimation
Graham C. Goodwin, María M. Seron and José A. De Doná

Randomized Algorithms for Analysis and Control of Uncertain Systems
Roberto Tempo, Giuseppe Calafiore and Fabrizio Dabbene

Switched Linear Systems
Zhendong Sun and Shuzhi S. Ge

Subspace Methods for System Identification
Tohru Katayama

Digital Control Systems
Ioan D. Landau and Gianluca Zito

Multivariable Computer-controlled Systems
Efim N. Rosenwasser and Bernhard P. Lampe

Dissipative Systems Analysis and Control (2nd Edition)
Bernard Brogliato, Rogelio Lozano, Bernhard Maschke and Olav Egeland

Algebraic Methods for Nonlinear Control Systems
Giuseppe Conte, Claude H. Moog and Anna M. Perdon

Polynomial and Rational Matrices
Tadeusz Kaczorek

Simulation-based Algorithms for Markov Decision Processes
Hyeong Soo Chang, Michael C. Fu, Jiaqiao Hu and Steven I. Marcus

Iterative Learning Control
Hyo-Sung Ahn, Kevin L. Moore and YangQuan Chen

Distributed Consensus in Multi-vehicle Cooperative Control
Wei Ren and Randal W. Beard

Control of Singular Systems with Random Abrupt Changes
El-Kébir Boukas

Alessandro Astolfi · Dimitrios Karagiannis
Romeo Ortega

Nonlinear and Adaptive Control with Applications

 Springer

Alessandro Astolfi, PhD
Department of Electrical and Electronic
  Engineering
Imperial College London
London SW7 2AZ
UK

and

Dipartimento di Informatica, Sistemi
  e Produzione
Università di Roma "Tor Vergata"
00133 Roma
Italy

Dimitrios Karagiannis, PhD
Department of Electrical and Electronic
  Engineering
Imperial College London
London SW7 2AZ
UK

Romeo Ortega, PhD
Centre National de la Recherche Scientifique
Laboratoire des Signaux et Systèmes
Supélec
91192 Gif-sur-Yvette
France

ISBN 978-1-84800-065-0

e-ISBN 978-1-84800-066-7

DOI 10.1007/978-1-84800-066-7

Communications and Control Engineering Series ISSN 0178-5354

British Library Cataloguing in Publication Data
A catalogue record for this book is available from the British Library

Library of Congress Control Number: 2007941070

© 2008 Springer-Verlag London Limited

MATLAB® and Simulink® are registered trademarks of The MathWorks, Inc., 3 Apple Hill Drive, Natick, MA 01760-2098, USA. http://www.mathworks.com

Apart from any fair dealing for the purposes of research or private study, or criticism or review, as permitted under the Copyright, Designs and Patents Act 1988, this publication may only be reproduced, stored or transmitted, in any form or by any means, with the prior permission in writing of the publishers, or in the case of reprographic reproduction in accordance with the terms of licences issued by the Copyright Licensing Agency. Enquiries concerning reproduction outside those terms should be sent to the publishers.

The use of registered names, trademarks, etc. in this publication does not imply, even in the absence of a specific statement, that such names are exempt from the relevant laws and regulations and therefore free for general use.

The publisher makes no representation, express or implied, with regard to the accuracy of the information contained in this book and cannot accept any legal responsibility or liability for any errors or omissions that may be made.

Cover design: LE-T_EX Jelonek, Schmidt & Vöckler GbR, Leipzig, Germany

Printed on acid-free paper

9 8 7 6 5 4 3 2 1

springer.com

To Elisabetta and the kids (A.A.)

To Lara (D.K.)

To the memory of my father (R.O.)

Preface

In the last few years we have witnessed the appearance of a series of challenging control engineering problems. Two common features of these new control problems are that the interesting range of operation of the system is not necessarily close to an equilibrium, hence *nonlinear* effects have to be explicitly taken into account for a successful controller design, and that, even though physical modelling allows to accurately identify certain well-defined nonlinear effects, the controller has to cope with a high level of *uncertainty*, mainly due to lack of knowledge on the system parameters and the inability to measure the whole system state.

This situation justifies the need for the development of tools for controller design for uncertain nonlinear systems, which is the main topic of this book.

Numerous theoretical control design methodologies for nonlinear systems have emerged over the last two decades. When viewed from a conceptual standpoint, they can be broadly classified into *analytically-oriented* and *computationally-oriented*. (The qualifiers analytical and computational are used to distinguish between symbolical analysis and numerical computations.)

The former approach, which is the one adopted in this book, proceeds from an analytical model of the system, and the controller design is the outcome of a systematic process that guarantees some specific behaviour. Since stability is a *sine qua non* condition, research following this approach usually runs under the heading *robust stabilisation*, and it includes Lyapunov-based methods, gain-assignment methods, and classical robust and adaptive tools.

Computationally-oriented techniques, on the other hand, do not necessarily require an analytical model, and they may be developed on the basis of a numerical model of the system to be controlled—obtained, for instance, by collecting large amounts of data to approximate its behaviour. Neural networks based control, fuzzy control and intelligent control are the more conspicuous representatives of this school. Recently, a second class of computationally-oriented techniques, that relies on analytical models of the system, has gained some popularity. In an attempt to mimic the developments of linear systems theory, piecewise linear (or linear parameter-varying) models are proposed to capture nonlinear effects. Typically some optimal control objective is formulated and the task of the controller design is to prove that, for the given numerical values of the system model, the optimisation is feasible, *e.g.*, it can be translated into linear matrix inequalities, and a control signal can be *numerically* computed. The optimal control approach suffers from two drawbacks. First, the solutions are fragile with respect to plant uncertainty, *e.g.*, lack of full state measurement and parametric uncertainty, which is the prevailing concern in many, if not all, practical applications. Second, computation of the optimal control law is feasible only for low-dimensional systems, which puts a serious question mark on the applicability of the method for nonlinear systems. In addition there is not always a clear reason, besides mathematical convenience, to express the desired behaviour of a dynamical system in terms of a scalar criterion to be optimised.

In summary, computationally-oriented approaches, while leveraging off a swiftly growing computer technology, provide solutions to some specific problems, but do not aim at explaining *why, how and when* these solutions indeed work. At a more philosophical level, it is the authors' opinion that controller design should not be reduced to the generation of numerical code that implements a control law, without any attempt to try to understand the underlying mechanism that makes it work. This information is encoded in the nonlinear system dynamics and revealed through a full-fledged nonlinear analysis.

We consider nonlinear control systems subject to various types of uncertainty, including lack of knowledge on the parameters, partial measurement of the system states and uncertainty on the system order and structure. To deal with all these situations we follow a common thread encrypted in the words *immersion* and *invariance* (I&I).

In the I&I approach we propose to capture the desired behaviour of the system to be controlled introducing a target dynamical system. The control problem is then reduced to the design of a control law which guarantees that the controlled system asymptotically behaves like the target system. More precisely, the I&I methodology relies on finding a manifold in state-space that can be rendered invariant and attractive, with internal dynamics a copy of the desired closed-loop dynamics, and on designing a control law that robustly *steers* the state of the system *sufficiently close* to this manifold.

I&I should be contrasted with the optimal control approach where the objective is captured by a scalar performance index to optimise. In addition, because of its two-step approach, it is conceptually different from existing (robust) stabilisation methodologies that rely on the use of control Lyapunov functions. However, it resembles the procedure used in sliding-mode control, where a given manifold—the sliding surface—is rendered attractive by a discontinuous control law. The key difference is that, while in sliding-mode control the manifold must be reached by the trajectories, in the proposed approach the manifold need not be reached. (This feature is essential in adaptive control and in output feedback design.)

The book is organised as follows. After a brief introduction where the main ideas of I&I are illustrated by means of examples (**Chapter 1**), in **Chapters 2 and 3** we introduce the I&I framework and show how it can be used to solve stabilisation and adaptive control problems.

In **Chapter 4** the method is applied to nonlinear systems with parametric uncertainties, where it is assumed that the full state is available for feedback.

In **Chapter 5** we show that I&I provides a natural framework for observer design for general nonlinear systems.

In **Chapter 6** the problem of output feedback stabilisation is solved for classes of nonlinear systems, which include systems with unstable zero dynamics and the well-known output feedback form. Furthermore, the method is extended to allow unstructured uncertainties to enter the system equations.

In **Chapter 7** the I&I approach is used to design a class of nonlinear *proportional-integral* controllers, where the gains of the controller are nonlinear functions that are chosen to guarantee stability for systems with unknown parameters and uncertain disturbances.

Chapters 8, **9** and **10** are devoted to applications from electrical, mechanical and electromechanical systems, including power converters, power systems, electric machines and autonomous aircraft.

Appendix A provides the basic definitions and recalls briefly results used throughout the book. In particular, characterisations of Lyapunov stability, input-to-state stability and a nonlinear version of the small-gain theorem are given along with some useful lemmas.

Acknowledgements

This book is the result of extensive research collaborations during the last five years. Some of the results of these collaborations have been reported in the papers [20, 19, 14, 15, 157, 162, 21, 99, 105, 16, 92, 186, 95, 106, 100, 103, 101, 93, 94, 102, 96, 39, 17, 97, 40, 104, 98, 18]. We are grateful to our co-authors, Nikita Barabanov, Daniele Carnevale, Gerardo Escobar, Mickaël Hilairet, Liu Hsu, Zhong-Ping Jiang, Eduardo Mendes, Mariana Netto, Hugo Rodríguez, and Aleksandar Stanković, for several stimulating discussions and for their hospitality while visiting their institutions. We also thank the research staff of the Laboratoire de Génie Électrique de Paris, for rendering available their experimental facilities.

Some of the topics of this book have been taught by the authors in a series of one-week graduate control courses offered in Paris for the last four years. These have been organised by the European Commission's Marie Curie Control Training Site (CTS) and by the European Embedded Control Institute (EECI) in the framework of the European Network of Excellence HYCON. We would like to thank Françoise Lamnabhi-Lagarrigue for giving us the oppurtunity to teach during these schools.

Workshops on the topics presented in this book were organised at the IEEE Conference on Decision and Control, Las Vegas, USA, 2002, and at the

XII Latin-American Congress on Automatic Control, Bahia, Brazil, 2006. A mini-tutorial was given at the European Control Conference, Kos, Greece, 2007. We have delivered lectures on selected topics of the book in the DISC Summer School, Eindhoven, The Netherlands, 2007, and in numerous research seminars.

Finally, a large part of this work would not have been possible without the financial support of several institutions. The first author would like to thank the Engineering and Physical Sciences Research Council (EPSRC) and the Leverhulme Trust. The second author's work was supported first by the European Commission's Training and Mobility of Researchers (TMR) programme through the Nonlinear and Adaptive Control (NACO2) network, then by BAE Systems and the EPSRC via the FLAVIIR Integrated Programme in Aeronautical Engineering, and finally by EPSRC via the Control and Power Portfolio Partnership. The third author would like to thank the European Network of Excellence HYCON for supporting part of his work.

Rome, London, Paris	*Alessandro Astolfi*
April 2007	*Dimitrios Karagiannis*
	Romeo Ortega

Contents

Notation xv

1 Introduction 1
   1.1 An I&I Perspective of Stabilisation 2
   1.2 Discussion and Literature Review 4
   1.3 Applications of I&I 6
      1.3.1 Robustification of Control Laws 6
      1.3.2 Underactuated Mechanical Systems 9
      1.3.3 Systems in Special Forms 9
      1.3.4 Adaptive Control 11
      1.3.5 Observer Design 12
      1.3.6 Nonlinear PI Control 13

2 I&I Stabilisation 15
   2.1 Main Stabilisation Result 15
   2.2 Systems with Special Structures 19
   2.3 Physical Systems 24

3 I&I Adaptive Control: Tools and Examples 33
   3.1 Introduction 33
   3.2 An Introductory Example 35
   3.3 Revisiting Classical Adaptive Control 38
      3.3.1 Direct Cancellation with Matching 38
      3.3.2 Direct Cancellation with Extended Matching 39
      3.3.3 Direct Domination 40
      3.3.4 Indirect Approach 40
      3.3.5 Discussion 40
   3.4 Linearly Parameterised Plant 42
   3.5 Linearly Parameterised Control 47
   3.6 Example: Visual Servoing 49

4 I&I Adaptive Control: Systems in Special Forms 55
- 4.1 Introduction 55
- 4.2 Systems in Feedback Form 57
  - 4.2.1 Adaptive Control Design 58
  - 4.2.2 Asymptotic Properties of Adaptive Controllers 59
  - 4.2.3 Unknown Control Gain 61
  - 4.2.4 Unmatched Uncertainties 63
- 4.3 Lower Triangular Systems 67
  - 4.3.1 Estimator Design 68
  - 4.3.2 Controller Design 70
  - 4.3.3 Estimator Design using Dynamic Scaling 74
- 4.4 Linear Systems 78
  - 4.4.1 Linear SISO Systems 79
  - 4.4.2 Linear Multivariable Systems 81

5 Nonlinear Observer Design 91
- 5.1 Introduction 92
- 5.2 Reduced-order Observers 93
- 5.3 Systems with Monotonic Nonlinearities Appearing in the Output Equation 100
- 5.4 Mechanical Systems with Two Degrees of Freedom 107
  - 5.4.1 Model Description 107
  - 5.4.2 Observer Design 108
  - 5.4.3 Non-diagonal Inertia Matrix with Bounded Entries 110
  - 5.4.4 Diagonal Inertia Matrix 111
- 5.5 Example: Ball and Beam 112

6 Robust Stabilisation via Output Feedback 115
- 6.1 Introduction 116
- 6.2 Linearly Parameterised Systems 117
- 6.3 Control Design Using a Separation Principle 121
- 6.4 Systems with Monotonic Nonlinearities Appearing in the Output Equation 124
- 6.5 Systems in Output Feedback Form 129
- 6.6 Robust Output Feedback Stabilisation 133
  - 6.6.1 Robust Observer Design 135
  - 6.6.2 Stabilisation via a Small-gain Condition 136
  - 6.6.3 Systems Without Zero Dynamics 140
  - 6.6.4 Systems with ISS Zero Dynamics 141
  - 6.6.5 Unperturbed Systems 141
  - 6.6.6 Linear Nonminimum-phase Systems 142
- 6.7 Example: Translational Oscillator/Rotational Actuator 145

Contents xiii

7 Nonlinear PI Control of Uncertain Systems 151
  7.1 Introduction 151
  7.2 Control Design 154
      7.2.1 Bounded Uncertainty 154
      7.2.2 Comparison with Adaptive Control 157
      7.2.3 Extensions 157
      7.2.4 Unbounded Uncertainty 159
  7.3 Unknown Control Direction 161
      7.3.1 State Feedback 161
      7.3.2 Observer-based Design 164
      7.3.3 Robustness 165
  7.4 Example: Visual Servoing 167

8 Electrical Systems 173
  8.1 Power Flow Control Using TCSC 173
      8.1.1 Problem Formulation 174
      8.1.2 Modified Model of the TCSC 175
      8.1.3 Controller Design 176
      8.1.4 Simulation Results 178
  8.2 Partial State Feedback Control of the DC–DC Ćuk Converter . 179
      8.2.1 Problem Formulation 181
      8.2.2 Full-information Controller 182
      8.2.3 I&I Adaptive Observer 184
      8.2.4 Partial State Feedback Controller 186
      8.2.5 Simulation Results 188
  8.3 Output Feedback Control of the DC–DC Boost Converter 189
      8.3.1 Problem Formulation 192
      8.3.2 Full-information Controller 193
      8.3.3 Output Feedback Controller 194
      8.3.4 Experimental Results 197
      8.3.5 A Remark on Robustness 198
  8.4 Adaptive Control of the Power Factor Precompensator 200
      8.4.1 Problem Formulation 201
      8.4.2 Full-information Controllers 203
      8.4.3 I&I Adaptation 205
      8.4.4 Experimental Results 207

9 Mechanical Systems 211
  9.1 Control of Flexible Joints Robots 211
      9.1.1 Control Design 213
      9.1.2 A 2-DOF Flexible Joints Robot 216
      9.1.3 A 3-DOF Flexible Joints Robot 217
  9.2 Position-feedback Control of a Two-link Manipulator 218
      9.2.1 Observer Design 219
      9.2.2 State Feedback Controller 220

		9.2.3 Output Feedback Design 222

        9.2.3 Output Feedback Design 222
        9.2.4 Simulation Results 223
  9.3 Adaptive Attitude Control of a Rigid Body 226
        9.3.1 Model Description 226
        9.3.2 Controller Design 227
        9.3.3 Estimator Design 228
        9.3.4 Simulation Results 229
  9.4 Trajectory Tracking for Autonomous Aerial Vehicles 229
        9.4.1 Model Description 230
        9.4.2 Controller Design 231
        9.4.3 Airspeed Regulation 233
        9.4.4 Simulation Results 234

10 Electromechanical Systems 237
  10.1 Observer Design for Single-machine Infinite-bus Systems...... 237
        10.1.1 Model Description 238
        10.1.2 Controller Design 239
        10.1.3 Observer Design 239
        10.1.4 Simulation Results 241
  10.2 Adaptive Control of Current-fed Induction Motors 243
        10.2.1 Problem Formulation 244
        10.2.2 Estimator Design 246
        10.2.3 Rotor Flux Estimation 247
        10.2.4 Controller Design 248
        10.2.5 Simulation Results 249
  10.3 Speed/Flux Tracking for Voltage-fed Induction Motors 252
        10.3.1 Problem Formulation 252
        10.3.2 Nonlinear Control Design 254
        10.3.3 Adaptive Control Design 258
        10.3.4 Simulations and Experimental Results 262

A Background Material 269
  A.1 Lyapunov Stability and Convergence 269
  A.2 Input-to-state Stability 271
  A.3 Invariant Manifolds and System Immersion 273

References 275

Index 289

Notation

We consider dynamical systems described by differential equations of the form[1]
$$\dot{x} = f(x, u, t),$$
where $x \in \mathbb{R}^n$ is the system state, $u \in \mathbb{R}^m$ is the input signal, t denotes time (in seconds), and the overdot "˙" denotes differentiation with respect to time. The solutions (or *trajectories*) of the above equation are denoted by $x(t)$, with $t \in [t_0, T)$ and $t_0 < T \leq \infty$, and $x(t_0)$ are the initial conditions. For simplicity we often assume $t_0 = 0$.

The transpose of a matrix A is denoted by $A^\top$ and I is the identity matrix; $e_i \in \mathbb{R}^n$ denotes a vector whose ith element is 1 and all other elements are zero. A diagonal $n \times n$ matrix is also written as $\mathrm{diag}(a_1, \ldots, a_n)$, where a_i are the diagonal elements. A column vector consisting of subvectors $x_1, \ldots, x_n$ is also written as $\mathrm{col}(x_1, \ldots, x_n)$.

The absolute value of a scalar y is denoted by $|y|$. The p-norm of a vector $x = [x_1, \ldots, x_n]^\top \in \mathbb{R}^n$, for $1 \leq p < \infty$, is defined as
$$|x|_p \triangleq \left(|x_1|^p + \cdots + |x_n|^p\right)^{1/p}.$$
Note that if $n = 1$ then $|x|_p = |x|$, for all $1 \leq p < \infty$. In this book we mainly use the 2-norm (or Euclidean norm) in which case the subscript p is often dropped, i.e., $|x| \triangleq |x|_2$. The induced p-norm of a matrix A, for $1 \leq p < \infty$, is defined as
$$|A|_p \triangleq \sup_{x \neq 0} \frac{|Ax|_p}{|x|_p},$$
where sup denotes the supremum, i.e., the least upper bound. For $p = 2$ the above definition yields $|A| \triangleq |A|_2 = \left(\lambda_{\max}(A^\top A)\right)^{1/2}$, where $\lambda_{\max}(A^\top A)$ is the maximum eigenvalue of $A^\top A$.

---

[1]This class obviously includes time-invariant systems and systems that are affine in the control.

The $\mathcal{L}_p$-norm of a (vector) signal $x(t)$, defined for all $t \geq 0$, for $1 \leq p < \infty$, is defined as

$$\|x\|_p \triangleq \left(\int_0^\infty (|x(t)|_p)^p \mathrm{d}t \right)^{1/p}.$$

$\mathcal{L}_p$ denotes the space of signals $x : \mathbb{R}_{\geq 0} \to \mathbb{R}^n$ such that $\|x\|_p$ exists and is finite; $\mathcal{L}_\infty$ denotes the space of signals $x : \mathbb{R}_{\geq 0} \to \mathbb{R}^n$ that are bounded, i.e., $\sup_{t \geq 0} |x(t)|$ exists and is finite. (Note that $\mathbb{R}_{\geq 0}$ denotes the set of nonnegative real numbers.)

We denote by $\mathcal{C}^n$ the class of functions or mappings that are n times differentiable, i.e., their partial derivatives up to order n exist and are continuous. Throughout the book we also use the term *smooth* to indicate functions or mappings that are $\mathcal{C}^k$, for some large k such that all required derivatives are well-defined.

For any $\mathcal{C}^1$ function $V : \mathbb{R}^n \to \mathbb{R}$ and any vector field $f : \mathbb{R}^n \times \mathbb{R}^m \times \mathbb{R} \to \mathbb{R}^n$, $L_f V(x)$ denotes the *Lie derivative* of V along f at x, i.e.,

$$L_f V(x) \triangleq \frac{\partial V}{\partial x} f(x, u, t).$$

A list of acronyms used in the text is given in the table below.

BIBS	Bounded-input bounded-state
I&I	Immersion and invariance
IFOC	Indirect field-oriented controller
IOS	Input-to-output stable
ISS	Input-to-state stable
LTI	Linear time-invariant
MIMO	Multi-input multi-output
MRAC	Model-reference adaptive control
PDE	Partial differential equation
PFP	Power factor precompensator
PI	Proportional-integral
PWM	Pulse-width modulation
RMS	Root mean square
SISO	Single-input single-output
SMIB	Single-machine infinite-bus
TCSC	Thyristor-controlled series capacitor
TORA	Translational oscillator with rotational actuator

1
Introduction

We present a new method for designing nonlinear and adaptive controllers for (uncertain) nonlinear systems. The method relies upon the notions of system immersion and manifold invariance, which are classical tools from nonlinear regulator theory and geometric nonlinear control, but are used in the present work from a new perspective. We call the new methodology *immersion and invariance* (I&I). The basic idea of the I&I approach is to achieve the control objective by *immersing*[1] the plant dynamics into a (possibly lower-order) target system that captures the desired behaviour.

The main features of I&I that will be illustrated throughout the book are as follows.

- It reduces the controller design problem to other subproblems which, in some instances, might be easier to solve.
- It differs from most of the existing controller design methodologies because it does not require, in principle, the knowledge of a (control) Lyapunov function.
- It is well-suited to situations where a stabilising controller for a nominal reduced-order model is known, and we would like to robustify it with respect to higher-order dynamics. This is achieved by designing a control law that asymptotically immerses the *full* system dynamics into the reduced-order ones.
- In adaptive control problems the method yields stabilising schemes that counter the effect of the uncertain parameters adopting a robustness perspective. This is in contrast with some of the existing adaptive designs that (relying on certain matching conditions) treat these terms as disturbances to be rejected. The I&I method does not invoke certainty equivalence, nor requires a linear parameterisation. Furthermore, it provides a procedure to add cross terms between the parameter estimates and the plant states in the Lyapunov function.

---
[1] A formal definition of immersion is given in Appendix A (see Definition A.11).

- It provides a natural framework for the formulation and solution of observer design and robust output feedback stabilisation problems—with state observation and parameter estimation treated in a unified setting.
- It allows to formulate and solve the problem of designing nonlinear proportional-integral controllers for a class of coarsely defined, low-order systems.

In this chapter we informally present the underlying concepts of I&I, briefly review how immersion and invariance are used in other control problems, comment on the relations between I&I and other controller design approaches—in particular, Lyapunov-based designs—and give some simple motivating examples.

1.1 An I&I Perspective of Stabilisation

We illustrate the I&I approach with the basic problem of stabilisation, by state feedback, of an equilibrium point of a nonlinear system. The state feedback stabilisation problem is chosen for ease of presentation but, as will become clear, the approach is applicable to a broad class of control problems, including tracking, parameter and state estimation, and stabilisation by output feedback.

Consider the system
$$\dot{x} = f(x) + g(x)u, \tag{1.1}$$
with $x \in \mathbb{R}^n$ and $u \in \mathbb{R}^m$, and the problem of finding, whenever possible, a state feedback control law $u = v(x)$ such that the closed-loop system has a (globally) asymptotically stable equilibrium at the origin. This problem can be addressed by considering the problem of finding a target dynamical system
$$\dot{\xi} = \alpha(\xi), \tag{1.2}$$
with $\xi \in \mathbb{R}^p$ and $p < n$, which has a (globally) asymptotically stable equilibrium at the origin, a smooth mapping $x = \pi(\xi)$, and a control law $v(x)$ such that
$$\pi(\xi(0)) = x(0), \tag{1.3}$$
$$\pi(0) = 0, \tag{1.4}$$
and
$$f(\pi(\xi)) + g(\pi(\xi))v(\pi(\xi)) = \frac{\partial \pi}{\partial \xi}\alpha(\xi). \tag{1.5}$$

If the above conditions hold, then any trajectory $x(t)$ of the closed-loop system
$$\dot{x} = f(x) + g(x)v(x) \tag{1.6}$$
is the image through the mapping $\pi(\cdot)$ of a trajectory $\xi(t)$ of the target system (1.2), as illustrated in Figure 1.1. From (1.4) and the fact that the zero

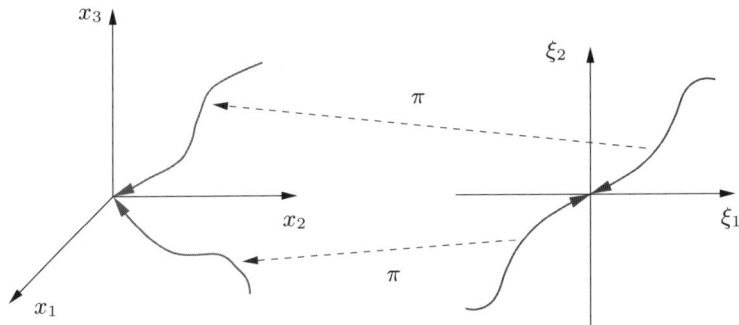

Fig. 1.1. Graphical illustration of the mapping between the trajectories of the system to be controlled and the target system for $p = 2$ and $n = 3$.

equilibrium of the target system is asymptotically stable, this implies that $x(t)$ converges to the origin. Thus the stabilisation problem for the zero equilibrium of the system (1.1) can be recast as the problem of solving the partial differential equation (1.5) with the boundary conditions (1.3) and (1.4).

A geometric interpretation of (1.3)–(1.5) is the following. Consider the closed-loop system (1.6) and a manifold in the n-dimensional state-space, defined by

$$\mathcal{M} = \{x \in \mathbb{R}^n \,|\, x = \pi(\xi),\ \xi \in \mathbb{R}^p\},$$

and such that (1.3) and (1.4) hold. From (1.5), the manifold $\mathcal{M}$ is invariant[2] with internal dynamics (1.2), hence all trajectories $x(t)$ that start on the manifold remain there and asymptotically converge to the point $\pi(0)$, which is the origin, by (1.4). Moreover, condition (1.3) guarantees that the initial state of (1.6) lies on $\mathcal{M}$.

The above formulation is impractical for two reasons. First, from (1.3) and (1.5), the mapping $\pi(\cdot)$ and the control $v(\cdot)$ depend, in general, on the initial conditions. Second, it may be impossible to find, for any $x(0) \in \mathbb{R}^n$, a mapping $\pi(\xi)$ such that (1.3), (1.4) and (1.5) hold.

These obstacles can be removed by determining a solution of (1.4) and (1.5) only and by modifying the control law $u = v(x)$ so that, for all initial conditions, the trajectories of the system (1.6) remain bounded and *asymptotically converge* to the manifold $x = \pi(\xi)$, i.e., $\mathcal{M}$ is rendered attractive. The attractivity of the manifold $\mathcal{M}$ can be expressed in terms of the distance

$$|\zeta| = \text{dist}(x, \mathcal{M}),$$

which should be driven to zero. Notice that the distance $|\zeta|$ is not uniquely defined. This provides an additional degree of freedom in the control design.

---

[2] A manifold $\mathcal{M} \subset \mathbb{R}^n$ is invariant if $x(0) \in \mathcal{M} \Rightarrow x(t) \in \mathcal{M}$, for all $t \geq 0$, see Definition A.10.

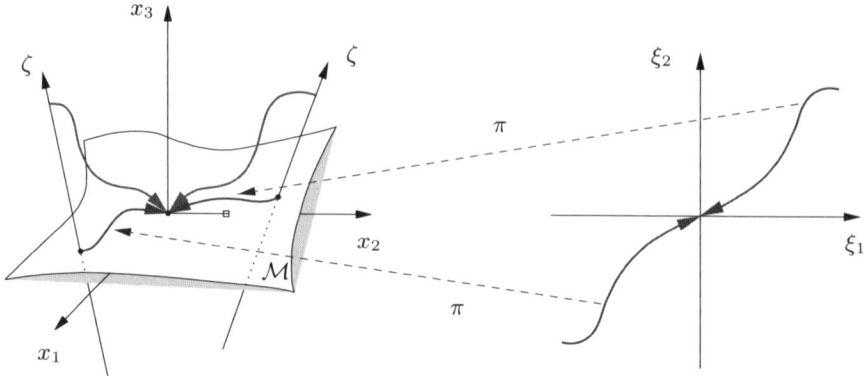

Fig. 1.2. Graphical illustration of the immersion and invariance approach.

A graphical illustration of the I&I approach for $p = 2$ and $n = 3$ is given in Figure 1.2. Observe that $\pi(\cdot)$ maps a trajectory on the ξ-space to a trajectory on the x-space, which is restricted to the manifold $\mathcal{M}$ containing the origin. Moreover, all trajectories starting outside $\mathcal{M}$ (*i.e.*, with $|\zeta| \neq 0$) converge to the origin.

1.2 Discussion and Literature Review

The concept of invariance has been widely used in control theory. The development of linear and nonlinear geometric control theory (see [228, 155, 78] for a comprehensive introduction) has shown that invariant subspaces, and their nonlinear counterpart, invariant distributions, play a fundamental role in the solution of many design problems. Slow and fast invariant manifolds, which naturally arise in singularly perturbed systems, have been used for stabilisation [114] and analysis of slow adaptation systems [184]. The notion of invariant manifolds is also crucial in the design of stabilising control laws for classes of nonlinear systems. More precisely, the theory of the centre manifold [41] has been instrumental in the design of stabilising control laws for systems with non-controllable linear approximation, see, *e.g.*, [3], whereas the concept of zero dynamics and the strongly related notion of zeroing manifold have been exploited in several local and global stabilisation methods, including passivity-based control [163], backstepping [123] and forwarding [194].

The notion of immersion has also a longstanding tradition in control theory. Its basic idea is to "transform" the system under consideration into a system with prespecified properties. For example, the classical problem of immersion of a generic nonlinear system into a linear and controllable system by means of static or dynamic state feedback has been extensively studied, see [155, 78] for further detail. State observation has traditionally being formu-

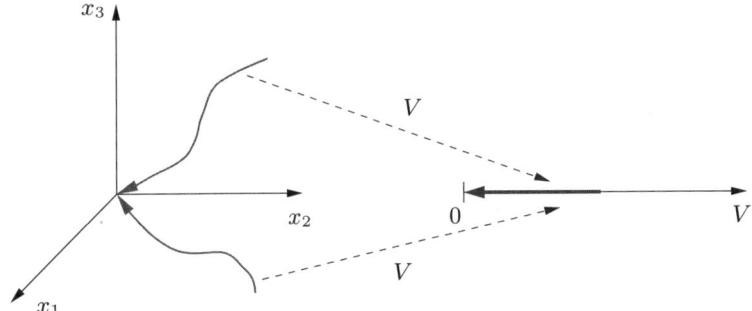

Fig. 1.3. Submersion interpretation of Lyapunov-based design methods.

lated in terms of system immersion, see [108]. More recently, immersion has been used in the nonlinear regulator theory to derive necessary and sufficient conditions for robust regulation. In particular, in [34, 79] it is shown that robust regulation is achievable provided that the exosystem can be immersed into a linear and observable system. Finally, in [144] it is shown that a dynamical system (possibly infinite dimensional) has a stable equilibrium if it can be immersed into another dynamical system with a stable equilibrium by means of a so-called *stability preserving* mapping. The definition of the latter given in [144] is related to some of the concepts used in this book.

Lyapunov-based design methods are somewhat dual to the approach (informally) described above. As a matter of fact, in Lyapunov design one seeks a function $V(x)$, which is positive-definite[3] (and proper[4], if global stability is sought after) and such that the system $\dot{V} = \alpha(V)$, for some function $\alpha(\cdot)$, has a (globally) asymptotically stable equilibrium at zero. Note that the function $V : x \to I$, where I is an interval of the real axis, is a *submersion* and the "target dynamics", namely the dynamics of the Lyapunov function, are one-dimensional, see Figure 1.3.

The I&I reformulation of the stabilisation problem is implicit in sliding mode control, where the target dynamics are the dynamics of the system on the sliding manifold, which is made attractive by a discontinuous control law, while $v(x)$ is the so-called equivalent control [223].

A procedure similar to I&I is proposed in [109], with the fundamental difference that $p = n$ and consequently $\pi(\cdot)$ is not an immersion but a change of co-ordinates.

Stabilisation via I&I is also related to passivity-based stabilisation methods, see [36, 163] and the survey paper [22]. In passivity-based control, stabilisation is achieved by finding an output such that the system is passive (with some suitable storage function, *e.g.*, with a minimum at the equilibrium to be stabilised). If the system is not open-loop passive, it is necessary

---
[3] See Definition A.1.
[4] See Definition A.4.

to find an output signal whose relative degree is one and whose associated zero dynamics are (weakly) minimum-phase. As shown in [36], under suitable assumptions, these are necessary and sufficient conditions for the system to be feedback equivalent to a passive system. As explained above, in I&I a stabilising control law is derived starting from the selection of a target dynamical system, $\dot{\xi} = \alpha(\xi)$, then computing the mapping $x = \pi(\xi)$, hence the manifold $\mathcal{M}$, and finally designing the control law that drives the distance $|\zeta|$ from the manifold to zero. It is clear that the system $\dot{x} = f(x) + g(x)u$ with output ζ is minimum-phase and its zero dynamics are precisely the dynamics of the target system. If the relative degree of ζ is one, then the system is feedback equivalent to a passive system.

1.3 Applications of I&I

From the discussion in Section 1.1 it is obvious that I&I requires the selection of a *target dynamical system*. This is in general a non-trivial task, as the solvability of the underlying control design problem depends upon such a selection. For general nonlinear systems the classical target dynamics are linear[5]. For physical systems the choice of linear target dynamics is not necessarily the most suitable one because, on one hand, workable designs should respect the constraints imposed by the physical structure and, on the other hand, it is well-known that most physical systems are not feedback linearisable.

In the remaining part of this section we present a series of control design problems for which it is possible to define some natural target dynamics and, therefore, exploit the I&I formulation. The problems are identified through representative examples that are solved in forthcoming chapters.

1.3.1 Robustification of Control Laws

In many cases of practical interest it is possible to design a control law for a reduced-order version of a given model. For instance, for systems admitting a slow/fast decomposition—which usually appears in applications where actuator dynamics or high-frequency (*e.g.*, bending) modes must be taken into account—it is sometimes possible to solve the control problem for the slow (*e.g.*, rigid) subsystem. A physically reasonable selection for the target dynamics is the latter subsystem placed in closed loop with a stabilising controller. In this case the application of I&I may be interpreted as a procedure to robustify, against higher-order dynamics, a controller derived from a low-order model. Finally, it is clear that the definition of the target system, which is not unique, allows to capture also performance objectives.

We now illustrate this application of I&I with two physical examples.

---

[5] A complete theory in this direction has been developed both for continuous and discrete time systems [155, 78].

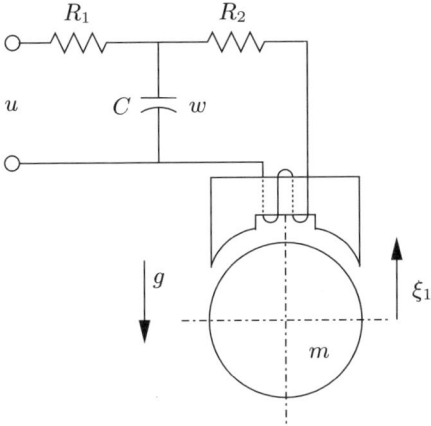

Fig. 1.4. Diagram of the levitated ball and the actuator.

Example 1.1 (Magnetic levitation). Consider a magnetic levitation system consisting of an iron ball in a vertical magnetic field created by a single electromagnet (see Figure 1.4). Adopting standard modelling assumptions for the electromagnetic coupling we obtain the model (with domain of validity[6] $-\infty < \xi_1 < 1$)

$$\dot{\xi}_1 = \frac{1}{m}\xi_2,$$
$$\dot{\xi}_2 = \frac{1}{2k}\xi_3^2 - mg, \quad (1.7)$$
$$\dot{\xi}_3 = -\frac{1}{k}R_2(1-\xi_1)\xi_3 + w,$$

where the state vector ξ consists of the ball position and its momentum and the flux in the inductance, w is the voltage applied to the electromagnet, m is the mass of the ball, g is the gravitational acceleration, R_2 is the coil resistance, and k is a positive constant that depends on the number of coil turns.

In low-power applications it is typical to neglect the dynamics of the actuator, hence it is assumed that w is the manipulated variable[7]. In medium-to-high power applications the voltage w is generated using a rectifier that includes a capacitance. The dynamics of this actuator can be described by the RC circuit shown in Figure 1.4, where the actual control voltage is u, while

---

[6] It is assumed that the ball position ξ_1 is scaled so that when $\xi_1 = 1$ the ball touches the electromagnet.

[7] In this case it is possible to asymptotically stabilise an equilibrium of the system (1.7) via a nonlinear static state feedback, derived using, for instance, feedback linearisation [78], backstepping [123], or passivity-based control [166], see [220] for a comparative study.

R_1 and C model the parasitic resistance and capacitance, respectively. The model of the levitated ball system, including the actuator dynamics, is given by the equations

$$\begin{aligned}
\dot{x}_1 &= \frac{1}{m} x_2, \\
\dot{x}_2 &= \frac{1}{2k} x_3^2 - mg, \\
\dot{x}_3 &= -\frac{1}{k} R_2 \left(1 - x_1\right) x_3 + x_4, \\
\dot{x}_4 &= -\frac{1}{Ck} \left(1 - x_1\right) x_3 - \frac{1}{R_1 C} x_4 + \frac{1}{R_1 C} u,
\end{aligned} \quad (1.8)$$

where $x_1 = \xi_1$, $x_2 = \xi_2$, $x_3 = \xi_3$ and we have added the co-ordinate $x_4 = w$ that represents the voltage across the capacitor. Note that (1.7) is the *slow* model of (1.8) taking as small parameter the parasitic time constant $R_1 C$. ■

Example 1.2 (Flexible joints robot). Consider the n-degrees-of-freedom (n-DOF) flexible joints robot model

$$\begin{aligned}
D(q)\ddot{q} + C(q, \dot{q})\dot{q} + G(q) + K(q - \theta) &= 0, \\
J\ddot{\theta} + K(\theta - q) &= u,
\end{aligned} \quad (1.9)$$

where $q \in \mathbb{R}^n$ and $\theta \in \mathbb{R}^n$ are the link and motor shaft angles, respectively, $D(q) = D(q)^\top > 0$ is the inertia matrix of the rigid arm, J is the constant diagonal inertia matrix of the actuators, $C(q, \dot{q})$ is the matrix related to the Coriolis and centrifugal terms, $G(q)$ is the gravity force vector of the rigid arm, $K = \mathrm{diag}(k_1, \ldots, k_n) > 0$ is the joint stiffness matrix, and u is the n-dimensional vector of torques applied to the motor shaft[8]. Taking the natural state-space realisation of (1.9), namely $x_1 = q$, $x_2 = \dot{q}$, $x_3 = \theta$, $x_4 = \dot{\theta}$, yields the system

$$\begin{aligned}
\dot{x}_1 &= x_2, \\
\dot{x}_2 &= -D(x_1)^{-1} \left(C(x_1, x_2)x_2 + G(x_1) + K(x_1 - x_3)\right), \\
\dot{x}_3 &= x_4, \\
\dot{x}_4 &= -J^{-1} \left(K(x_3 - x_1) - u\right).
\end{aligned} \quad (1.10)$$

Neglecting the flexible modes of the robot results in the reduced-order model

$$\begin{aligned}
\dot{\xi}_1 &= \xi_2, \\
\dot{\xi}_2 &= -D(\xi_1)^{-1} \left(C(\xi_1, \xi_2)\xi_2 + G(\xi_1) - w(\xi_1, \xi_2)\right),
\end{aligned} \quad (1.11)$$

where $w(\cdot)$ is the vector of torques applied to the links and defined as a function of ξ_1 and ξ_2. Note that the system (1.10) with output $y = x_3 - x_1 - K^{-1} w(x_1, x_2)$ has a well-defined vector relative degree $(r_1, \ldots, r_n) = (2, \ldots, 2)$ and the zero dynamics are exactly given by the target dynamics (1.11). ■

---
[8] See, e.g., [163] for further detail on the model.

Designing new controllers for the models (1.8) and (1.10) could be a time-consuming task, hence we might want to simply modify the ones for models (1.7) and (1.11) to make them robust against the actuator dynamics. This is a typical scenario where the I&I approach provides a useful tool.

1.3.2 Underactuated Mechanical Systems

For underactuated mechanical systems a sensible target system is the unactuated part of the mechanism, possibly with a feedback control. In this situation the definition of the target system, which is not unique and provides a degree of freedom to the design, can be used to trade-off solvability of the I&I problem with achievable performance as shown in the following example.

Example 1.3 (Cart and pendulum system). Consider the cart–pendulum system, depicted in Figure 1.5, and assume that a partial feedback linearisation stage has been applied. After normalisation this yields the model

$$\begin{aligned}\dot{x}_1 &= x_2, \\ \dot{x}_2 &= \sin(x_1) - u\cos(x_1), \\ \dot{x}_3 &= u,\end{aligned} \qquad (1.12)$$

where $(x_1, x_2) \in \mathcal{S}^1 \times \mathbb{R}$ are the pendulum angle (with respect to the upright vertical) and its velocity, $x_3 \in \mathbb{R}$ is the velocity of the cart, and $u \in \mathbb{R}$ is the input. The equilibrium to be stabilised is the upward position of the pendulum with the cart stopped, which corresponds to $x^* = 0$.

The key idea is to immerse the third-order system into a simple second-order pendulum, whose potential energy and damping functions are left to be designed (see Figure 1.5). Accordingly, we define the target dynamics as

$$\begin{aligned}\dot{\xi}_1 &= \xi_2, \\ \dot{\xi}_2 &= -V'(\xi_1) - R(\xi_1, \xi_2)\xi_2,\end{aligned} \qquad (1.13)$$

which are the equations of a (fully actuated) pendulum with energy function $H(\xi_1, \xi_2) = \frac{1}{2}\xi_2^2 + V(\xi_1)$ and, possibly nonlinear, damping function $R(\cdot)$ that, for generality, we have defined as a function of ξ_1 and ξ_2. To ensure that the target system has a locally asymptotically stable equilibrium at the origin, we specify that the potential energy function $V(\xi_1)$ satisfies $V'(0) = 0$ and $V''(0) > 0$, and the damping function is such that $R(0,0) > 0$. ∎

1.3.3 Systems in Special Forms

I&I is also applicable to two classes of systems with special structures that have attracted a great deal of attention from researchers: systems in feedback and feedforward forms.

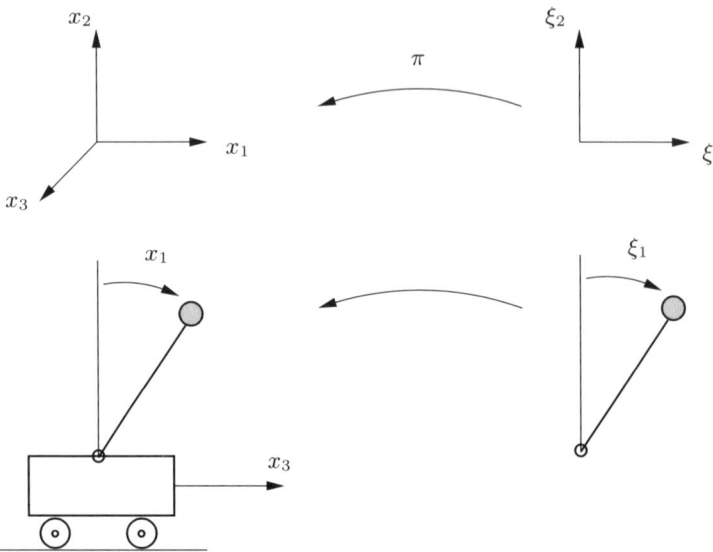

Fig. 1.5. Pendulum on a cart and target dynamics.

Example 1.4 (Systems in feedback form). Consider the class of systems in feedback form described by the equations[9]

$$\dot{x}_1 = f(x_1, x_2),$$
$$\dot{x}_2 = u, \qquad (1.14)$$

with $x = \text{col}(x_1, x_2) \in \mathbb{R}^n \times \mathbb{R}$, $u \in \mathbb{R}$ and $f(0,0) = 0$, where the system

$$\dot{x}_1 = f(x_1, 0)$$

has a globally asymptotically stable equilibrium at the origin. A sensible choice for the target dynamics is therefore given by

$$\dot{\xi} = f(\xi, 0), \qquad (1.15)$$

and this implies that the mapping $\pi(\cdot) = \text{col}(\pi_1(\cdot), \pi_2(\cdot))$ is such that $x_1 = \pi_1(\xi) = \xi$. ∎

Example 1.5 (Systems in feedforward form). Consider the class of systems in feedforward form described by the equations[10]

---

[9] This class of systems can be stabilised using the *backstepping* method, see [123] for an in-depth description.

[10] Globally asymptotically stabilising control laws for such systems can be constructed using the so-called *forwarding* method, as described in [194, 143, 181, 89].

$$\dot{x}_1 = h(x_2),$$
$$\dot{x}_2 = f(x_2) + g(x_2)u, \qquad (1.16)$$

with $x = \mathrm{col}(x_1, x_2) \in \mathbb{R} \times \mathbb{R}^n$, $u \in \mathbb{R}$, $h(0) = 0$ and $f(0) = 0$, where it is assumed that the zero equilibrium of the system $\dot{x}_2 = f(x_2)$ is globally asymptotically stable. In view of this assumption, the natural choice for the target dynamics is

$$\dot{\xi} = f(\xi), \qquad (1.17)$$

and this implies that the mapping $\pi(\cdot) = \mathrm{col}(\pi_1(\cdot), \pi_2(\cdot))$ is such that $x_2 = \pi_2(\xi) = \xi$. ∎

1.3.4 Adaptive Control

In adaptive control a natural candidate for the target dynamics is the plant placed in closed loop with a stabilising known-parameters controller. It is interesting to note that, even though the target dynamics are not known (since they depend on the unknown parameters), I&I is still applicable.

Example 1.6 (Adaptive controller design). Consider the first-order nonlinear system

$$\dot{x} = \theta x^2 + u, \qquad (1.18)$$

where $\theta \in \mathbb{R}$ is an unknown constant parameter. Note that if θ were known, the zero equilibrium of the system (1.18) would be globally asymptotically stabilised by the control law

$$u = -kx - \theta x^2, \qquad (1.19)$$

with $k > 0$. The corresponding *certainty-equivalent* controller is then given by

$$u = -kx - \hat{\theta}x^2,$$
$$\dot{\hat{\theta}} = w,$$

where w is the *update law*, which is typically chosen to cancel all parameter-dependent terms from the time-derivative of the Lyapunov function

$$V(x, \hat{\theta}) = \frac{1}{2}x^2 + \frac{1}{2\gamma}(\hat{\theta} - \theta)^2,$$

with $\gamma > 0$, which is given by

$$\dot{V}(x, \hat{\theta}) = -kx^2 - (\hat{\theta} - \theta)x^3 + \frac{1}{\gamma}(\hat{\theta} - \theta)w.$$

Selecting $w = \gamma x^3$ yields $\dot{V}(x, \hat{\theta}) = -kx^2$. This establishes boundedness of x and $\hat{\theta}$, and convergence of x to zero (see Theorem A.2). However, no conclusion can be drawn about the behaviour of the "estimation error" $\hat{\theta}-\theta$, except that it converges to a constant value.

Alternatively, the adaptive control problem can be placed in the I&I framework by considering the augmented system

$$\dot{x} = \theta x^2 + u,$$
$$\dot{\hat{\theta}} = w,$$

with control inputs u and w, and the target system

$$\dot{\xi} = -k\xi,$$

which is obtained by applying the known-parameters controller (1.19) to the system (1.18). Note that the selection of the target system imposes that the mapping $\pi(\cdot) = \text{col}(\pi_1(\cdot), \pi_2(\cdot))$ should be such that $\pi_1(\xi) = \xi$. ∎

1.3.5 Observer Design

We now show that the problem of observer design can be formulated using the I&I perspective. In this particular application of I&I the target dynamics are not given *a priori* but "define" the observer to be constructed.

Example 1.7 (Observer design). Consider a nonlinear system described by equations of the form

$$\begin{aligned}\dot{\eta} &= f_1(\eta, y),\\ \dot{y} &= f_2(\eta, y),\end{aligned} \qquad (1.20)$$

where $\eta \in \mathbb{R}^n$ is the unmeasured part of the state and $y \in \mathbb{R}^m$ is the measurable part of the state. Assume that there exists a dynamical system

$$\dot{\xi} = \alpha(y, \xi), \qquad (1.21)$$

with $\xi \in \mathbb{R}^p$, $p \geq n$, and two mappings $\beta : \mathbb{R}^p \times \mathbb{R}^m \to \mathbb{R}^p$ and $\phi : \mathbb{R}^n \times \mathbb{R}^m \to \mathbb{R}^p$, that are left-invertible with respect to their first argument and such that the manifold

$$\mathcal{M} = \{\, (\eta, y, \xi) \in \mathbb{R}^n \times \mathbb{R}^m \times \mathbb{R}^p \;:\; \beta(\xi, y) = \phi(\eta, y) \,\}$$

is attractive and invariant. Then the dynamical system (1.21) is an *observer* for the system (1.20) and an asymptotically convergent estimate of the state η is given by

$$\hat{\eta} = \phi^{\text{L}}(\beta(\xi, y)),$$

where $\phi^{\text{L}}(\cdot)$ is a left-inverse of $\phi(\cdot)$ with respect to its first argument. As a matter of fact, the state estimation error is zero on the manifold $\mathcal{M}$ and by rendering the manifold attractive, $\beta(\xi(t), y(t))$ tends to $\phi(\eta(t), y(t))$, which consequently implies that $\hat{\eta}(t)$ tends to $\eta(t)$. ∎

1.3.6 Nonlinear PI Control

The I&I approach allows to design nonlinear PI controllers with gains that are *nonlinear functions* of the measurable signals, thus avoiding the re-tuning stage which is necessary in classical PI control. We present the basic idea with a simple example.

Example 1.8 (Systems with matched uncertainty). Consider the simple scalar system
$$\dot{y} = \varphi(y) + u, \tag{1.22}$$
where $\varphi(y)$ is an unknown continuous function that ranges in the interval $\varphi_m < \varphi(y) < \varphi_M$. The control objective is to drive the state of the system to zero despite this reduced prior knowledge. Towards this end, we select a nonlinear PI controller structure of the form
$$\begin{aligned}\dot{\beta}_I &= w_I(y),\\ u &= v(\beta_P(y) + \beta_I, y),\end{aligned} \tag{1.23}$$
where the functions $\beta_P(\cdot)$, $w_I(\cdot)$ and $v(\cdot)$ are to be defined. Note that the classical linear PI scheme corresponds to the selection
$$\begin{aligned}\beta_P(y) &= k_P y,\\ w_I(y) &= k_I y,\\ u &= -\beta_P(y) - \beta_I,\end{aligned}$$
where the positive constants k_P and k_I are the proportional and integral tuning gains, respectively.

Select the target dynamics to be $\dot{\xi} = -\lambda \xi$, with $\lambda > 0$. Defining the signal
$$z = \beta_P(y) + \beta_I \tag{1.24}$$
and closing the loop with the nonlinear PI controller (1.23), we can write the system dynamics in *perturbed* form as
$$\dot{y} = -\lambda y + [\varphi(y) + v(z,y) + \lambda y]. \tag{1.25}$$
The control objective is then to keep all trajectories bounded and to drive the term in brackets asymptotically to zero, that in I&I parlance is tantamount to making the manifold
$$\mathcal{M} = \{\,(y,z) \in \mathbb{R} \times \mathbb{R} \;:\; \varphi(y) + v(z,y) + \lambda y = 0\,\}$$
attractive and invariant. The problem is then to find functions $\beta_P(\cdot)$, $w_I(\cdot)$ and $v(\cdot)$ such that, for each y, there exists (at least one) $\bar{z}_y$ such that[11] $(0, \bar{z}_y) \in \mathcal{M}$

---

[11] We use the sub-index $(\cdot)_y$ to underscore the fact that, in general, the quantity depends on y.

and z converges asymptotically towards $\bar{z}_y$. Note that on the manifold $\mathcal{M}$ the dynamics of the system coincide with the target dynamics.

Consider now the dynamics of z, which are described by the equation

$$\dot{z} = w_I(y) + \frac{\partial \beta_P}{\partial y}[\varphi(y) + v(z,y)]$$

and assume that $\mathcal{M}$ is not empty. To ensure that $\bar{z}_y$ is an equilibrium of the z dynamics, it is sufficient to select the integral parameter of the nonlinear PI controller as

$$w_I(y) = \frac{\partial \beta_P}{\partial y}\lambda y,$$

which yields

$$\dot{z} = \frac{\partial \beta_P}{\partial y}[\varphi(y) + v(z,y) + \lambda y]. \tag{1.26}$$

To complete the description of the nonlinear PI controller, the functions $\beta_P(y)$ and $v(z,y)$ must be such that all trajectories of the closed-loop system (1.25), (1.26) converge to the equilibrium $(y,z) = (0, \bar{z}_y)$. ■

2
I&I Stabilisation

This chapter contains the basic result for I&I stabilisation, namely a set of sufficient conditions for the construction of globally asymptotically stabilising, static, state feedback control laws for general, control affine, nonlinear systems. Note, however, that similar considerations can be given to non-affine systems and to tracking problems, while local versions follow *mutatis mutandis*.

To illustrate this result and some of the extensions cited above, we provide several academic and physically motivated examples. The former include singularly perturbed systems and systems in feedback and feedforward forms, while the latter consist of a mechanical system with flexibility modes, an electromechanical system with parasitic actuator dynamics, and an underactuated mechanical system.

2.1 Main Stabilisation Result

Theorem 2.1. *Consider the system*

$$\dot{x} = f(x) + g(x)u, \qquad (2.1)$$

with $x \in \mathbb{R}^n$, $u \in \mathbb{R}^m$, and an equilibrium point $x^* \in \mathbb{R}^n$ to be stabilised. Assume that there exist smooth mappings $\alpha : \mathbb{R}^p \to \mathbb{R}^p$, $\pi : \mathbb{R}^p \to \mathbb{R}^n$, $\phi : \mathbb{R}^n \to \mathbb{R}^{n-p}$, $c : \mathbb{R}^p \to \mathbb{R}^m$ and $v : \mathbb{R}^{n \times (n-p)} \to \mathbb{R}^m$, with $p < n$, such that the following hold.

(A1) The target system

$$\dot{\xi} = \alpha(\xi), \qquad (2.2)$$

with $\xi \in \mathbb{R}^p$ has a globally asymptotically stable equilibrium at $\xi^* \in \mathbb{R}^p$ and

$$x^* = \pi(\xi^*).$$

(A2) For all $\xi \in \mathbb{R}^p$,
$$f(\pi(\xi)) + g(\pi(\xi))c(\pi(\xi)) = \frac{\partial \pi}{\partial \xi}\alpha(\xi). \qquad (2.3)$$

(A3) The set identity
$$\{x \in \mathbb{R}^n \,|\, \phi(x) = 0\} = \{x \in \mathbb{R}^n \,|\, x = \pi(\xi),\ \xi \in \mathbb{R}^p\} \qquad (2.4)$$
holds.

(A4) All trajectories of the system
$$\dot{z} = \frac{\partial \phi}{\partial x}\left(f(x) + g(x)v(x,z)\right), \qquad (2.5)$$
$$\dot{x} = f(x) + g(x)v(x,z), \qquad (2.6)$$
are bounded and (2.5) has a uniformly globally asymptotically stable equilibrium at $z = 0$.

Then x^* is a globally asymptotically stable equilibrium of the closed-loop system
$$\dot{x} = f(x) + g(x)v(x, \phi(x)). \qquad (2.7)$$

Proof. We establish the claim in two steps. First, it is shown that the equilibrium x^* is globally attractive, then that it is Lyapunov stable. Let $z(0) = \phi(x(0))$ and note that the right-hand side of (2.5) is $\dot{\phi}$, hence by (A4) any trajectory of the closed-loop system (2.7) is bounded and it is such that $\lim_{t \to \infty} z(t) = 0$, *i.e.*, it converges towards the manifold $\phi(x) = 0$, which is well defined by (A3). Moreover, by (A1) and (A2) the manifold is invariant and internally asymptotically stable, hence all trajectories of the closed-loop system converge to the equilibrium x^*.

Note now that, since (2.5) has a stable equilibrium at the origin, we have that, for any $\epsilon_1 > 0$, there exists $\delta_1 > 0$ such that $|\phi(x(0))| < \delta_1$ implies $|\phi(x(t))| < \epsilon_1$, for all $t \geq 0$. By (A1) and (A3) we also have that $\phi(x^*) = 0$. Consider now a projection of $x(t)$ on the manifold $\mathcal{M} = \{x \in \mathbb{R}^n \,|\, \phi(x) = 0\}$, denoted by $x_P(t)$, such that $|x(t) - x_P(t)| = \gamma(|\phi(x(t))|)$, for some class-$\mathcal{K}$ function $\gamma(\cdot)$, and such that $x_P(t) = \pi(\xi(t))$. By (A1) we have that, for any $\epsilon_2 > 0$, there exists $\delta_2 > 0$ such that $|x_P(0) - x^*| < \delta_2$ implies $|x_P(t) - x^*| < \epsilon_2$, for all $t \geq 0$. Selecting $\epsilon_1 = \epsilon_2 = \frac{1}{2}\epsilon$, and using the triangle inequality
$$|x(t) - x^*| \leq \gamma(|\phi(x(t))|) + |x_P(t) - x^*|,$$
it follows that, for any $\epsilon > 0$, there exists $\delta > 0$ (dependent on δ_1 and δ_2) such that $|x(0) - x^*| < \delta$ implies $|x(t) - x^*| < \epsilon$, for all $t \geq 0$, which proves the claim. □

The result summarised in Theorem 2.1 lends itself to the following interpretation. Given the system (2.1) and the target dynamical system (2.2), the

2.1 Main Stabilisation Result

goal is to find a manifold $\mathcal{M}$, described implicitly by $\{x \in \mathbb{R}^n \mid \phi(x) = 0\}$, and in parameterised form by $\{x \in \mathbb{R}^n \mid x = \pi(\xi), \xi \in \mathbb{R}^p\}$, which can be rendered invariant and asymptotically stable, and such that the (well-defined) restriction of the closed-loop system to $\mathcal{M}$ is described by $\dot{\xi} = \alpha(\xi)$.

Notice that the control u that renders the manifold invariant is not unique, since it is uniquely defined only on $\mathcal{M}$, i.e., $v(\pi(\xi), 0) = c(\pi(\xi))$. From all possible controls we select one that drives to zero the *off-the-manifold* coordinates z and keeps the system trajectories bounded, i.e., such that (A4) holds.

The following observations concerning the assumptions (A1)–(A4) of Theorem 2.1 are in order.

1. In most applications of Theorem 2.1 described in this chapter, the target system is *a priori* defined, hence condition (A1) is automatically satisfied.
2. If the control objective is to track a given trajectory, then Theorem 2.1 has to be rephrased in terms of tracking errors and the target system should *generate* the reference trajectory.
3. Given the target system (2.2), equation (2.3) of condition (A2) defines a partial differential equation (PDE) in the unknown $\pi(\cdot)$, where $c(\cdot)$ is a free parameter[1]. Finding the solution of this equation is (in general) a difficult task. Despite this fact, as shown by numerous examples in the book, a suitable selection of the target dynamics, i.e., following physical and system theoretic considerations, allows to simplify this task. In some cases (see Example 2.6) it is possible to interlace the steps of definition of the target dynamics (2.2) and generation of the manifold (2.4) by viewing the PDE (2.3) as an algebraic equation relating $\alpha(\cdot)$ with $\pi(\cdot)$ (and its partial derivatives) and then selecting suitable expressions for $\pi(\cdot)$ that ensure the desired stability properties of the target dynamics.
4. Assumption (A3) states that the image of the mapping $\pi(\cdot)$ can be expressed as the zero level set of a (smooth) function $\phi(\cdot)$. Roughly speaking, this is a condition on the invertibility of the mapping that translates into a rank restriction on $\frac{\partial \pi}{\partial \xi}$. In the linear case, where $\pi(\xi) = T\xi$ with T some constant $(n \times p)$ matrix, we have $\phi(x) = T^\perp x$, where $T^\perp T = 0$, and (A3) holds if and only if T is full rank. In general, if $\pi : \mathbb{R}^p \to \mathbb{R}^n$ is an injective and proper[2] immersion then the image of $\pi(\cdot)$ is a submanifold of $\mathbb{R}^n$. (A3) thus requires that such a submanifold can be described (globally) as the zero level set of the function $\phi(\cdot)$. Note, finally, that if there exists a

---

[1]Note that, if the linearisation of (2.1) at $x = x^*$ is controllable and all functions are locally analytic, it has been shown in [109], using Lyapunov's auxiliary theorem and under some *non-resonance* conditions, that there always exists $c(\cdot)$ such that the solution exists locally.

[2]An immersion is a mapping $\pi(\cdot) : \mathbb{R}^p \to \mathbb{R}^n$, with $p < n$. It is injective if rank$(\pi) = p$, and it is proper if the inverse image of any compact set is also compact [1, Chapter 3].

partition of $x = \mathrm{col}(x_1, x_2)$, with $x_1 \in \mathbb{R}^p$ and $x_2 \in \mathbb{R}^{n-p}$, and a corresponding partition of $\pi(\xi) = \mathrm{col}(\pi_1(\xi), \pi_2(\xi))$ such that $\pi_1(\xi)$ is a global diffeomorphism, then the function $\phi(x) = x_2 - \pi_2(\pi_1^{-1}(x_1))$ is such that (A3) holds.

5. In many cases of practical interest, to have asymptotic convergence of $x(t)$ to x^* it is sufficient to require that the system (2.5) has a uniformly globally stable equilibrium at $z = 0$ and

$$\lim_{t \to \infty} g(x(t))(v(x(t), z(t)) - v(x(t), 0)) = 0, \qquad (2.8)$$

i.e., it is not necessarily required that the manifold is reached. This fact, which distinguishes the present approach from others, such as sliding mode, is instrumental to the development of the adaptive and output feedback control theory in Chapters 3, 4 and 6.

6. In Theorem 2.1 a stabilising control law is derived starting from the selection of a target dynamical system. A different perspective can be taken: given the mapping $x = \pi(\xi)$, hence the mapping $z = \phi(x)$, find (if possible) a control law which renders the manifold $z = 0$ invariant and asymptotically stable, and a vector field $\dot{\xi} = \alpha(\xi)$, with a globally asymptotically stable equilibrium ξ^*, such that equation (2.3) holds. If this goal is achieved then the system (2.1) with output $z = \phi(x)$ is (globally) minimum-phase and its zero dynamics—i.e., the dynamics on the output zeroing manifold $\mathcal{M}$—are given by (2.2). In this respect, the result in Theorem 2.1 can be regarded as a *dual* of the classical stabilisation methods based on the construction of passive or minimum-phase outputs[3].

Note 2.1. The partial differential equation (2.3) also arises in nonlinear regulator theory[4]. However, equation (2.3) and its solution are used in the present context in a new form. First of all, in classical regulator theory, the system $\dot{\xi} = \alpha(\xi)$ is assumed Poisson stable [34], whereas in the I&I framework it is required to have an asymptotically stable equilibrium. Second, while in regulator theory the mapping $\pi(\xi)$ is needed to define a controlled invariant manifold for the system composed of the plant and the exosystem, in the present approach the mapping $\pi(\xi)$ is used to define a parameterised controlled invariant manifold which is a submanifold of the state-space of the system to be stabilised. Finally, in regulator theory the exosystem $\dot{\xi} = \alpha(\xi)$ is driving the plant to be controlled, whereas in the I&I approach the closed-loop system contains a *copy* of the dynamics $\dot{\xi} = \alpha(\xi)$. Finally, equation (2.3) arises also in the feedback linearisation problem studied in [109]. Therein, the goal is to obtain a closed-loop system which is locally equivalent to a target linear system. Unlike the present context, the target system in [109] has the same dimension as the system to be controlled, i.e., the mapping $\pi(\cdot)$ is a (local) diffeomorphism rather than an immersion. ◁

---

[3] See [36, 163] and the survey paper [22].

[4] To be precise, two equations arise in nonlinear regulator theory. The first one is equation (2.3), the second is an equation expressing the fact that the tracking error is zero on the invariant manifold defined via the solution of equation (2.3), see [34].

The following definition is used in the rest of the book to provide concise statements.

Definition 2.1. *A system described by equations of the form (2.1) with target dynamics (2.2) is said to be (locally) I&I stabilisable if the assumptions (A1)–(A4) of Theorem 2.1 hold (locally).*

2.2 Systems with Special Structures

In this section we show that some of the systems with special structures considered in the literature are I&I stabilisable. In particular we consider singularly perturbed systems and systems in feedback and feedforward forms.

Example 2.1 (Singularly perturbed systems). We present an academic example, with the twofold objective of putting in perspective the I&I formulation and giving the flavour of the required computations. Consider the two-dimensional system[5]

$$\begin{aligned} \dot{x}_1 &= x_1 x_2^3, \\ \epsilon \dot{x}_2 &= x_2 + u, \end{aligned} \quad (2.9)$$

with $x_1(0) \geq 0$[6], and the target dynamics

$$\dot{\xi} = -\xi^5, \quad (2.10)$$

where $\xi \in \mathbb{R}$. Fixing $\pi_1(\xi) = \xi$, equations (2.3) become

$$\begin{aligned} \xi \pi_2(\xi)^3 &= -\xi^5, \\ \pi_2(\xi) + c(\xi) &= -\epsilon \frac{\partial \pi_2}{\partial \xi} \xi^5. \end{aligned} \quad (2.11)$$

From the first equation we obtain $\pi_2(\xi) = -\xi^{4/3}$, which is defined for $\xi \geq 0$, while the mapping $c(\cdot)$ is defined by the second equation. The manifold $x = \pi(\xi)$ can be implicitly described by $\phi(x) = x_2 + x_1^{4/3} = 0$ and the off-the-manifold dynamics (2.5) are given by

$$\epsilon \dot{z} = v(x, z) + x_2 + \frac{4}{3} \epsilon x_1^{4/3} x_2^3.$$

The I&I design is completed by choosing $v(x, z) = -x_2 - \frac{4}{3} \epsilon x_1^{4/3} x_2^3 - z$, which yields the closed-loop dynamics

$$\begin{aligned} \epsilon \dot{z} &= -z, \\ \dot{x}_1 &= x_1 x_2^3, \\ \epsilon \dot{x}_2 &= -\frac{4}{3} \epsilon x_1^{4/3} x_2^3 - z. \end{aligned} \quad (2.12)$$

---

[5]This example has been adopted from [114], where the composite control approach has been used, see Note 2.2.
[6]Similarly to [114], we consider the system on the invariant set $x_1 \geq 0$.

From the first equation in (2.12) it is clear that the z-subsystem has a globally asymptotically stable equilibrium at zero, hence, to complete the proof, it only remains to show that all trajectories of the system (2.12) are bounded. To this end, consider the (partial) change of co-ordinates $\eta = x_2 + x_1^{4/3}$ yielding

$$\epsilon \dot{z} = -z,$$
$$\dot{x}_1 = x_1 \left(\eta - x_1^{4/3} \right)^3,$$
$$\epsilon \dot{\eta} = -z,$$

and note that $z(t)$ and $\eta(t)$ are bounded for all t. Finally, boundedness of $x_1(t)$ can be proved observing that the dynamics of x_1 can be expressed in the form

$$\dot{x}_1 = -x_1^5 + \rho(x_1, \eta),$$

for some function $\rho(\cdot)$ satisfying $|\rho(x_1, \eta)| \leq k(\eta)|x_1|^4$, for some $k(\eta) > 0$ and for $|x_1| > 1$. The control law is obtained as[7]

$$u = v(x, \phi(x)) = -2x_2 - x_1^{4/3} \left(1 + \frac{4}{3}\epsilon x_2^3 \right).$$

As a result, the system (2.9) is I&I stabilisable with target dynamics (2.10) in its domain of definition.

A *global* stabilisation result can be obtained if instead of (2.10) we choose the target dynamics $\dot{\xi} = -|\xi|^3 \xi$. Then the solution of (2.11) is $\pi_1(\xi) = \xi$ and $\pi_2(\xi) = -|\xi|$. As a result, the manifold takes the form $\phi(x) = x_2 + |x_1| = 0$ and it can be rendered globally attractive, while keeping all trajectories bounded, by the control law $u = -2x_2 - \epsilon|x_1|x_2^3 - |x_1|$. ∎

Note 2.2. The composite control approach of [114] is applicable for singularly perturbed systems of the form

$$\dot{x}_1 = f(x_1, x_2, u),$$
$$\epsilon \dot{x}_2 = g(x_1, x_2, u),$$

for which a slow manifold, defined by the function $x_2 = h(x_1, u, \epsilon)$, exists and results from the solution of the PDE

$$\epsilon \left(\frac{\partial h}{\partial x_1} + \frac{\partial h}{\partial u} \frac{\partial u}{\partial x_1} \right) f(x_1, h(x_1, u, \epsilon), u) = g(x_1, h(x_1, u, \epsilon), u).$$

In [114] it is proposed to expand $h(x_1, u, \epsilon)$ and u in a power series of ϵ, namely $h(\cdot) = \bar{h} + \mathcal{O}(\epsilon^k)$, where $\bar{h} = h_0 + \epsilon h_1 + \cdots + \epsilon^k h_k$, and $u = \bar{u} + \mathcal{O}(\epsilon^k)$, where $\bar{u} = u_0 + \epsilon u_1 + \cdots + \epsilon^k u_k$, taking as u_0 the control law that stabilises some equilibrium of the slow subsystem. Collecting the terms with the same powers of ϵ yields equations relating the terms h_i and u_i, which can be iteratively solved to approximate (up to any order $\mathcal{O}(\epsilon^k)$) the solution of the PDE. The control law is then constructed as $u = \bar{u} - K(x_2 - \bar{h})$, where the last term, with $K > 0$, is a fast control that *steers* the trajectories onto the slow manifold. ◁

---

[7] For comparison we note that the composite controller obtained in [114] is $u = -2x_2 - x_1^{4/3}(1 + \frac{4}{3}\epsilon x_1^4)$.

2.2 Systems with Special Structures

We conclude this section by showing that for the systems in feedback and feedforward forms presented in Examples 1.4 and 1.5, respectively, the I&I method can be used to construct (globally) stabilising control laws that are different from those obtained by the standard (backstepping and forwarding) approaches.

Example 2.2 (Systems in feedback form). Consider the class of systems in feedback form introduced in Example 1.4. The following statement summarises the application of the I&I approach to this special case.

Proposition 2.1. *Consider a system described by equations of the form (1.14) with $f(0,0) = 0$ and suppose that the system $\dot{x}_1 = f(x_1, 0)$ has a globally asymptotically stable equilibrium at zero. Then the system (1.14) is (globally) I&I stabilisable with target dynamics (1.15).*

Proof. To establish the claim we need to prove that conditions (A1)–(A4) of Theorem 2.1 hold. To begin with, note that (A1) is trivially satisfied, whereas the mappings

$$\begin{bmatrix} \pi_1(\xi) \\ \pi_2(\xi) \end{bmatrix} = \begin{bmatrix} \xi \\ 0 \end{bmatrix}, \qquad c(\pi) = 0, \qquad \phi(x_1, x_2) = x_2$$

are such that conditions (A2) and (A3) hold.

Note now that the off-the-manifold variable $z = x_2$ can be used as a partial co-ordinate, hence instead of verifying (A4) we simply need to show that it is possible to select u such that the trajectories of the closed-loop system are bounded and $z = x_2$ converges to zero.

To this end, let $u = -K(x_1, x_2)x_2$, with $K(x_1, x_2) \geq k > 0$ for any (x_1, x_2) and for some k, and consider the system

$$\dot{x}_1 = f(x_1, x_2),$$
$$\dot{x}_2 = -K(x_1, x_2)x_2.$$

Note that x_2 converges to zero. To prove boundedness of x_1 pick any $M > 0$ and let $V(x_1)$ be a positive-definite and proper function such that

$$\frac{\partial V}{\partial x_1} f(x_1, 0) < 0, \tag{2.13}$$

for all $|x_1| > M$. Note that such a function $V(x_1)$ exists, by global asymptotic stability of the zero equilibrium of the system $\dot{x}_1 = f(x_1, 0)$, but $V(x_1)$ is not necessarily a Lyapunov function for $\dot{x}_1 = f(x_1, 0)$. Consider now the positive-definite and proper function $W(x_1, x_2) = V(x_1) + \frac{1}{2}x_2^2$ and note that, for some function $F(x_1, x_2)$ and for any smooth function $\gamma(x_1) > 0$, one has

$$\dot{W} = \frac{\partial V}{\partial x_1} f(x_1, 0) + \frac{\partial V}{\partial x_1} F(x_1, x_2)x_2 - K(x_1, x_2)x_2^2$$

$$\leq \frac{\partial V}{\partial x_1} f(x_1, 0) + \frac{1}{\gamma(x_1)} \left| \frac{\partial V}{\partial x_1} \right|^2 + \gamma(x_1)|F(x_1, x_2)|^2 x_2^2 - K(x_1, x_2)x_2^2.$$

As a result, setting $\gamma(x_1)$ such that

$$\frac{\partial V}{\partial x_1}f(x_1,0) + \frac{1}{\gamma(x_1)}\left|\frac{\partial V}{\partial x_1}\right|^2 < 0,$$

for all $|x_1| > M$, and selecting

$$K(x_1,x_2) > \gamma(x_1)|F(x_1,x_2)|^2$$

yields the claim. $\square$

It is worth noting that although system (1.14) is stabilisable using standard backstepping arguments[8], the control law obtained using backstepping is different from the control law suggested by Proposition 2.1. The former requires the knowledge of a Lyapunov function for the system $\dot{x}_1 = f(x_1,0)$ and it is such that, in closed loop, the manifold $x_2 = 0$ is not invariant, whereas the latter requires only the knowledge of the function $V(x_1)$ satisfying equation (2.13) for sufficiently large $|x_1|$ and renders the manifold $x_2 = 0$ invariant and globally attractive.

To illustrate the result in Proposition 2.1 and compare it with the control law resulting from standard backstepping, consider the two-dimensional system

$$\dot{x}_1 = -x_1 + \lambda x_1^3 x_2,$$
$$\dot{x}_2 = u.$$

A backstepping-based stabilising control law is

$$u = -\lambda x_1^4 - x_2,$$

whereas a direct application of the procedure described in the proof of Proposition 2.1 shows that the I&I stabilising control law is

$$u = -\left(2 + x_1^8\right)x_2.$$

The latter does not require the knowledge of the parameter λ, however, it is in general more *aggressive* because of the higher power in x_1.

As a second example consider the system

$$\dot{x}_1 = A(\theta)x_1 + F(x_1,x_2)x_2,$$
$$\dot{x}_2 = u,$$

with $x_1 \in \mathbb{R}^n$ and $\theta \in \mathbb{R}^s$ an unknown constant vector. Assume that the matrix $A(\theta)$ is Hurwitz for all θ. Then the system is I&I stabilisable with target dynamics $\dot{\xi} = A(\theta)\xi$. A simple computation shows that an I&I stabilising control law is

$$u = -\left(1 + x_1^2\right)|F(x_1,x_2)|^2 x_2 - x_2,$$

and this does not require the knowledge of a Lyapunov function for the system $\dot{x}_1 = A(\theta)x_1$, nor the knowledge of the parameter θ. $\blacksquare$

---
[8] See [123].

2.2 Systems with Special Structures

Example 2.3 (Systems in feedforward form). We now show that, using the approach pursued in this chapter, a new class of control laws can be constructed for the special class of systems in feedforward form given in Example 1.5.

Proposition 2.2. *Consider a system described by equations of the form (1.16) and suppose that the zero equilibrium of the system $\dot{x}_2 = f(x_2)$ is globally asymptotically stable. Assume, moreover, that there exists a smooth function $M(x_2)$ such that, for all x_2, $L_f M(x_2) = h(x_2)$, the set $\mathcal{S} = \{x_2 \in \mathbb{R}^n \mid L_g M(x_2) = 0\}$ is composed of isolated points, and $0 \notin \mathcal{S}$. Then the system (1.16) is globally I&I stabilisable with target dynamics (1.17).*

Proof. To begin with note that (A1) in Theorem 2.1 is trivially satisfied and that the mappings

$$\begin{bmatrix} \pi_1(\xi) \\ \pi_2(\xi) \end{bmatrix} = \begin{bmatrix} M(\xi) \\ \xi \end{bmatrix}, \qquad c(\xi) = 0$$

are such that condition (A2) holds. The implicit description of the manifold in (A3) is $z = \phi(x) = x_1 - M(x_2)$ and the off-the-manifold dynamics are

$$\dot{z} = -L_g M(x_2) u.$$

To complete the proof it remains to verify condition (A4), or alternatively the weaker condition (2.8). To this end, let

$$u = \epsilon \frac{1}{1 + |g(x_2)|} \frac{L_g M(x_2)}{1 + |L_g M(x_2)|} \frac{z}{1 + |z|},$$

with $\epsilon = \epsilon(x_2) > 0$, and consider the closed-loop system

$$\dot{z} = -\epsilon \frac{1}{1 + |g(x_2)|} \frac{(L_g M(x_2))^2}{1 + |L_g M(x_2)|} \frac{z}{1 + |z|},$$
$$\dot{x}_1 = h(x_2),$$
$$\dot{x}_2 = f(x_2) + \epsilon \frac{g(x_2)}{1 + |g(x_2)|} \frac{L_g M(x_2)}{1 + |L_g M(x_2)|} \frac{z}{1 + |z|}.$$

Note now that, from the first equation, z is bounded, hence x_2 is bounded, provided that ϵ is sufficiently small. It follows that $L_g M(x_2) z$, and hence x_2, converge to zero. Note now that, if ϵ is sufficiently small, z converges exponentially to zero. As a result, $\eta = x_1 - M(x_2)$ is bounded, hence x_1 is also bounded for all t, which proves the claim. □

To illustrate Proposition 2.2 consider the system

$$\dot{x}_1 = x_{21}^3,$$
$$\dot{x}_{21} = x_{22}^3, \qquad (2.14)$$
$$\dot{x}_{22} = v,$$

and let $v = -x_{21}^3 - x_{22}^3 + u$. The system (2.14) satisfies all the assumptions of Proposition 2.2, with $M(x_2) = M(x_{21}, x_{22}) = -x_{21} - x_{22}$. As a result its zero equilibrium is stabilised by the control law[9] $u = -z$ with $z = x_1 + x_{21} + x_{22}$.

Consider now the system

$$\dot{x}_1 = x_2^3,$$
$$\dot{x}_2 = -x_2^3 + (1 - x_2^2)u.$$

A simple computation shows that $M(x_2) = -x_2$ satisfies the assumptions of Proposition 2.2 and that $\mathcal{S} = \{x_2 = -1, x_2 = 1\}$. Hence, a straightforward application of the procedure outlined in the proof of Proposition 2.2 shows that the control law

$$u = -\frac{1 - x_2^2}{1 + x_2^4}(x_1 + x_2),$$

renders the zero equilibrium of the closed-loop system globally asymptotically stable. ∎

Note 2.3. The system (2.14) has been studied in several papers. In [35] a globally stabilising control law has been designed through the construction of a minimum-phase relative degree one output map, whereas in [168] a globally stabilising control law has been obtained using a control Lyapunov function. The control law in [168] is $u = -x_{21}^3 - x_{22}^3 - x_1 - x_{22}$ and it is similar to the one proposed above. Note that the former requires the knowledge of a (control) Lyapunov function, whereas the latter does not. Finally, the zero equilibrium of the system (2.14) can be stabilised with the modified version of forwarding proposed in [142], which is again a Lyapunov-based design methodology, and with homogeneous feedback laws. ◁

2.3 Physical Systems

In this section the I&I stabilisation method is applied to the three physical examples introduced in Section 1.3, namely the magnetically levitated ball with actuator dynamics, the robot manipulator with joint flexibilities, and the cart and pendulum system. The first example is presented to highlight the connections between I&I and other existing techniques, in particular we show how the I&I framework allows to recover the composite control and backstepping solutions, while for the robot problem we prove that with I&I we can generate a novel family of global tracking controllers under standard assumptions. In both examples we start from the assumption that we know a stabilising controller for a nominal reduced-order model, and we robustify it with respect to some higher-order dynamics.

Finally, the objective of the cart and pendulum example is to suggest a procedure to overcome the need to solve the PDE of the immersion condition— i.e., the computation of the function $\pi(\cdot)$ that defines the manifold $\mathcal{M}$. This

---

[9] Due to the special form of the system (2.14), the control law applied here is simpler than the one given in the proof of Proposition 2.2.

computation is unquestionably the main difficulty for the application of the I&I methodology. Towards this end, we propose to transform the PDE into an algebraic equation where the target dynamics are viewed as a function of $\pi(\cdot)$ (and its partial derivatives), and then select $\pi(\cdot)$ to ensure that the target dynamics have the desired stability properties.

Example 2.4 (Magnetic levitation). Consider the magnetic levitation system (1.8) of Example 1.1 and the problem of (locally) stabilising the ball around a given position. A solution to this problem is expressed by the following proposition.

Proposition 2.3. *The full-order model of the levitated ball system (1.8) is I&I stabilisable with target dynamics (1.7), where $w = w(\xi)$ is any stabilising state feedback such that (1.7) in closed loop with $w(\xi(t)) + c(t)$, where $c(t)$ is a bounded signal, has bounded trajectories.*

Proof. As (A1) is automatically satisfied, we only verify the remaining conditions (A2)–(A4) of Theorem 2.1. First, simple calculations show that a solution of equation (2.3) is given by the mapping $\pi(\xi) = \text{col}(\xi_1, \xi_2, \xi_3, w(\xi))$. This solution can be easily obtained by fixing $\pi_1(\xi) = \xi_1$ and $\pi_3(\xi) = \xi_3$, a choice which captures the control objective. The parameterised manifold $x = \pi(\xi)$ can be implicitly defined as $\phi(x) = x_4 - w(x_1, x_2, x_3) = 0$, hence condition (A3) is also satisfied. Finally, we have to choose a function $v(x, z)$ that preserves boundedness of trajectories and asymptotically stabilises the zero equilibrium of the off-the-manifold dynamics (2.5), which in this example are given by

$$\dot{z} = \frac{1}{Ck}(1 - x_1)x_3 - \frac{1}{R_1 C}(x_4 - v(x, z)) - \dot{w}, \qquad (2.15)$$

where $\dot{w}$ is evaluated on the manifold, hence it is computable from the full-order dynamics (1.8). An obvious simple selection is

$$v(x, z) = x_4 - z + R_1 C \dot{w} + \frac{R_1}{k}(1 - x_1)x_3,$$

which yields the system

$$\dot{z} = -\frac{1}{R_1 C} z,$$
$$\dot{x}_1 = \frac{1}{m} x_2,$$
$$\dot{x}_2 = \frac{1}{2k} x_3^2 - mg, \qquad (2.16)$$
$$\dot{x}_3 = -\frac{R_2}{k}(1 - x_1)x_3 + x_4,$$
$$\dot{x}_4 = -\frac{1}{R_1 C} z + \dot{w}.$$

To complete the proof of the proposition it only remains to prove that, for the given $w(\cdot)$, the trajectories of (2.16) are bounded. To this end, note that in the co-ordinates (z, x_1, x_2, x_3, η), with $\eta = x_4 - w(x_1, x_2, x_3)$, the system is described by the equations

$$\dot{z} = -\frac{1}{R_1 C} z,$$

$$\dot{x}_1 = \frac{1}{m} x_2,$$

$$\dot{x}_2 = \frac{1}{2k} x_3^2 - mg,$$

$$\dot{x}_3 = -\frac{R_2}{k}(1 - x_1)x_3 + w(x_1, x_2, x_3) + \eta,$$

$$\dot{\eta} = -\frac{1}{R_1 C} z,$$

from which we conclude that z converges to zero exponentially, hence η is bounded. The proof is completed by invoking the assumed robustness property with respect to bounded input disturbances. The control law is finally obtained as

$$u = w(x_1, x_2, x_3) + R_1 C \dot{w} + \frac{R_1}{k}(1 - x_1)x_3.$$

□

We use this example to compare the I&I formulation with the composite control approach. Taking x_4 as the slow variable and $R_1 C$ as the small parameter the problem reduces to finding a function $h(\cdot)$ and a control u such that $x_4 = h(x_1, x_2, x_3, u)$ describes an invariant manifold. This requires the solution of a PDE of the form

$$-\frac{1}{Ck}(1-x_1)x_3 + \frac{1}{R_1 C}(u - h) = \frac{\partial h}{\partial x_3}\left(-\frac{R_2}{k}(1-x_1)x_3 + h\right)$$

$$+ \frac{1}{m}\frac{\partial h}{\partial x_1}x_2 + \frac{\partial h}{\partial x_2}\left(\frac{1}{2k}x_3^2 - mg\right) + \frac{\partial h}{\partial u}\dot{u}.$$

In this simple case an exact solution is possible setting $h = w$ and

$$u_1 = \dot{w} + \frac{1}{Ck}(1-x_1)x_3.$$

This choice ensures $h_k = 0$ for all $k \geq 1$. The resulting invariant manifold $(x_4 = w)$ and the controller are the same obtained via I&I stabilisation. However, while in the composite control approach the function u_1 that determines the controller is essentially imposed, in I&I we have some freedom in the choice of the function $v(x, z)$ to stabilise (2.15). Finally, notice that this controller also results from direct application of backstepping to system (1.8). ■

Example 2.5 (Flexible joints robot). Consider the problem of global tracking for the n-DOF flexible joints robot model introduced in Example 1.2. We present a procedure to robustify an arbitrary global state feedback *tracking* controller designed for the rigid robot.

Proposition 2.4. *The flexible joints robot model (1.10) is globally I&I stabilisable with target dynamics (1.11), where $w = w(\xi, t)$ is any time-varying state feedback that ensures that the solutions $\xi_1(t)$ of (1.11) globally track any bounded, four times differentiable trajectory $\xi_1^*(t)$, and with the additional property that in closed loop with $w(\xi,t) + c(t)$, where $c(t)$ is a bounded signal, trajectories remain bounded*[10].

Proof. As discussed in Section 2.1, since the control objective is to track a reference trajectory, (A1) is replaced by a condition on the trajectory of the target system (1.11). Hence, to establish the claim we verify that the conditions (A2)–(A4) of Theorem 2.1 are satisfied. First, it is easy to see that a solution of equations (2.3) is given by the mapping

$$\pi(\xi, t) = \begin{bmatrix} \xi_1 \\ \xi_2 \\ \xi_1 + K^{-1} w(\xi, t) \\ \pi_4(\xi, t) \end{bmatrix},$$

where $\pi_4(\xi, t) = \frac{\partial \pi_3}{\partial \xi} \dot{\xi}$, with $\dot{\xi}$ as defined in (1.11). This solution follows immediately considering (1.10) and fixing $\pi_1(\xi) = \xi_1$, as required by the control objective. An implicit definition of the manifold[11] $\phi(x, t) = 0$ is obtained selecting

$$\phi(x,t) = \begin{bmatrix} x_3 - x_1 - K^{-1} w(x,t) \\ x_4 - x_2 - K^{-1} \left(\frac{\partial w}{\partial x_1} x_2 - \frac{\partial w}{\partial x_2} D^{-1}(x_1) \left(C(x_1, x_2) x_2 \right. \right. \\ \left. \left. + g(x_1) + K(x_1 - x_3) \right) - \frac{\partial w}{\partial t} \right) \end{bmatrix}.$$

It is important to underscore that, while $\phi_1(x,t)$ is obtained from the obvious choice

$$\phi_1(x,t) = x_3 - \pi_3(\xi, t)|_{\xi_1 = x_1, \xi_2 = x_2},$$

the term $\phi_2(x,t)$ is not defined likewise. However, the set identity (2.4) is satisfied for the definition above as well. To verify this observe that

---
[10] A possible selection is the Slotine and Li controller [200, 163], which is given by

$$w(\xi,t) = D(\xi_1)(\ddot{\xi}_1^* - \Lambda \dot{\tilde{\xi}}_1) + C(\xi_1, \xi_2)(\dot{\xi}_1^* - \Lambda \tilde{\xi}_1) - K_p(\dot{\tilde{\xi}} + \Lambda \tilde{\xi}_1) + g(\xi_1),$$

where $K_p = K_p^\top > 0$, $\Lambda = \Lambda^\top > 0$ and $\tilde{\xi}_1 = \xi_1 - \xi_1^*$.

[11] Note that in this case the target dynamics and the equations defining the invariant manifold depend explicitly on t.

$$x_4 - \pi_4(\xi,t)|_{\xi_1=x_1,\xi_2=x_2} = \phi_2(x,t)$$
$$+ K^{-1}\frac{\partial w}{\partial x_2}D^{-1}(x_1)K\left(x_3 - x_1 - K^{-1}w(x,t)\right),$$

but the last right-hand term in brackets is precisely $\phi_1(x,t)$. The interest in defining $\phi_2(x,t)$ in this way is that $\dot{\phi}_1 = \phi_2$, simplifying the task of stabilising the zero equilibrium of the off-the-manifold dynamics, which are given by

$$\dot{z}_1 = z_2,$$
$$\dot{z}_2 = m(x,t) + J^{-1}v(x,z,t),$$

where $m(x,t)$ can be computed via differentiation of $\phi_2(x,t)$. It is then a trivial task to select a control law $v(x,z,t)$ that asymptotically stabilises the zero equilibrium of the z system, an obvious simple selection being

$$v(x,z,t) = -J\left(m(x,t) + K_1 z_1 + K_2 z_2\right),$$

with K_1, K_2 arbitrary positive-definite matrices. To complete the proof it is necessary to show that all trajectories of the closed-loop system with state (x, z_1, z_2) are bounded. To this end, it suffices to rewrite the system in the co-ordinates $(x_1, x_2, \phi_1, \phi_2, z_1, z_2)$ and use arguments similar to those in the proof of Proposition 2.3. □

It must be stressed that the target dynamics (1.11) are *not* the dynamics of the rigid model obtained from a singular perturbation reduction of the full model (1.10) with small parameters $1/k_i$. In the latter model the inertia matrix is $D+J$ and not simply D as in the present case[12]. The motivation to choose these target dynamics is clear noting that, if we take the rigid model resulting from a singular perturbation reduction, the solution of equations (2.3) leads to $\pi_3(\xi) = \xi_1 + K^{-1}(w - \dot{\xi}_2)$, complicating the subsequent analysis. ∎

Note 2.4. In [207] the composite control approach is used to derive approximate, feedback linearising, asymptotically stabilising controllers for the full-inertia model. In the case of the block-diagonal inertia matrix considered here the slow manifold equations can be solved exactly and the stabilisation is global. ◁

Example 2.6 (Cart and pendulum system). Consider the problem of upward stabilisation of the underactuated cart and pendulum system (1.12) with target dynamics the fully actuated pendulum (1.13). The following statement describes a procedure to generate I&I stabilising feedback laws without the need to solve the PDE (2.3).

Proposition 2.5. *Let $\pi_3(\cdot) : \mathcal{S}^1 \times \mathbb{R} \to \mathbb{R}$ be such that $\frac{\partial \pi_3}{\partial \xi_2}$ is a function of ξ_1 only and let*

$$V'(\xi_1) = -\frac{\sin(\xi_1)}{\Delta(\xi_1)}, \qquad R(\xi_1,\xi_2) = \frac{\cos(\xi_1)}{\Delta(\xi_1)}\frac{\partial \pi_3}{\partial \xi_1}, \qquad (2.17)$$

---
[12] See the discussion in [207].

with
$$\Delta(\xi_1) = 1 + \cos(\xi_1)\frac{\partial \pi_3}{\partial \xi_2}. \tag{2.18}$$

Assume that the functions $\Delta(\xi_1)$ and $R(\xi_1, \xi_2)$ are such that $\Delta(0) < 0$ and $R(0,0) > 0$. Then the cart–pendulum system (1.12) in closed loop with the I&I controller

$$v(x, \phi(x)) = \frac{1}{\Delta(x_1)} \left(-\gamma\phi(x) + \frac{\partial \pi_3}{\partial x_1}x_2 + \frac{\partial \pi_3}{\partial x_2}\sin(x_1) \right), \tag{2.19}$$

with $\gamma > 0$ and $\phi(x) = x_3 - \pi_3(x_1, x_2)$, has a locally asymptotically stable equilibrium at zero.

Proof. We proceed to verify the hypotheses (A1)–(A4) of Theorem 2.1. By assumption the target system (1.13), namely

$$\dot{\xi}_1 = \xi_2,$$
$$\dot{\xi}_2 = -V'(\xi_1) - R(\xi_1, \xi_2)\xi_2,$$

has a locally asymptotically stable equilibrium at the origin, therefore (A1) is satisfied. Given the control objectives and our choice of target dynamics, a natural selection of the mapping $\pi(\cdot)$ is

$$\pi(\xi) = \begin{bmatrix} \xi_1 \\ \xi_2 \\ \pi_3(\xi_1, \xi_2) \end{bmatrix}, \tag{2.20}$$

where $\pi_3(\cdot)$ is a function to be defined. With this choice of $\pi(\cdot)$ and of the target dynamics the PDE (2.3) reduces to

$$\sin(\xi_1) - \cos(\xi_1)c(\pi(\xi)) = -V'(\xi_1) - R(\xi_1, \xi_2)\xi_2, \tag{2.21}$$
$$c(\pi(\xi)) = \frac{\partial \pi_3}{\partial \xi_1}\xi_2 - \frac{\partial \pi_3}{\partial \xi_2}\left(V'(\xi_1) + R(\xi_1, \xi_2)\xi_2\right). \tag{2.22}$$

Replacing $c(\cdot)$ from (2.22) in (2.21) yields the PDE

$$\left(\cos(\xi_1)\frac{\partial \pi_3}{\partial \xi_1} - R(\xi_1, \xi_2)\Delta(\xi_1)\right)\xi_2 = \sin(\xi_1) + \Delta(\xi_1)V'(\xi_1), \tag{2.23}$$

where $\Delta(\cdot)$ is defined in (2.18). Clearly, (2.23) holds by (2.17), hence (A2) holds.

The implicit manifold condition (A3) is verified noting that the manifold $\mathcal{M}$ can be implicitly described by $\mathcal{M} = \{x \in \mathbb{R}^3 \mid \phi(x) = 0\}$, with $\phi(x) = x_3 - \pi_3(x_1, x_2)$. Finally, we prove condition (A4). The off-the-manifold dynamics are

$$\dot{z} = -\frac{\partial \pi_3}{\partial x_1}x_2 - \frac{\partial \pi_3}{\partial x_2}\sin(x_1) + \Delta(x_1)v(x, z).$$

2 I&I Stabilisation

Selecting

$$v(x,z) = \frac{1}{\Delta(x_1)}\left(-\gamma z + \frac{\partial \pi_3}{\partial x_1}x_2 + \frac{\partial \pi_3}{\partial x_2}\sin(x_1)\right),$$

which is well-defined locally around zero by assumption, yields $\dot{z} = -\gamma z$. Finally, replacing $z = \phi(x)$ in the foregoing equation yields the controller (2.19). To complete the proof we show that, for a set of initial conditions containing a neighbourhood of the origin, the trajectories of the system (2.5), (2.6), namely

$$\begin{aligned}\dot{z} &= -\gamma z,\\ \dot{x}_1 &= x_2,\\ \dot{x}_2 &= \sin(x_1) - \cos(x_1)v(x,z),\\ \dot{x}_3 &= v(x,z),\end{aligned} \quad (2.24)$$

with $v(\cdot)$ given by (2.19), are bounded. Towards this end, note that from (2.18) and (2.17) there exist $\epsilon_1 > 0$ and $\epsilon_2 > 0$ such that, locally around zero,

$$\Delta(x_1) \leq -\epsilon_1, \qquad R(x_1, x_2) \geq \epsilon_2. \quad (2.25)$$

Note now that, using the functions $V'(\cdot)$ and $R(\cdot)$ defined in (2.17), the first three equations of (2.24) can be rewritten in the form

$$\begin{aligned}\dot{z} &= -\gamma z,\\ \dot{x}_1 &= x_2,\\ \dot{x}_2 &= -V'(x_1) - R(x_1, x_2)x_2 + \frac{\gamma \cos(x_1)}{\Delta(x_1)}z.\end{aligned} \quad (2.26)$$

Consider the function $H(x_1, x_2) = \frac{1}{2}x_2^2 + V(x_1) + \frac{\gamma}{2\epsilon_2 \epsilon_1^2}z^2$ and note that, from (2.17), $V'(0) = 0$ and $V''(0) = -\frac{1}{\Delta(0)} > 0$, hence the function $H(x_1, x_2)$ is positive-definite with a local minimum at zero. Differentiating along the trajectories of (2.26) yields

$$\begin{aligned}\dot{H} &= -R(x_1, x_2)x_2^2 + \frac{\gamma \cos(x_1)}{\Delta(x_1)}x_2 z - \frac{\gamma^2}{\epsilon_2 \epsilon_1^2}z^2\\ &\leq -\epsilon_2 x_2^2 + \frac{\epsilon_2}{2}x_2^2 + \frac{\gamma^2}{2\epsilon_2 \epsilon_1^2}z^2 - \frac{\gamma^2}{\epsilon_2 \epsilon_1^2}z^2\\ &\leq -\frac{\epsilon_2}{2}x_2^2 - \frac{\gamma^2}{2\epsilon_2 \epsilon_1^2}z^2,\end{aligned}$$

where we have used (2.25) and Young's inequality. It follows that there exists a neighbourhood of the origin such that all trajectories $(x_1(t), x_2(t), z(t))$ starting in this set remain bounded and asymptotically converge to $(0, 0, 0)$. Finally, boundedness of x_3 follows from the fact that $x_3(t) = z(t) + \pi_3(x_1(t), x_2(t)) - z(0) + x_3(0)$ and the right-hand side is bounded. □

To complete the design we propose functions $\pi_3(\cdot)$ that verify the assumptions $\Delta(0) < 0$ and $R(0,0) > 0$ which, in view of (2.17) and (2.18), impose that $\frac{\partial \pi_3}{\partial x_1}(0) < 0$ and $1 + \frac{\partial \pi_3}{\partial x_2}(0) < 0$. Two simple candidates are

$$\pi_3(x_1, x_2) = -k_1 x_1 - k_2 x_2$$

and

$$\pi_3(x_1, x_2) = -k_1 x_1 - \frac{k_2}{\cos(x_1)} x_2,$$

where $k_1 > 0$ and $k_2 > 1$ are tuning gains.

Alternatively, to enforce a particular behaviour on the target dynamics, we can also proceed dually, that is, fix the desired potential energy $V(\cdot)$ and then work backwards to compute $\pi_3(\cdot)$, $\Delta(\cdot)$ and $R(\cdot)$. A particularly interesting choice is $V(x_1) = \frac{k_1}{2} \tan^2(x_1)$, with $k_1 > 0$, which has a unique minimum at zero and is radially unbounded on the interval $(-\frac{\pi}{2}, \frac{\pi}{2})$. Replacing in (2.17) yields $\Delta(x_1) = -\frac{1}{k_1} \cos^3(x_1)$, which satisfies the assumption $\Delta(0) < 0$. From (2.18), and after some simple calculations, we obtain

$$\pi_3(x_1, x_2) = -\left(\frac{1}{\cos(x_1)} + \frac{1}{k_1} \cos^2(x_1)\right) x_2 + \psi(x_1),$$

where $\psi(\cdot)$ is a free function. As it can be easily shown, $R(0,0) = -k_1 \psi'(0)$, hence $\psi(\cdot)$ must be such that $\psi'(0) < 0$ to ensure the damping is positive, e.g., $\psi(x_1) = -k_2 x_1$, with $k_2 > 0$.

Simulations have been carried out with $\pi_3(x_1, x_2) = -k_1 x_1 - \frac{k_2}{\cos(x_1)} x_2$, yielding[13]

$$u = \frac{1}{k_2 - 1} \left[\gamma \left(x_3 + k_1 x_1 + \frac{k_2}{\cos(x_1)} x_2 \right) + k_1 x_2 + k_2 \tan(x_1) \left(\frac{x_2^2}{\cos(x_1)} + 1 \right) \right].$$

Note, however, that this controller is not globally defined because $\pi_3(\cdot)$ has a singularity at $\pm \frac{\pi}{2}$. Simulation results are shown in Figures 2.1 and 2.2 for the controller gains $k_1 = 3$, $k_2 = 4$ and for two different values of γ. The initial conditions are set to $x(0) = (\frac{\pi}{2} - 0.1, 0, 0)$, which correspond to zero velocities and with the pendulum close to horizontal. The results clearly show the desired closed-loop behaviour: first, convergence towards the manifold $z = 0$ at a speed determined by γ, and then, once close to the manifold, where the cart–pendulum system behaves like a simple pendulum, convergence towards the equilibrium. Note from Figure 2.1 that increasing γ, hence increasing the speed of convergence to the manifold, does not necessarily lead to a faster overall transient response. This is due to the fact that, even though the closed-loop system (2.26) is the cascade connection of an exponentially stable and an asymptotically stable system, the peaking phenomenon[14] appears when we increase the rate of convergence of the former. ∎

---

[13]The simplicity of this control law should be contrasted with other schemes proposed in the literature, e.g., [31, 30, 2].

[14]See [115].

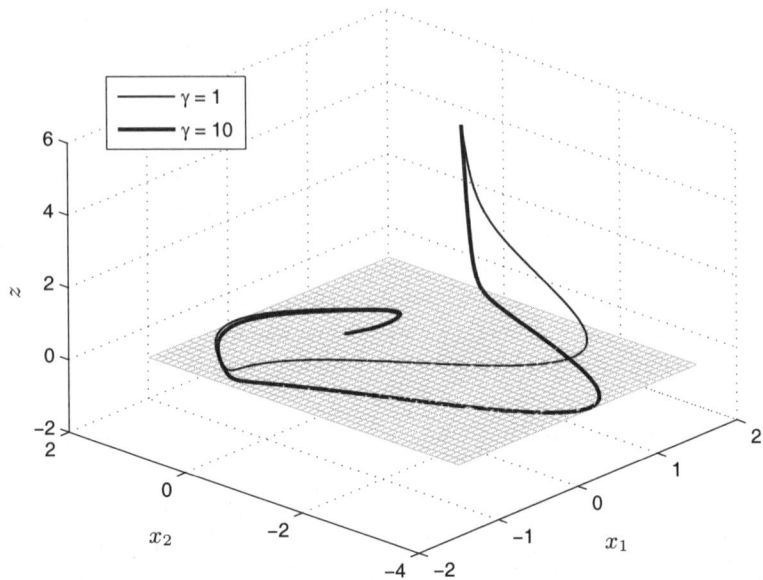

Fig. 2.1. Trajectories of the cart–pendulum system—the shaded area corresponds to the invariant manifold $z = 0$.

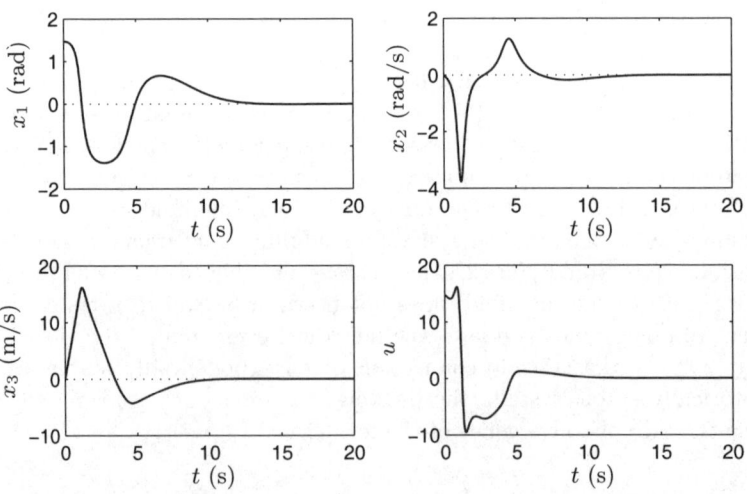

Fig. 2.2. Time histories of the states and the control input of the cart–pendulum system for $\gamma = 1$, $k_1 = 3$ and $k_2 = 4$.

3
I&I Adaptive Control: Tools and Examples

In this chapter it is shown how the I&I methodology can be used to develop a novel framework for adaptive stabilisation of nonlinear systems with parametric uncertainties that is not certainty equivalent and does not rely on the assumption of linear parameterisation. A key step is to add to the classical certainty-equivalent control law a new term that, together with the parameter update law, is designed to achieve adaptive stabilisation. The role of this new term is to *shape the manifold* into which the adaptive system is immersed.

We first address the problem of (adaptive) stabilisation via state feedback for general nonlinear systems. After giving the problem formulation, to illustrate the main features of the method, we study a simple nonlinear system with a single unknown parameter. To put the adaptive I&I in perspective we briefly review the existing techniques to solve adaptive control problems for nonlinear systems. We then provide two general theorems. The first one applies to linearly parameterised plants with nonlinearly parameterised controllers, while the second one applies to the case in which the controller depends linearly on the parameters and the plant satisfies a matching assumption[1].

We close the chapter with the solution of an adaptive control problem for a nonlinearly parameterised system, which describes a prototype robotic vision problem.

3.1 Introduction

Consider again the system (2.1), namely

$$\dot{x} = f(x) + g(x)u, \tag{3.1}$$

with an equilibrium x^* to be stabilised, where the functions $f(\cdot)$ and $g(\cdot)$ depend on an unknown parameter vector $\theta \in \mathbb{R}^q$, and the problem of finding,

---

[1] If both the plant and the controller are linearly parameterised, it is possible to combine the procedures described in Sections 3.4 and 3.5.

whenever possible, an adaptive state feedback control law of the form

$$\dot{\hat{\theta}} = w(x, \hat{\theta}),$$
$$u = v(x, \hat{\theta}),$$
(3.2)

such that all trajectories of the closed-loop system (3.1), (3.2) are bounded and $\lim_{t\to\infty} x(t) = x^*$.

To this end, it is natural to assume that a *full-information* control law (that depends on θ) is known, *i.e.*, that the following stabilisability condition holds.

Assumption 3.1 (Stabilisability). There exists a function $v(x, \theta)$, where $\theta \in \mathbb{R}^q$, such that the system

$$\dot{x} = f^*(x) \triangleq f(x) + g(x)v(x, \theta)$$
(3.3)

has a globally asymptotically stable equilibrium at $x = x^*$.

The I&I adaptive control problem is then formulated as follows.

Definition 3.1. *The system (3.1) with Assumption 3.1 is said to be adaptively I&I stabilizable if there exist $\beta(\cdot)$ and $w(\cdot)$ such that all trajectories of the extended system*

$$\dot{x} = f(x) + g(x)v(x, \hat{\theta} + \beta(x)),$$
$$\dot{\hat{\theta}} = w(x, \hat{\theta})$$
(3.4)

are bounded and satisfy

$$\lim_{t \to \infty} [g(x(t))v(x(t), \hat{\theta}(t) + \beta(x(t))) - g(x(t))v(x(t), \theta)] = 0.$$
(3.5)

Note that for all trajectories staying on the manifold

$$\mathcal{M} = \{(x, \hat{\theta}) \in \mathbb{R}^n \times \mathbb{R}^q \mid \hat{\theta} - \theta + \beta(x) = 0\}$$

condition (3.5) holds. Moreover, by Definition 3.1 and Assumption 3.1, adaptive I&I stabilisability implies that

$$\lim_{t \to \infty} x(t) = x^*.$$
(3.6)

From the first equation in (3.4) we see that the I&I approach departs from the certainty equivalence philosophy. In other words, we do not apply directly the estimate coming from the update law in the controller.

It is important to recall that, in general, $f(\cdot)$, and possibly $g(\cdot)$, depends on the unknown θ, therefore, it is not reasonable to require that $\hat{\theta}$ converges to any particular equilibrium, but merely that it remains bounded. However, in many cases it is possible to also establish global stability of the equilibrium $(x, \hat{\theta}) = (x^*, \theta)$, for instance, see the example of Section 3.2 and the results in Chapter 4.

Note 3.1. This formulation of the I&I adaptive stabilization problem is in contrast with the more restrictive formulation adopted in [21], where it is required that the extended system (3.4) is I&I stabilisable with target dynamics $\dot{\xi} = f^*(\xi)$. It is interesting to note, however, that in [21] it is shown that the conditions (A1)–(A3) of Theorem 2.1 hold under quite general assumptions and the off-the-manifold coordinate used in the sequel is precisely the one derived adopting this perspective. We refer the interested reader to [21] for details. ◁

3.2 An Introductory Example

Consider the first-order nonlinear system (1.18), which for ease of reference we rewrite, namely
$$\dot{x} = \theta x^2 + u. \tag{3.7}$$
In Example 1.6 a classical adaptive controller has been derived. In the following we show that, adopting an I&I approach, we avoid the (non-robust) cancellation and provide a means of shaping the dynamic response of the estimation error.

Consider the augmented system
$$\begin{aligned} \dot{x} &= \theta x^2 + u, \\ \dot{\hat{\theta}} &= w, \end{aligned} \tag{3.8}$$
where w is the update law to be determined, and define in the extended space $(x, \hat{\theta})$ the one-dimensional manifold
$$\mathcal{M} = \{(x, \hat{\theta}) \in \mathbb{R}^2 \,|\, \hat{\theta} - \theta + \beta(x) = 0\},$$
where $\beta(\cdot)$ is a continuous function yet to be specified. The motivation for this definition is that the dynamics of the system (3.8) restricted to the manifold $\mathcal{M}$ (provided it is invariant) are described by the equation
$$\dot{x} = \left(\hat{\theta} + \beta(x)\right) x^2 + u,$$
hence they are completely known and the equilibrium $x = 0$ is asymptotically stabilised by the control law
$$u = -kx - \left(\hat{\theta} + \beta(x)\right) x^2, \tag{3.9}$$
with $k > 0$. For the above design to be feasible, the first step in the I&I approach consists in finding an update law w that renders the manifold $\mathcal{M}$ invariant. To this end, consider the dynamics of the off-the-manifold co-ordinate
$$z \triangleq \hat{\theta} - \theta + \beta(x), \tag{3.10}$$
which are given by the equation

$$\dot{z} = w + \frac{\partial \beta}{\partial x}\left((\hat{\theta} + \beta(x) - z)x^2 + u\right),$$

and note that the update law

$$w = -\frac{\partial \beta}{\partial x}\left((\hat{\theta} + \beta(x))x^2 + u\right) = \frac{\partial \beta}{\partial x}kx \qquad (3.11)$$

is such that the manifold $\mathcal{M}$ is indeed invariant and the off-the-manifold dynamics are described by the equation

$$\dot{z} = -\frac{\partial \beta}{\partial x}x^2 z.$$

Selecting the function $\beta(\cdot)$ as

$$\beta(x) = \gamma\frac{x^3}{3}, \qquad (3.12)$$

with $\gamma > 0$, yields the system

$$\dot{z} = -\gamma x^4 z. \qquad (3.13)$$

Consider now the Lyapunov function $V = \frac{1}{2}z^2$, whose time-derivative along the trajectories of (3.13) satisfies $\dot{V} = -\gamma x^4 z^2 \leq 0$, hence the system (3.13) has a globally stable equilibrium at zero. By integrating $\dot{V}$ it follows that $V(\infty) - V(0) = -\gamma \int_0^\infty |x^2(t)z(t)|^2 dt$, hence $x^2(t)z(t) \in \mathcal{L}_2$, for all $x(t)$. Using (3.12) and the definition of z, the closed-loop system (3.8), (3.9), (3.11) can be written in the (z, x) co-ordinates as

$$\begin{aligned}\dot{z} &= -\gamma x^4 z, \\ \dot{x} &= -kx - x^2 z,\end{aligned} \qquad (3.14)$$

and this has a globally stable equilibrium at the origin and x converges to zero. Moreover, as is shown in Section 4.2.2, the extra term $\beta(x)x^2$ in the control law (3.9) renders the closed-loop system (3.8), (3.9), (3.11) *input-to-state stable* (ISS) with respect to $\hat{\theta} - \theta$.

A comparative graph of the responses of the I&I adaptive controller and the classical controller (see Example 1.6), for $\theta = 2$, $k = 1$ and $\gamma = 1$, is shown in Figure 3.1. A phase-plane diagram of the controlled system (3.8), which shows the convergence of the trajectories to $x = 0$, is given in Figure 3.2.

It must be noted that the selection (3.12) is not unique. For instance, an alternative choice is

$$\beta(x) = \begin{cases} \frac{\gamma}{3}(x - \epsilon)^3, & x > \epsilon, \\ 0, & |x| \leq \epsilon, \\ \frac{\gamma}{3}(x + \epsilon)^3, & x < -\epsilon, \end{cases}$$

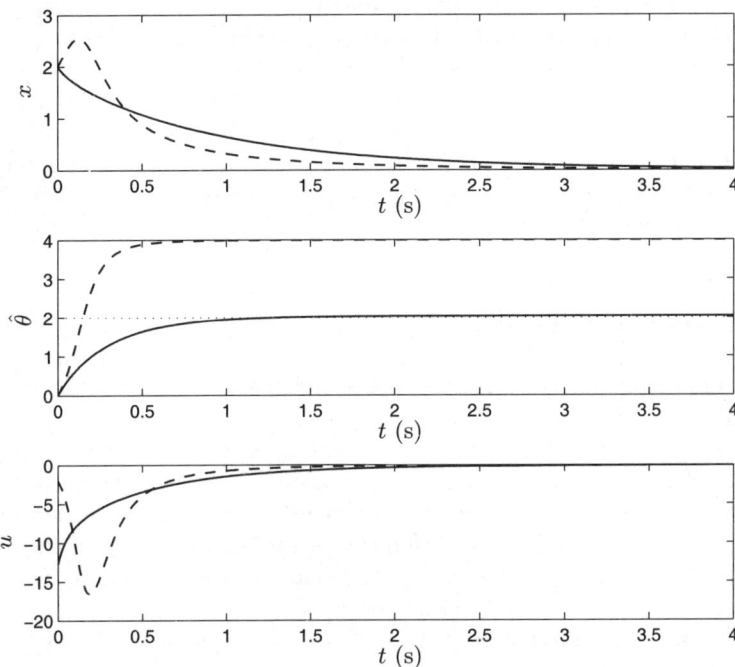

Fig. 3.1. Time histories of the states and the control input for the controlled system (3.8). Dashed line: Classical controller. Solid line: Proposed controller.

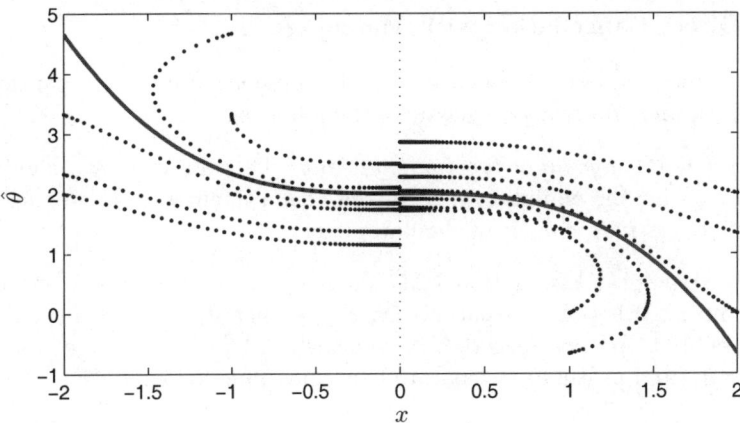

Fig. 3.2. Phase-plane diagram of the controlled system (3.8)—the solid line corresponds to the invariant manifold (3.10).

which introduces a dead-zone, *i.e.*, the adaptation is switched off when $-\epsilon \leq x \leq \epsilon$. This is useful to ensure boundedness of $\hat{\theta}$ in the presence of noise.

The closed-loop system (3.14) can be regarded as a cascaded interconnection between two stable systems whose "gains" can be tuned via the arbitrary constants k and γ. This *modularity* property is one of the prominent features of the I&I approach and is exploited in Chapter 4 to obtain adaptive state feedback controllers that are easier to tune than their classical counterparts. It also plays an essential role in achieving the (robust) output feedback stabilisation results of Chapter 6.

3.3 Revisiting Classical Adaptive Control

To put in perspective the I&I approach—underscoring its new features—it is convenient to recall first the "classical" procedures to address the problem of adaptive control for nonlinear systems. The main difference with respect to the case of linear plants is that the system in closed loop with the known-parameters controller still depends, in general, on the unknown parameters. Consequently, the Lyapunov function that establishes stability of the desired equilibrium for this system is also a function of the parameters, hence it is unknown.

There are four different ways to try to bypass this difficulty[2], which can be classified as follows:

1. direct cancellation with matching;
2. direct cancellation with extended matching;
3. direct domination;
4. indirect approach.

3.3.1 Direct Cancellation with Matching

The first and simplest way to deal with the presence of unknown parameters in the Lyapunov function is to assume the following.

Assumption 3.2 (Lyapunov function matching). There exists a Lyapunov function $V(x, \theta)$ for the equilibrium x^* of the ideal system $\dot{x} = f^*(x)$ such that $\frac{\partial V}{\partial x}(x, \theta)g(x)$ is *independent* of the parameters[3].

If the structural Assumption 3.2 holds and if, in addition, the function $v(\cdot)$ of Assumption 3.1 is linearly parameterised, *i.e.*, $v(x, \theta) = v_0(x) + v_1(x)\theta$, then it is possible to propose a certainty-equivalent adaptive control of the form $u = v(x, \hat{\theta})$, and postulate a separate Lyapunov function candidate

---

[2] A fifth alternative is the supervisory control of Morse [149, 150], which is however formulated under different assumptions.

[3] The most notable example where this assumption is satisfied is in the Slotine and Li controller for robot manipulators, see, *e.g.*, [163].

$$W(x,\tilde{\theta}) = V(x,\theta) + \frac{1}{2}\tilde{\theta}^\top \Gamma^{-1}\tilde{\theta},$$

where $\Gamma = \Gamma^\top > 0$ and $\tilde{\theta} = \hat{\theta} - \theta$ is the parameter estimation error. Choosing the estimator that cancels the $\tilde{\theta}$-dependent term in $\dot{W}$ yields the error dynamics

$$\begin{aligned}\dot{x} &= f^*(x) + g(x)v_1(x)\tilde{\theta},\\ \dot{\tilde{\theta}} &= -\Gamma v_1^\top(x)\left(\frac{\partial V}{\partial x}g(x)\right)^\top\end{aligned} \quad (3.15)$$

and $\dot{W} = \frac{\partial V}{\partial x}f^*(x) \leq 0$. If $\frac{\partial V}{\partial x}f^*(x) < 0$, for all $x \neq x^*$, then LaSalle's invariance principle yields $\lim_{t\to\infty} x(t) = x^*$. Otherwise, it is necessary to add the following, rather restrictive, detectability assumption[4].

Assumption 3.3 (Detectability). The trajectories of the system (3.15) are such that $\frac{\partial V}{\partial x}f^*(x(t)) \equiv 0$ implies $\lim_{t\to\infty} x(t) = x^*$.

It is important to underscore that Assumption 3.3 may hold in the known-parameters case, *i.e.*, when $\tilde{\theta} = 0$, but be violated when considering the system with state $(x, \hat{\theta})$. (This is illustrated in Example 3.3.)

3.3.2 Direct Cancellation with Extended Matching

When Assumption 3.2 is not satisfied, the Lyapunov function is expressed with the parameter estimates, instead of the actual unknown parameters, but this brings along a new term to the derivative, namely $\frac{\partial V}{\partial \hat{\theta}}\dot{\hat{\theta}}$, that has to be countered.

The second approach assumes that the vector field $f(x)$ can be written as

$$f(x) = f_0(x) + f_1(x)\theta, \quad (3.16)$$

and that the effect of the parameters can be *rejected* when considered as a disturbance with known derivative. The key assumption introduced in this framework is the following condition[5].

Assumption 3.4 (Extended matching). There exists a function $\psi(x, \hat{\theta})$ such that

$$\frac{\partial V}{\partial x}g(x)\psi(x,\hat{\theta}) + \frac{\partial V}{\partial \hat{\theta}}\Gamma\left(\frac{\partial V}{\partial x}f_1(x)\right)^\top = 0.$$

Note that such a function is unlikely to exist since $\frac{\partial V}{\partial x}g(x)$ is, in general, zero at some points. However, Assumption 3.4 holds for systems in *lower triangular* forms, and this is at the basis of adaptive backstepping algorithms[6].

---
[4] A procedure that combines direct and indirect adaptive control to overcome, in some cases, this obstacle has been proposed in [170].
[5] See [123, Section 4.1.1].
[6] See [123, 137].

3.3.3 Direct Domination

In the direct domination approach, robustness, instead of disturbance rejection, is utilised, thus matching conditions are replaced by growth conditions—either on the Lyapunov function or on the system equation[7]. Unfortunately, it is difficult to characterise the class of systems for which the required Lyapunov function growth conditions hold[8].

3.3.4 Indirect Approach

Finally, the adaptive stabilization problem can also be approached, as done in indirect adaptive control, adopting an identification perspective. Namely, if $f(x)$ is as in equation (3.16), it is easy to show[9] that using the filtered signals

$$\dot{r} = -\lambda(x+r) - (f_0(x) + g(x)u),$$
$$\dot{\Phi} = -\lambda\Phi(x) + f_1(x),$$

where $\lambda > 0$ is a design parameter, the identification error $e_I = \Phi(x)\hat{\theta} - (x+r)$ exponentially converges to $\Phi(x)\tilde{\theta}$, at a rate determined by λ. This suggests the update law $\dot{\hat{\theta}} = -\Phi(x)^\top e_I$, which ensures that the parameter estimation error is nonincreasing.

Since the control is, in general, nonlinearly parameterised, the parameter estimates may cross to regions that contain control singularities that may be difficult to avoid. Further, as the estimation of the parameters is decoupled from the control, the only way to guarantee that the "quality of the control" improves with time is by ensuring that the parameter estimation error actually decreases. This, in its turn, imposes an excitation restriction on the regressor matrix $\Phi(x(t))$ that is hard to verify *a priori*.

3.3.5 Discussion

For the purpose of comparison with the I&I approach the following remarks are in order.

1. The analysis of the standard schemes invariably relies on the construction of error equations (similar to (3.15)), where the x-dynamics are decomposed into a stable portion and a perturbation due to the parameter mismatch. In the I&I formulation we also write the x-dynamics in an error equation form but with the ideal system perturbed by the off-the-manifold co-ordinates z.

---

[7] See [175] and [192].

[8] It should be pointed out, however, that systems in so-called strict feedback form with polynomial nonlinearities satisfy the Lyapunov growth condition of [175]. See also [38] for an interesting application to a nonlinearly parameterised nonlinear system.

[9] See, *e.g.*, [170].

2. The first and second procedures are based on the cancellation of a perturbation term in the Lyapunov function derivative, which is a fragile operation that generates a manifold of equilibria and is at the core of the robustness problems of classical adaptive control[10]. In the third approach this term is not cancelled but dominated. In the I&I formulation we do not try to cancel the perturbation term coming from z but—rendering the manifold attractive—only make it sufficiently small. In this respect I&I resembles the third approach as well as indirect adaptive control.
3. Compared to the indirect method, I&I provides additional flexibility to reparameterise the plant so as to avoid controller singularities. This feature is instrumental in Section 4.4.2 to reduce the prior knowledge on the high frequency gain for linear multivariable plants.
4. The assumption that the *controller* is linearly parameterised is critical in some of the classical procedures. (In some cases it may be replaced by a convexity condition, but this is very hard to verify for more than one uncertain parameter.) In adaptive I&I we do not, *a priori*, require this assumption.

Example 3.1. We now illustrate, with an example, some of the difficulties encountered by standard adaptive techniques, that can be overcome with adaptive I&I.

Consider the (normalised) averaged model of a thyristor-controlled series capacitor (TCSC) system[11]

$$\dot{x}_1 = x_2 u,$$
$$\dot{x}_2 = -x_1 u - \theta_1 x_2 - \theta_2, \tag{3.17}$$

where $x_1 \in \mathbb{R}$, $x_2 \in \mathbb{R}$ are the dynamic phasors of the capacitor voltage, $u > 0$ is the control signal which is directly related to the thyristor firing angle, and $\theta = [\theta_1, \theta_2]^\top$ are unknown positive parameters representing the nominal action of the control and one component of the phasor of the line current, respectively. The control objective is to drive the state to the equilibrium $x^* = [-1, 0]^\top$ with a positive control action.

Note that the known-parameters controller $u = \theta_2$ achieves the desired objective, as seen from the Lyapunov function $V(x) = \frac{1}{2}(x_1 + 1)^2 + \frac{1}{2}x_2^2$, whose time-derivative along the trajectories of (3.17) is $\dot{V} = -\theta_1 x_2^2$. To make this controller adaptive we first try the direct approach and, as Assumption 3.2 is verified, propose $u = \hat{\theta}_2$, which, following the calculations above with the Lyapunov function $V(x) + \frac{1}{2}(\tilde{\theta} - \theta)$, suggests the estimator $\dot{\hat{\theta}} = -x_2$. It is easy to show that the detectability Assumption 3.3 is not satisfied and the closed-loop system has a manifold of equilibria, described by $x_1(\tilde{\theta}_2 + \theta_2) + \theta_2 = 0$, which contains the desired equilibrium, hence it is not asymptotically stable.

---
[10] See [165, Section 2.2.2] and [131, Chapter 5] for additional details.
[11] This device is used in flexible AC transmission systems (FACTS) to regulate the power flow in a distribution line, see Section 8.1 for a detailed design.

The second approach does not apply either because Assumption 3.4 is not satisfied.

The indirect approach is also hampered by the detectability obstacle. Indeed, reparameterising the second equation in (3.17) with the regressor $\Phi(x) = -[x_2, 1]^\top$ and implementing the estimator

$$\begin{aligned} \dot{\hat{x}}_2 &= -x_1 u + \hat{\theta}^\top \Phi(x) - (\hat{x}_2 - x_2), \\ \dot{\hat{\theta}} &= -\Phi(x)(\hat{x}_2 - x_2), \end{aligned} \qquad (3.18)$$

ensures that the prediction error $\hat{x}_2 - x_2$ converges to zero, but, again, the overall system is not detectable with respect to this output. Furthermore, as $\Phi(x)$ contains a constant term, it cannot be persistently exciting and parameter convergence is not achieved. (The same scenario appears if we use filtered signals in the estimator as suggested in Section 3.3.4.) Finally, a combined direct–indirect scheme[12] results in

$$\dot{\hat{\theta}} = -\Phi(x)(\hat{x}_2 - x_2) - \begin{bmatrix} 0 \\ x_2 \end{bmatrix},$$

and the derivative of the Lyapunov function $V(x) + \frac{1}{2}(\hat{x}_2 - x_2)^2 + \frac{1}{2}|\tilde{\theta}|^2$ is $-(\hat{x}_2 - x_2)^2 - \theta_1 x_2^2$, but this signal still does not satisfy Assumption 3.3.

In summary, to the best of the authors' knowledge, none of the standard methods provides a solution to the considered adaptive control problem for system (3.17). ∎

3.4 Linearly Parameterised Plant

Consider the stabilisation problem for systems of the form (3.1) under the assumption that the plant is linearly parameterised, *i.e.*, it can be described by equations of the form

$$\dot{x} = f_0(x) + f_1(x)\theta + g(x)u, \qquad (3.19)$$

with state $x \in \mathbb{R}^n$ and input $u \in \mathbb{R}^m$, where $\theta \in \mathbb{R}^q$ is an unknown constant vector. Our main tool is a generalisation of the approach of Section 2.1, which allows the design of the control law to be *decoupled* from the design of the adaptation law (in the spirit of indirect adaptive control), thus leading to more flexible control schemes[13].

The first step is to assume that a stabilising *full-information* control law (*i.e.*, a control law that depends on the unknown parameters) is available.

---

[12] See [170].

[13] For linear plants, although the construction is totally different, the resulting error equations are similar to the ones obtained with indirect adaptive control [77], but obviating the need for the linear filters.

3.4 Linearly Parameterised Plant

Then the problem is reduced to finding an adaptation law such that the closed-loop system is immersed into the system that would result from applying the full-information controller. This two-step approach is outlined in the following theorem.

Theorem 3.1. *Consider the system (3.19) with an equilibrium point x^* to be stabilised and assume that the following hold.*

(A1) There exists a full-information control law $u = v(x, \theta)$ such that the closed-loop system

$$\dot{x} = f^*(x) \triangleq f_0(x) + f_1(x)\theta + g(x)v(x, \theta)$$

has a globally asymptotically stable equilibrium at x^*.

(A2) There exists a mapping $\beta : \mathbb{R}^n \to \mathbb{R}^q$ such that all trajectories of the system

$$\dot{z} = -\left[\frac{\partial \beta}{\partial x} f_1(x)\right] z$$
$$\dot{x} = f^*(x) + g(x)\left(v(x, \theta + z) - v(x, \theta)\right)$$
(3.20)

are bounded and satisfy

$$\lim_{t \to \infty} \left[g(x(t))\left(v(x(t), \theta + z(t)) - v(x(t), \theta)\right)\right] = 0.$$

Then the system (3.19) is adaptively I&I stabilisable.

Proof. Consider the extended system

$$\dot{x} = f_0(x) + f_1(x)\theta + g(x)u,$$
$$\dot{\theta} = w,$$
(3.21)

and an adaptive state feedback control law of the form

$$\dot{\hat{\theta}} = w(x, \hat{\theta}),$$
$$u = v(x, \hat{\theta} + \beta(x)),$$
(3.22)

where $v(\cdot)$ and $\beta(\cdot)$ satisfy (A1) and (A2), respectively. Define the manifold

$$z \triangleq \hat{\theta} - \theta + \beta(x) = 0$$
(3.23)

and note that the dynamics of the off-the-manifold co-ordinates z are given by the equation

$$\dot{z} = w(x, \hat{\theta}) + \frac{\partial \beta}{\partial x}\left[f_0(x) + f_1(x)(\hat{\theta} + \beta(x) - z) + g(x)v(x, \hat{\theta} + \beta(x))\right].$$

Selecting the function $w(\cdot)$ as

$$w(x,\hat{\theta}) = -\frac{\partial \beta}{\partial x}\left[f_0(x) + f_1(x)(\hat{\theta} + \beta(x)) + g(x)v(x,\hat{\theta} + \beta(x))\right]$$

yields the first equation in (3.20), while the second equation is obtained from (3.19) using (A1) and the definition of z, and adding and subtracting $g(x)v(x,\theta)$. Hence, by assumptions (A1) and (A2), all trajectories of the closed-loop system (3.21), (3.22) are bounded and satisfy (3.5), which proves the claim. □

The fact that the system (3.19) is linearly parameterised is essential for obtaining the cascade form (3.20) and for ensuring that the off-the-manifold dynamics are independent of the unknown parameter vector θ. Note that, in the case of nonlinear parameterisation, the off-the-manifold dynamics generally depend on the unknown parameters. However, a solution using the I&I approach may still be feasible as shown in Section 3.6.

At first glance, assumption (A2) may seem quite restrictive in the sense that, in general, there is no systematic way of treating the cascade (3.20). However, explicit conditions in terms of Lyapunov functions can be obtained when the control law $v(x,\theta)$ is linear in θ or, more generally, when it satisfies a Lipschitz condition, as the following corollary shows.

Corollary 3.1. *Consider the system (3.19) and assume that condition (A1) of Theorem 3.1 holds with a control law satisfying the Lipschitz condition*

$$|v(x,\theta + z) - v(x,\theta)| \leq M(x)|z|, \qquad (3.24)$$

for all $z \in \mathbb{R}^q$ and for some function $M : \mathbb{R}^n \to \mathbb{R}_{>0}$. Moreover, assume that the following hold.

(A2′) There exists a mapping $\beta : \mathbb{R}^n \to \mathbb{R}^q$ such that

$$\frac{\partial \beta}{\partial x}f_1(x) + \left[\frac{\partial \beta}{\partial x}f_1(x)\right]^\top \geq M(x)^2 I > 0.$$

(A2″) There exists a radially unbounded function $V : \mathbb{R}^n \to \mathbb{R}_{\geq 0}$ such that

$$\frac{\partial V}{\partial x}f^*(x) \leq 0, \qquad \limsup_{|x|\to\infty} \frac{\left|\frac{\partial V}{\partial x}f^*(x)\right|}{\left|\frac{\partial V}{\partial x}g(x)\right|^2} \leq K,$$

for some $0 < K < \infty$. Then the system (3.19) is adaptively I&I stabilisable.

Proof. It suffices to show that assumptions (A2′) and (A2″) together with the Lipschitz condition (3.24) imply that (A2) holds, hence Theorem 3.1 applies. To establish this, consider the function $W(x,z) = V(x) + \frac{1}{2}\rho|z|^2$, with $\rho > 0$, where $V(x)$ is a Lyapunov function for the system $\dot{x} = f^*(x)$ that satisfies (A2″). Invoking assumption (A2′) and using Young's inequality, the time-derivative of $W(x,z)$ along the trajectories of (3.20) can be bounded, for any $\alpha > 0$, as

$$\dot{W} \leq \frac{\partial V}{\partial x} f^*(x) + \frac{1}{2\alpha} \left| \frac{\partial V}{\partial x} g(x) \right|^2 + \frac{\alpha}{2} M(x)^2 |z|^2 - \rho z^\top \frac{\partial \beta}{\partial x} f_1(x) z.$$

The conclusion then follows from (A2′) and (A2″) by selecting $\rho > \alpha$ and α sufficiently large and noting that by (A2′) z is bounded and converges to zero. □

From inspection of (3.20) it is clear that if $|z|$ decreases then, under some weak conditions on $v(x,\theta)$, the *disturbance* term $|v(x,\theta+z) - v(x,\theta)|$ also decreases, and if it is eventually "dominated" by the stability margin of the ideal system $\dot{x} = f^*(x)$ then stabilisation is achieved. It is therefore evident that ensuring stability of the zero equilibrium of the z-subsystem in (3.20), uniformly in x, is the main constraint imposed by the theorem.

A sufficient condition for the z-subsystem in (3.20) to have a globally stable equilibrium at zero is

$$\frac{\partial f_1}{\partial x}(x) = \left[\frac{\partial f_1}{\partial x}(x) \right]^\top,$$

which, from Poincaré's Lemma, is a necessary and sufficient condition for $f_1(x)$ to be a gradient function, that is, for the existence of a function $h : \mathbb{R}^n \to \mathbb{R}$ such that $f_1(x) = \frac{\partial h}{\partial x}$. In this case, we simply take $\beta(x) = h(x)$ and the z-dynamics become[14]

$$\dot{z} = -|f_1(x)|^2 z.$$

The fact that the Lyapunov function $V(x)$ for the ideal system verifies the second condition in (A2″) is not very restrictive. (Recall that the Lyapunov function does not need to be known.) For example, if $\frac{\partial V}{\partial x} f^*(x) = 0$ implies $\frac{\partial V}{\partial x} g(x) = 0$, then it is possible[15] to define a new function $U(V(x))$ such that the second condition in (A2″) holds for $U(V(x))$.

From a Lyapunov analysis perspective, the I&I procedure automatically creates cross terms between plant states and estimated parameters in the Lyapunov function, as suggested in the proof of Corollary 3.1. Also, we reiterate the fact that the stabilisation mechanism does not rely on cancellation of the perturbation term $g(x)(v(x,\theta+z) - v(x,\theta))$ in (3.20).

Note 3.2. It is clear from the construction of the adaptive I&I control laws that, besides the classical "integral action" of the parameter estimator, through the action of $\beta(x)$ we have introduced in the control law a "proportional" term. (See [175] for an

---

[14] In this respect, it is worth comparing (3.20) with the error system that results from application of indirect adaptive control, which is given by $\dot{\tilde{\theta}} = -\Phi(x)^\top \Phi(x) \tilde{\theta}$, $\dot{x} = f^*(x) + g(x)(v(x,\theta+\tilde{\theta}) - v(x,\theta))$, where $\Phi(x)$ is the regressor matrix, see [170] for details.

[15] See the discussion, in a different context, in Section 9.5 of [78].

adaptive algorithm that also includes this term.) This kind of parameter update law was called in early references *PI adaptation* [126]. Although it was recognised that PI update laws were potentially superior to purely integral adaptation, except for the notable exception of output error identification, their performance improvement was never clearly established, see, *e.g.*, [165] for a tutorial account of these developments. The contribution of the I&I method in this direction is to show that, as discussed in more detail in Chapter 4, the cascaded forms, such as (3.20), that are obtained in the I&I framework, allow to recover the performance of the full-information controller.◁

Example 3.2. We now look back at the TCSC system (3.17) of Example 3.1 and show how the adaptive stabilisation problem can be solved using Theorem 3.1.

Proposition 3.1. *The system (3.17) is adaptively I&I stabilisable at the equilibrium point* $x^* = [-1, 0]^\top$.

Proof. Direct application of the construction in the proof of Theorem 3.1 with

$$\dot{\hat{\theta}}_2 = \frac{\partial \beta}{\partial x_2} \left(x_1 u + \hat{\theta}_2 + \beta(x_2) \right),$$

$$u = \hat{\theta}_2 + \beta(x_2),$$

$\beta(x_2) = -\lambda x_2$ and $\lambda > 0$, yields the error equations

$$\dot{x}_1 = x_2(z + \theta_2),$$
$$\dot{x}_2 = -x_1(z + \theta_2) - \theta_2 - \theta_1 x_2,$$
$$\dot{z} = -\lambda(z - \theta_1 x_2).$$

Notice that, due to the presence of the term $\theta_1 x_2$, in this case we do not obtain a cascade system as in Theorem 3.1. (This is due to the fact that we are only estimating θ_2 and not the entire vector θ.) However, we can still carry out the stability analysis with the Lyapunov function $W(x, z) = \frac{1}{2}(x_1 + 1)^2 + \frac{1}{2}x_2^2 + \frac{1}{2\lambda\theta_1}z^2$, whose derivative is

$$\dot{W} = -\theta_1 \left(x_2 - \frac{1}{\theta_1} z \right)^2,$$

which establishes Lyapunov stability of the equilibrium (x^*, θ_2) and global boundedness of solutions. The proof is completed verifying that (x^*, θ_2) is an isolated stable equilibrium and all trajectories converge towards an equilibrium. □

It must be pointed out that the closed-loop system also admits the equilibrium manifold $\{(x, z) \mid x_2 = -\frac{\theta_2}{\theta_1}, z = -\theta_2\}$, which does not contain the equilibrium x^*. One way to ensure this manifold is never reached is to enforce the constraint $u > 0$ with a saturation function (see Section 8.1). Another way is to estimate both θ_1 and θ_2 and apply a control law of the form $u = (\hat{\theta}_1 + \beta_1(x_2) - k)x_2 + (\hat{\theta}_2 + \beta_2(x_2))$, with $k > 0$. The details of this second solution are left to the reader. ■

3.5 Linearly Parameterised Control

Consider the case of linearly parameterised control, that is, assume that the stabilising, known-parameters, state feedback control law has the form

$$v(x, \theta) = v_0(x) + v_1(x)\theta. \tag{3.25}$$

We present a stabilisation result, similar in spirit to the one in Section 3.4, in which we do not make any explicit assumption on the dependence of the system on the unknown parameters, but assume instead a "realisability" condition for the adaptation law. The result is summarised in the following statement.

Theorem 3.2. *Consider the system (3.1) with an equilibrium point x^* to be stabilised and assume that the following hold.*

(B1) There exists a full-information control law $u = v(x, \theta)$ of the form (3.25) such that the closed-loop system

$$\dot{x} = f^*(x) \triangleq f(x) + g(x)v(x, \theta)$$

has a globally asymptotically stable equilibrium at x^*.

(B2) There exists a mapping $\beta : \mathbb{R}^n \to \mathbb{R}^q$, with $\frac{\partial \beta}{\partial x} f^*(x)$ independent of the unknown parameters, such that all trajectories of the system

$$\begin{aligned} \dot{z} &= \left[\frac{\partial \beta}{\partial x} g(x) v_1(x) \right] z \\ \dot{x} &= f^*(x) + g(x) v_1(x) z \end{aligned} \tag{3.26}$$

are bounded and satisfy

$$\lim_{t \to \infty} [g(x(t))v_1(x(t))z(t)] = 0.$$

Then the system (3.1) is adaptively I&I stabilisable.

Proof. In view of the derivations in the proof of Theorem 3.1, we only need to establish that under the foregoing assumptions we can obtain the error equations (3.26). The second equation follows immediately from (B1), (3.25) and (3.23), while the first error equation is obtained from

$$\dot{z} = w(x, \hat{\theta}) + \frac{\partial \beta}{\partial x}(f^*(x) + g(x)v_1(x)z)$$

and selecting the parameter update law as

$$w(x, \hat{\theta}) = -\frac{\partial \beta}{\partial x} f^*(x).$$

Hence, by assumptions (B1) and (B2), all trajectories of the closed-loop system (3.4) are bounded and satisfy (3.5). □

Similarly to the discussion in Section 3.4, it is clear that the success of the proposed design hinges upon the ability to assign the sign of (the symmetric part of) the matrix $\frac{\partial \beta}{\partial x} g(x) v_1(x)$. Comparing with the corresponding matrix in (3.20) we see that $g(x) v_1(x)$ and $f_1(x)$ play the same role, and the discussion about solvability of the problem carried out in Section 3.4 applies here as well. In particular, for the case of single-input systems with one uncertain parameter, assumption (B2) is easily satisfied by picking $\beta(x)$ such that

$$\frac{\partial \beta}{\partial x} = -\gamma \left(g(x) v_1(x) \right)^\top$$

with $\gamma > 0$. However, $\beta(x)$ should also ensure that $\frac{\partial \beta}{\partial x} f^*(x)$ is independent of the unknown parameters[16].

Finally, condition (B2) can also be expressed in terms of Lyapunov functions as the following corollary shows.

Corollary 3.2. *Consider the system (3.1) and assume that condition (B1) of Theorem 3.2 and the following condition hold.*

(B2′) For all $x \in \mathbb{R}^n$ and $z \in \mathbb{R}^q$ we have

$$Q(x, z) \triangleq z^\top \left[\frac{\partial \beta}{\partial x} g(x) v_1(x) + \left(\frac{\partial \beta}{\partial x} g(x) v_1(x) \right)^\top \right] z$$
$$+ \frac{\partial V}{\partial x} \left(f^*(x) + g(x) v_1(x) z \right) \leq 0,$$

where $V(x)$ is a Lyapunov function for the system $\dot{x} = f^*(x)$. Moreover, $Q(x, z) = 0$ implies (3.5).

Then the system (3.1) is adaptively I&I stabilisable.

Proof. In view of Theorem 3.2 we only need to show that condition (B2′) implies (B2). To this end, consider the system (3.26) and the Lyapunov function $W(x, z) = V(x) + |z|^2$, whose time-derivative satisfies $\dot{W} = Q(x, z) \leq 0$, where we have used (B2′). Hence, by Theorem A.3, all trajectories are bounded and satisfy (3.5). □

Note that the key requirement in Corollary 3.2 is the attractivity of the manifold $z = 0$, which is ensured by the (restrictive) condition that the matrix in square brackets in $Q(x, z)$ is negative-semidefinite (uniformly in x).

Example 3.3. Consider the problem of adaptive I&I stabilisation to a nonzero position $(x_1^*, 0)$ of the pendulum system

---

[16] This realisability assumption is weaker than the strict matching assumption discussed in, e.g., [123], which requires that the uncertain parameters enter in the image of $g(x)$. It is clear that, in this case, we can always find $v_1(x)$ such that $f^*(x)$ is independent of the parameters.

$$\dot{x}_1 = x_2,$$
$$\dot{x}_2 = -\theta \sin(x_1) - (x_1 - x_1^*) - x_2 + u, \qquad (3.27)$$

where we assume that a PD controller has already been applied[17]. The target dynamics can be chosen independent of θ, say as

$$f^*(x) = \begin{bmatrix} x_2 \\ -(x_1 - x_1^*) - x_2 \end{bmatrix},$$

hence the adaptive I&I controller becomes $u = \sin(x_1)(\hat{\theta} + \beta(x))$. The error equations (3.26) take the form

$$\dot{z} = -\frac{\partial \beta}{\partial x_2} \sin(x_1) z,$$
$$\dot{x} = f^*(x) + \begin{bmatrix} 0 \\ \sin(x_1) \end{bmatrix} z, \qquad (3.28)$$

which immediately suggests $\beta(x) = -x_2 \sin(x_1)$, yielding $\dot{z} = -\sin^2(x_1) z$. The resulting adaptive control law is

$$\dot{\hat{\theta}} = -\big((x_1 - x_1^*) + x_2\big)\sin(x_1) - x_2^2 \cos(x_1),$$
$$u = \hat{\theta} \sin(x_1) - x_2 \sin^2(x_1).$$

As in the example of Section 3.2, the foregoing selection for $\beta(x)$ ensures that the system (3.28) has a globally stable equilibrium at $(x, z) = (x^*, 0)$ and $\sin(x_1(t))z(t) \in \mathcal{L}_2$, hence x converges to the desired equilibrium. ∎

3.6 Example: Visual Servoing

In this section we illustrate with a visual servoing problem how the adaptive I&I methodology can be applied in the nonlinearly parameterised case. Consider the visual servoing of a planar two-link robot manipulator in the so-called fixed-camera configuration, where the camera orientation and scale factor are unknown[18]. The control goal is to place the robot end-effector in some desired constant position, or to make it track a (slowly moving) trajectory, by using a vision system equipped with a fixed camera that is perpendicular to the plane where the robot evolves, as depicted in Figure 3.3.

---

[17] As discussed in [163] the classical schemes with the pendulum energy as Lyapunov function are inefficient because of the detectability obstacle. This difficulty can be overcome introducing cross terms in the Lyapunov function.

[18] We refer the interested reader to [74, 110, 15] for further detail on this problem. The idea presented in this section has been applied in [229] to an n-link manipulator and validated experimentally.

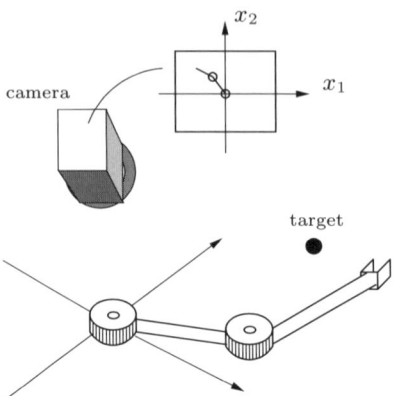

Fig. 3.3. Diagram of the visual servoing problem.

We model the action of the camera as a static mapping from the joint positions $q \in \mathbb{R}^2$ to the position (in pixels) of the robot tip in the image output, denoted $x \in \mathbb{R}^2$. This mapping is described by

$$x = ae^{J\theta}(k(q) - \vartheta_1) + \vartheta_2, \tag{3.29}$$

where θ is the orientation of the camera with respect to the robot frame, $a \geq a_m > 0$ and ϑ_1, ϑ_2 denote intrinsic camera parameters (scale factors, focal length and centre offset, respectively), $k : \mathbb{R}^2 \to \mathbb{R}^2$ defines the robot direct kinematics, and

$$J = \begin{bmatrix} 0 & -1 \\ 1 & 0 \end{bmatrix}, \quad e^{J\theta} = \begin{bmatrix} \cos(\theta) & -\sin(\theta) \\ \sin(\theta) & \cos(\theta) \end{bmatrix}.$$

Invoking standard time-scale separation arguments and assuming an inner fast loop for the robot velocity control, we concentrate on the kinematic problem to generate the references for the robot velocities. The robot dynamics are then described by a simple integrator $\dot{q} = v$, where $v \in \mathbb{R}^2$ are the joint velocities. The direct kinematics yield $\dot{k} = \mathcal{J}(q)\dot{q}$, where $\mathcal{J}(q) = \frac{\partial k}{\partial q}$ is the analytic robot Jacobian, which is assumed nonsingular. Differentiating (3.29) and replacing the latter expression yields the dynamic model of the overall system of interest, namely

$$\dot{x} = ae^{J\theta}u, \tag{3.30}$$

where $u = \mathcal{J}(q)v$ is a new input. The problem is to find a control law u such that $x(t)$ asymptotically tracks a reference trajectory $x^* = x^*(t)$ in spite of the lack of knowledge of a and θ.

Note that, if θ were known, a stabilising control law for system (3.30) could be obtained, under some assumptions, without the knowledge of the uncertain parameter a. Indeed, the feedback

$$v(x,\theta) = -\frac{1}{a_m}e^{-J\theta}\left(\tilde{x} - \dot{x}^*\right),$$

where $\tilde{x} = x - x^*$ is the tracking error, yields the target closed-loop dynamics

$$\dot{\tilde{x}} = -\frac{a}{a_m}\left(\tilde{x} - \dot{x}^*\right) - \dot{x}^*,$$

whose trajectories converge to zero if either $a/a_m = 1$ or $|\dot{x}^*(t)|$ goes to zero. Motivated by this property, the adaptive I&I methodology yields the following result.

Proposition 3.2. *Consider the system (3.30) and a bounded reference trajectory x^*, with bounded first- and second-order derivatives $\dot{x}^*$, $\ddot{x}^*$, and assume that a positive lower bound on the scale factor a is known, i.e., $a \geq a_m > 0$. Then the adaptive I&I controller*

$$\begin{aligned}u &= -\frac{1}{a_m}e^{-J\left(\hat{\theta}+\frac{1}{2}|s|^2\right)}s,\\ \dot{\hat{\theta}} &= s^T\left(s + \dot{x}^* + \ddot{x}^*\right),\end{aligned} \qquad (3.31)$$

where $s = \tilde{x} - \dot{x}^*$, is such that all trajectories of the closed-loop system (3.30), (3.31) are bounded and the tracking error either satisfies

$$\lim_{t\to\infty}|\tilde{x}(t) - w(t)| = 0, \qquad (3.32)$$

where $w(t)$ is the solution of

$$\dot{w} = Rw - (R+I)\dot{x}^*, \qquad (3.33)$$

with[19] *$R = -\frac{a}{a_m}e^{-J\arccos(a_m/a)}$, or*

$$\lim_{t\to\infty}|s(t)| = 0, \quad \lim_{t\to\infty}\tilde{x}(t) = 0. \qquad (3.34)$$

In particular, if either $a_m = a$ or $\lim_{t\to\infty}|\dot{x}^*(t)| = 0$, then $\lim_{t\to\infty}\tilde{x}(t) = 0$.

Proof. Consider the co-ordinate

$$z = \hat{\theta} - \theta + \frac{1}{2}|s|^2$$

and note that

$$\dot{\tilde{x}} = -\frac{a}{a_m}e^{-Jz}\tilde{x} + \left(\frac{a}{a_m}e^{-Jz} - I\right)\dot{x}_*. \qquad (3.35)$$

Comparing with (3.33), it is clear that the control objective is achieved if $\lim_{t\to\infty}z(t) = \arccos(a_m/a)$, which is established in the sequel.

---

[19] Since $0 < a_m/a \leq 1$, the matrix R is Hurwitz. Furthermore, for all initial conditions, the solutions $w(t)$ of (3.33) converge to zero, if either $a/a_m = 1$ or $\lim_{t\to\infty}|\dot{x}^*(t)| = 0$.

The dynamics of z are given by

$$\dot{z} = -|s|^2 \left(\frac{a}{a_m}\cos(z) - 1\right), \tag{3.36}$$

where we have used (3.31) and the fact that $s^\top e^{-Jz} s = |s|^2 \cos(z)$. From the shape of the function $\frac{a}{a_m}\cos(z) - 1$ and the fact that $a/a_m > 1$, we see that all trajectories of the z subsystem are bounded (uniformly in s). Since z is bounded, it remains to prove that s, and therefore $\tilde{x}$, are also bounded. To this end, rewrite the $\tilde{x}$ dynamics in terms of s as

$$\dot{s} = -\frac{a}{a_m} e^{-Jz} s - \dot{x}_* - \ddot{x}_* \tag{3.37}$$

and consider the function

$$W(s,z) = \frac{1}{2}|s|^2 - z. \tag{3.38}$$

Taking the derivative along the trajectories of (3.36), (3.37) yields

$$\dot{W} = -|s|^2 - s^\top (\dot{x}_* + \ddot{x}_*) \leq -\frac{1}{2}|s|^2 + d_1,$$

where $d_1 \geq \frac{1}{2}|\dot{x}_* + \ddot{x}_*|^2$, and we have used the inequality $2|s||\dot{x}_* + \ddot{x}_*| \leq |s|^2 + |\dot{x}_* + \ddot{x}_*|^2$. From boundedness of z, there exists a positive constant c_1 such that $W \leq \frac{1}{2}|s|^2 + c_1$. As a result, $\dot{W} \leq -W + d_1 + c_1$, from which we immediately conclude that W, and consequently s and $\tilde{x}$, are bounded.

We now establish the convergence results (3.32) and (3.34). Consider first the case $s(t) \notin \mathcal{L}_2$. It follows from (3.36) that, for all initial conditions, the trajectories of the z subsystem converge towards the points $z = \arccos(a_m/a)$. This means that the dynamics (3.35) can be written in the form

$$\dot{\tilde{x}} = [R + B_1(t)]\tilde{x} - [R + B_1(t) + I]\dot{x}_*,$$

for some bounded matrix $B_1(t)$, with $\lim_{t\to\infty} |B_1(t)| = 0$. The convergence result (3.32) then follows subtracting (3.33) from the equation above.

Consider now the case $s(t) \in \mathcal{L}_2$. It follows from (3.37) and boundedness of s, $\dot{x}_*$ and $\ddot{x}_*$, that $\dot{s}$ is also bounded. Hence, by Lemma A.2, we conclude that $\lim_{t\to\infty} |s(t)| = 0$, which implies that $\lim_{t\to\infty} |\tilde{x}(t) - \dot{x}_*(t)| = 0$. It follows from (3.35) that $\lim_{t\to\infty} |\tilde{x}(t)| = 0$. This completes the proof of the proposition. □

The adaptive I&I controller (3.31) has been tested through simulations[20]. The two-link robot direct kinematics transformation $y = k(q)$ is given by

$$y_1 = L_1 \cos(q_1) + L_2 \cos(q_1 + q_2) + O_1,$$
$$y_2 = L_1 \sin(q_1) + L_2 \sin(q_1 + q_2) + O_2,$$

---

[20] The simulations have been carried out in the same conditions as [70].

3.6 Example: Visual Servoing

where y_1 and y_2 are the end-effector Cartesian co-ordinates, L_1 and L_2 are the link lengths, and O_1 and O_2 are the robot base co-ordinates in the workspace frame. The parameter values are $L_1 = 0.8$ m, $L_2 = 0.5$ m, $O_1 = -0.666$ m, and $O_2 = -0.333$ m. For simplicity, the image co-ordinates (originally in pixels) are considered to have been normalised, i.e., x_1 and x_2 are nondimensional. A case of large disorientation is taken into consideration: $\theta = 1$ rad, with $\hat{\theta}(0) = 0$. The scaling factor a and its lower bound a_m are chosen as $a = 0.7$ and $a_m = 0.5$. The initial conditions of the manipulator are $q_1(0) = 1.3$ rad and $q_2(0) = -1.3$ rad.

For the set-point control case, with $x_1^* = x_2^* = 0.1$, the convergence is observed in 6 seconds (within 4% of the final value) and the control velocities are not greater than 0.9 rad/s (Figure 3.4). The tracking case has also been tested with the reference trajectory generated by a first-order filter

$$\dot{x}^* = -\lambda x^* + r,$$

where $r = [r_1, r_2]^\top$ is a vector of reference signals given by

$$r_1 = a_1 \sin(w_r t) + c + d \sin(1.5 w_r t),$$
$$r_2 = a_2 \sin(w_r t + \psi) + c + d \sin(1.5 w_r t + \psi),$$

with $a_1 = a_2 = d = 0.04$, $c = 0.1$, $\psi = 1$ rad, and $\lambda = 1$, and with w_r assuming two different values, i.e., $w_r = 0.07$ and $w_r = 0.03$. The simulations have been carried out for 80 seconds. Figure 3.5 shows the time histories of the errors between the normalised image co-ordinates x_1 and x_2 and their corresponding references x_1^* and x_2^*. Observe that the tracking is good, with a small steady-state error. As expected, by reducing the speed of the reference trajectory (that is, $w_r = 0.03$), the tracking error becomes smaller. The control effort for the I&I controller is not higher than 1.1 rad/s for both reference trajectories.

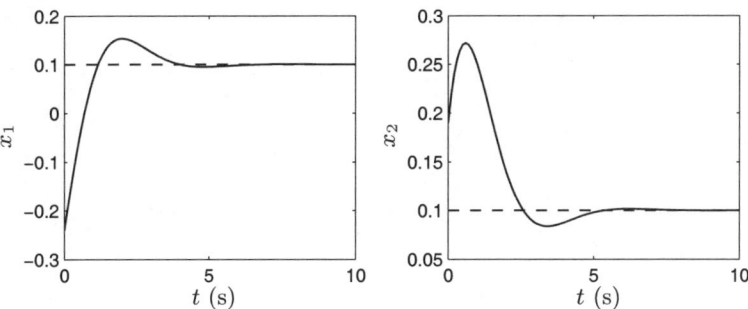

Fig. 3.4. Time histories of the normalised image output signals $x_1(t)$ and $x_2(t)$ for $x_1^* = x_2^* = 0.1$.

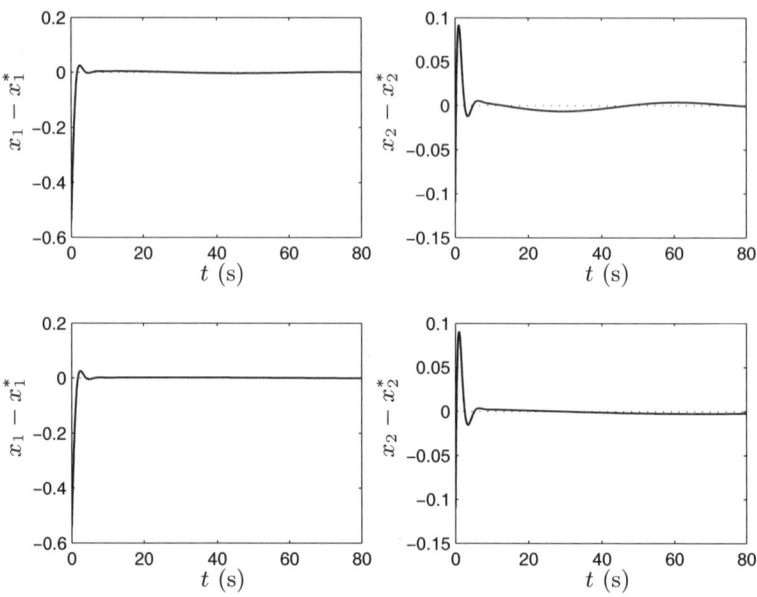

Fig. 3.5. Time histories of the tracking errors $x_1(t) - x_1^*(t)$ and $x_2(t) - x_2^*(t)$ for $w_r = 0.07$ (top) and for $w_r = 0.03$ (bottom).

4

I&I Adaptive Control: Systems in Special Forms

Chapters 2 and 3 have addressed the problem of (adaptive) stabilisation via state feedback for general nonlinear systems, based on the solution of a set of partial differential equations. In this chapter these ideas are exploited to develop constructive adaptive control algorithms for specific classes of nonlinear systems with parametric uncertainties. In particular, the proposed approach is applied to the class of systems in so-called parametric feedback form, yielding new adaptive control laws with interesting properties. These are compared with the classical adaptive (feedback linearising) controllers by means of a practical example.

4.1 Introduction

A considerable amount of nonlinear adaptive control methods deal with dynamical systems with parametric uncertainties. The most widely studied among these is the class of feedback linearisable systems that depend linearly on the unknown parameters[1]. As discussed in Chapter 3, the standard methodology for dealing with this class of systems is based on the so-called certainty equivalence principle and relies on finding a parameter update law such that a (quadratic) function of some functions of the state and the parameter estimation error becomes a Lyapunov function for the closed-loop system. This is achieved by cancelling out the parameter-dependent terms from the derivative of the Lyapunov function, thus ensuring global stability and boundedness of all closed-loop signals.

A typical example where this cancellation is straightforward is the class of systems

$$\dot{x}_1 = x_2, \quad \cdots \quad \dot{x}_{n-1} = x_n, \quad \dot{x}_n = u + \varphi(x)^\top \theta, \tag{4.1}$$

which satisfy the so-called *matching condition, i.e.,* the control input u enters through the same "integrator" as the unknown parameter vector θ. In this

---

[1] See [191, 137, 123] and references therein.

case the certainty equivalence control law is given by

$$u = -K^\top x - \varphi(x)^\top \hat{\theta}, \qquad (4.2)$$

where $\hat{\theta}$ is the estimate of θ and the vector $K = [k_1, \ldots, k_n]^\top \in \mathbb{R}^n$ is such that the matrix

$$A = \begin{bmatrix} 0 & 1 & \cdots & 0 \\ \vdots & & \ddots & \\ 0 & 0 & \cdots & 1 \\ -k_1 & -k_2 & \cdots & -k_n \end{bmatrix}$$

is Hurwitz, i.e., it satisfies the Lyapunov equation $A^\top P + PA = -Q$, for some positive-definite matrices P and Q. Consider now the candidate Lyapunov function $V = x^\top P x + \gamma^{-1} |\hat{\theta} - \theta|^2$ with $\gamma > 0$, whose time-derivative along the trajectories of the closed-loop system (4.1), (4.2) is given by

$$\dot{V} = -x^\top Q x - 2x^\top P e_n \varphi(x)^\top (\hat{\theta} - \theta) + 2\gamma^{-1}(\hat{\theta} - \theta)^\top \dot{\hat{\theta}},$$

and note that the update law (P is independent of θ)

$$\dot{\hat{\theta}} = \gamma \varphi(x) e_n^\top P x \qquad (4.3)$$

is such that the last two terms are cancelled out, rendering $\dot{V}$ negative semidefinite[2].

A shortcoming of the classical adaptive approach, briefly described above, is that the estimation error $\hat{\theta} - \theta$ is only guaranteed to be bounded (and converging to an unknown constant). However, little can be said about its dynamical behaviour, which may be unacceptable in terms of the transient response of the closed-loop system. Moreover, due to the strong coupling between the plant and the estimator dynamics, increasing the adaptation gain, i.e., the parameter γ in (4.3), does not necessarily speed-up the response of the system. In other words, there is an inherent limitation on the achievable performance.

Another issue that arises in classical adaptive control is that the corresponding non-adaptive *asymptotic* controller, which is obtained from the adaptive controller by fixing the parameter estimate to its limit value, is not necessarily stabilising[3]. In other words, if we let the adaptive controller run for *infinite* time and then freeze the parameter estimate, the resulting controller may be *destabilising*.

In this chapter a new algorithm for the stabilisation via state feedback of a class of linearly parameterised systems in feedback form is proposed, which

---

[2] For more general systems in feedback (lower triangular) form [137, 123], that do not satisfy the matching condition, this cancellation must be carried out in a number of steps equal to the order of the system, a procedure commonly known as *adaptive backstepping* [91].

[3] See the counter-example in [221].

addresses the two foregoing issues. The construction relies on the nonlinear adaptive stabilisation tools described in Chapter 3 and allows for prescribed (uniformly stable) dynamics to be assigned to the estimation error, thus leading to a *modular* scheme which is easier to tune than the one obtained from Lyapunov redesign. As a result, the performance of the adaptive system can be significantly improved. In addition, the respective non-adaptive controller ensures that trajectories remain bounded *for any* fixed estimate of the unknown parameter vector, provided that the "regressor" (*i.e.*, the vector multiplying the unknown parameters) satisfies a structural condition (see Proposition 4.2).

4.2 Systems in Feedback Form

Consider a class of single-input nonlinear systems that can be described in suitable co-ordinates by equations of the form

$$\begin{aligned} \dot{x}_1 &= f(x_1) + g(x_1)x_2, \\ \dot{x}_2 &= u + \varphi(x)^\top \theta, \end{aligned} \quad (4.4)$$

where $x = [x_1^\top, x_2]^\top \in \mathbb{R}^n \times \mathbb{R}$ is the state, $u \in \mathbb{R}$ is the control input, $f(\cdot)$, $g(\cdot)$ and $\varphi(\cdot)$ are smooth mappings, and $\theta \in \mathbb{R}^p$ is an unknown constant vector.

Assumption 4.1. The system $\dot{x}_1 = f(x_1)$ has a globally asymptotically stable equilibrium at zero, *i.e.*, there exists a positive-definite and proper Lyapunov function $V_1(x_1)$ satisfying

$$L_f V_1(x_1) \leq -\kappa(x_1),$$

where $\kappa(\cdot)$ is a positive-definite function.

Note that the above assumption is trivially satisfied for the class of systems (4.1) by redefining the state x_n and the control input u.

The control objective is to find a continuous adaptive state feedback control law of the form

$$\begin{aligned} \dot{\hat{\theta}} &= w(x, \hat{\theta}), \\ u &= v(x, \hat{\theta}), \end{aligned} \quad (4.5)$$

such that the closed-loop system (4.4), (4.5) has a globally stable equilibrium at $(x, \hat{\theta}) = (0, \theta)$ and

$$\lim_{t \to \infty} x(t) = 0. \quad (4.6)$$

The proposed approach follows the ideas in Chapter 3, where the adaptive stabilisation problem is formulated as a problem of asymptotically *immersing* the closed-loop system into the system that would result from applying the non-adaptive controller[4].

---

[4] See also [73, 5] for two related design examples.

4.2.1 Adaptive Control Design

Proposition 4.1. *Consider the system (4.4) and the adaptive state feedback control law*

$$\dot{\hat{\theta}} = -\frac{\partial \beta}{\partial x_1}\left(f(x_1) + g(x_1)x_2\right) - \frac{\partial \beta}{\partial x_2}\left(u + \varphi(x)^\top(\hat{\theta} + \beta(x))\right), \quad (4.7)$$

$$u = -kx_2 - \epsilon L_g V_1(x_1) - \varphi(x)^\top(\hat{\theta} + \beta(x)), \quad (4.8)$$

where

$$\beta(x) = \gamma \int_0^{x_2} \varphi(x_1, \chi) d\chi \quad (4.9)$$

and $k > 0$, $\epsilon > 0$, $\gamma > 0$ are constants. Then the closed-loop system (4.4), (4.7), (4.8) has a globally stable equilibrium at $(x, \hat{\theta}) = (0, \theta)$ and (4.6) holds.

Proof. Define the "estimation error" $z = \hat{\theta} - \theta + \beta(x)$, where $\beta(x)$ is given by (4.9), and note that the "error dynamics" are given by

$$\dot{z} = \dot{\hat{\theta}} + \frac{\partial \beta}{\partial x_1}\left(f(x_1) + g(x_1)x_2\right) + \frac{\partial \beta}{\partial x_2}\left(u + \varphi(x)^\top(\hat{\theta} + \beta(x) - z)\right). \quad (4.10)$$

Substituting (4.7) into (4.10) and exploiting (4.9) yields

$$\dot{z} = -\frac{\partial \beta}{\partial x_2}\varphi(x)^\top z = -\gamma\varphi(x)\varphi(x)^\top z. \quad (4.11)$$

To establish the dynamic properties of the above system, consider the function $V_2(z) = \gamma^{-1}z^\top z$ whose time-derivative along the trajectories of (4.11) satisfies $\dot{V}_2(z) = -2(\varphi(x)^\top z)^2 \leq 0$, which implies that $z(t) \in \mathcal{L}_\infty$ and $\varphi(x(t))^\top z(t) \in \mathcal{L}_2$. Note that the above holds *for any u*.

Assume now that u is given by (4.8) and note that the system (4.4) can be rewritten as

$$\begin{aligned}\dot{x}_1 &= f(x_1) + g(x_1)x_2, \\ \dot{x}_2 &= -kx_2 - \epsilon L_g V_1(x_1) - \varphi(x)^\top z,\end{aligned} \quad (4.12)$$

i.e., as an asymptotically stable system perturbed by an $\mathcal{L}_2$ signal. Consider now the function $W(x, z) = 2\epsilon V_1(x_1) + x_2^2 + k^{-1}V_2(z)$ and note that

$$\dot{W}(x, z) = 2\epsilon L_f V_1(x_1) - 2kx_2^2 - 2x_2\varphi(x)^\top z - 2k^{-1}\left(\varphi(x)^\top z\right)^2$$

which implies

$$\dot{W}(x, z) \leq -2\epsilon\kappa(x_1) - kx_2^2 - k^{-1}\left(\varphi(x)^\top z\right)^2 \leq 0,$$

where we have used Assumption 4.1. It follows that the system (4.11), (4.12) has a globally stable equilibrium at the origin, hence the system (4.4), (4.7), (4.8) has a globally stable equilibrium at $(0, \theta)$. Moreover, by LaSalle's invariance principle (see Theorem A.2), all trajectories converge to the invariant set $E = \{(x, z) \in \mathbb{R}^{n+1} \times \mathbb{R}^p : x = 0, \varphi(x)^\top z = 0\}$, hence condition (4.6) holds, which concludes the proof. □

The following remarks regarding the properties of the proposed adaptive controller are in order.

1. The introduction of the function $\beta(x)$ in the definition of the estimation error z allows construction of the error system (4.10), whose zero equilibrium is rendered uniformly stable, for any x and u and for any $\gamma > 0$, by the update law given in (4.7) and by the selection (4.9).
2. Integrating the system (4.11) when $z \in \mathbb{R}$ yields the solution

$$z(t) = z(0) \exp(-\gamma \int_0^t \varphi(x(\tau))^2 \mathrm{d}\tau), \qquad (4.13)$$

which implies that the convergence of the estimation error to its limit value can be *arbitrarily* increased simply by increasing the parameter γ. Moreover, from (4.13) it is straightforward to derive a "persistent excitation" condition which would guarantee asymptotic convergence of the estimation error to zero.
3. The control law given in (4.8) differs from the certainty equivalence control law (4.2) in that it contains an additional "feedback" expressed by the term $\varphi(x)^\top \beta(x)$. As is shown in Section 4.2.2, when $\varphi(x)$ is a polynomial function in x_2, this term renders the closed-loop system *input-to-state stable* (ISS) with respect to the error $\hat{\theta} - \theta$.
4. For simplicity we have used a linear feedback term, namely $-kx_2$, in the control law (4.8). As shown in the proof of Proposition 4.1, this is enough to ensure that the system (4.12) is $\mathcal{L}_2$ stable with respect to the perturbation $\varphi(x)^\top z$, which belongs to $\mathcal{L}_2$ from (4.11). However, a nonlinear feedback would accommodate a wider class of perturbations, thus allowing more flexibility in selecting the function $\beta(\cdot)$.

4.2.2 Asymptotic Properties of Adaptive Controllers

As stated in the introduction, a drawback of classical adaptive control is that, although the state $\hat{\theta}$ is guaranteed to converge to a constant, substituting this constant in (4.2) may result in a *destabilising* controller. In this section we show that for the control law given in (4.8) this does not hold. In particular, the closed-loop system (4.4), (4.8) has bounded trajectories and its zero equilibrium can be rendered globally practically stable *for any* fixed $\hat{\theta}$, provided a structural condition on the mapping $\varphi(x)$ holds and a bound on θ is known.

Proposition 4.2. *Consider the system (4.4) and the control law*

$$u = -kx_2 - \epsilon L_g V_1(x_1) - \varphi(x)^\top \left(\bar{\theta} + \beta(x) \right), \qquad (4.14)$$

where $\bar{\theta}$ is a constant, $\beta(x)$ is given by (4.9) with $\gamma > 0$, and $k > 0$. Suppose that $\varphi(x)$ can be expressed as[5]

---

[5] If $\varphi(x)$ is a polynomial in x_2 we can always redefine $\varphi(x)$ and θ so that (4.15) holds.

$$\varphi(x) = \left[\psi_1(x_1)x_2^{q_1}, \cdots, \psi_p(x_1)x_2^{q_p}\right]^\top, \qquad (4.15)$$

where $q_1, \ldots, q_p$ are nonnegative constants. Then all trajectories of the closed-loop system (4.4), (4.14) are bounded. Moreover, if θ belongs to a known compact set Θ, then for any $\delta > 0$ there exists γ, which depends on Θ, $\bar{\theta}$ and δ, such that all trajectories converge to the set $\Omega = \{x \in \mathbb{R}^{n+1} : |x| \leq \delta\}$.

Proof. Integrating (4.15) according to (4.9) yields

$$\beta(x) = \gamma x_2 \left[\psi_1(x_1)\frac{x_2^{q_1}}{q_1+1}, \cdots, \psi_p(x_1)\frac{x_2^{q_p}}{q_p+1}\right]^\top = \gamma x_2 \Gamma \varphi(x), \qquad (4.16)$$

where

$$\Gamma = \mathrm{diag}(\frac{1}{q_1+1}, \ldots, \frac{1}{q_p+1}).$$

Consider the function $V(x) = 2\epsilon V_1(x_1) + x_2^2$, whose time-derivative along the trajectories of (4.4), (4.14) is given by

$$\dot{V}(x) = 2\epsilon L_f V_1(x_1) - 2kx_2^2 - 2x_2\varphi(x)^\top (\bar{\theta} - \theta) - 2x_2\varphi(x)^\top \beta(x).$$

Substituting (4.16) in the above equation yields

$$\dot{V}(x) = 2\epsilon L_f V_1(x_1) - 2kx_2^2 - 2x_2\varphi(x)^\top (\bar{\theta} - \theta) - 2\gamma x_2^2 \varphi(x)^\top \Gamma \varphi(x)$$

$$\leq -2\epsilon\kappa(x_1) - 2kx_2^2 + \frac{2\gamma}{\bar{q}+1}x_2^2|\varphi(x)|^2 + \frac{\bar{q}+1}{2\gamma}|\bar{\theta}-\theta|^2$$

$$\quad - \frac{2\gamma}{\bar{q}+1}x_2^2|\varphi(x)|^2$$

$$= -2\epsilon\kappa(x_1) - 2kx_2^2 + \frac{\bar{q}+1}{2\gamma}|\bar{\theta}-\theta|^2$$

$$\leq -\epsilon\kappa(x_1) - kx_2^2 - \rho(|x|) + \frac{\bar{q}+1}{2\gamma}|\bar{\theta}-\theta|^2,$$

where $\bar{q} = \max(q_1, \ldots, q_p)$ and $\rho(\cdot)$ is a $\mathcal{K}_\infty$ function such that[6]

$$\rho(|x|) \leq \epsilon\kappa(x_1) + kx_2^2.$$

Defining $\mathcal{K}_\infty$ functions $\sigma_1(\cdot)$ and $\sigma_2(\cdot)$ such that $\sigma_1(|x|) \leq V(x) \leq \sigma_2(|x|)$ we obtain

$$\dot{V}(x) \leq -\epsilon\kappa(x_1) - kx_2^2 - \rho(\sigma_2^{-1}(V(x))) + \frac{\bar{q}+1}{2\gamma}|\bar{\theta}-\theta|^2,$$

hence $\dot{V}(x)$ is negative-definite whenever

$$V(x) \geq \sigma_2(\rho^{-1}(\frac{\bar{q}+1}{2\gamma}|\bar{\theta}-\theta|^2)),$$

---

[6] See Lemma A.1 or [112, Lemma 4.3].

which implies that all closed-loop trajectories are bounded and converge to the set

$$\bar{\Omega} = \{\, x \in \mathbb{R}^{n+1} : V(x) \le \sigma_2(\rho^{-1}(\frac{\bar{q}+1}{2\gamma}|\bar{\theta}-\theta|^2))\,\}$$
$$\subseteq \{\, x \in \mathbb{R}^{n+1} : |x| \le \sigma_1^{-1}(\sigma_2(\rho^{-1}(\frac{\bar{q}+1}{2\gamma}|\bar{\theta}-\theta|^2)))\,\}.$$

Suppose now that θ belongs to a known compact set Θ and let

$$c = \max_{\theta \in \Theta} |\bar{\theta}-\theta|^2.$$

Note that $\bar{\Omega} \subseteq \Omega = \{x \in \mathbb{R}^{n+1} : |x| \le \delta\}$, where

$$\delta = \sigma_1^{-1}(\sigma_2(\rho^{-1}(\frac{\bar{q}+1}{2\gamma}c))). \tag{4.17}$$

The proof is completed by noting that, given any δ (arbitrarily small) and for any $\bar{q}$ and c, there exists a constant γ (sufficiently large) such that (4.17) holds. □

Although the regressor $\varphi(x)$ is restricted to have the form (4.15), the same result holds for any vector $\varphi(x)$ with elements of the form $\varphi_i(x) = \psi_i(x_1)\rho_i(x_2)$, provided that the function $\rho_i(\cdot)$ satisfies the differential inequality

$$-cx_2 \frac{\partial \rho_i}{\partial x_2}\rho_i(x_2) + x_2^2 \left(\frac{\partial \rho_i}{\partial x_2}\right)^2 \le 0$$

for some constant $c > 0$. A solution to the above inequality is given by

$$\rho_i(x_2) = \rho_0 \exp(\int_0^{x_2} \frac{c\chi}{\chi^2 + h(\chi)} d\chi),$$

where ρ_0 is a constant and $h(\cdot)$ is a nonnegative function.

4.2.3 Unknown Control Gain

In this section the result in Proposition 4.1 is extended to the case in which the control input is multiplied by an *unknown* parameter. This simple extension can be used, for instance, to deal with modelling errors in the actuation or loss of control effectiveness due to actuator wear or damage.

Consider the class of systems given in (2.1) where the control input u is multiplied by an unknown constant parameter, namely

$$\begin{aligned} \dot{x}_1 &= f(x_1) + g(x_1)x_2, \\ \dot{x}_2 &= \theta_2 u + \varphi(x)^\top \theta_1, \end{aligned} \tag{4.18}$$

where $\theta_1 \in \mathbb{R}^p$, $\theta_2 \in \mathbb{R}$, and suppose that the sign of θ_2 (the "control direction") is known. Without loss of generality suppose that $\theta_2 > 0$.

62   4 I&I Adaptive Control: Systems in Special Forms

Proposition 4.3. *Consider the system (4.18) and the adaptive state feedback control law*

$$\dot{\hat{\theta}} = -\left(I + \frac{\partial \beta}{\partial \hat{\theta}}\right)^{-1} \left(\frac{\partial \beta}{\partial x_1}(f(x_1) + g(x_1)x_2) \right.$$
$$\left. + \frac{\partial \beta}{\partial x_2}(-kx_2 - \epsilon L_g V_1(x_1))\right), \quad (4.19)$$

$$u = -\left(\hat{\theta}_2 + \beta_2(x, \hat{\theta}_1)\right)\left(kx_2 + \epsilon L_g V_1(x_1) + \varphi(x)^\top (\hat{\theta}_1 + \beta_1(x))\right), \quad (4.20)$$

where $\beta(\cdot) = [\beta_1(\cdot)^\top, \beta_2(\cdot)]^\top$, *with*

$$\beta_1(x) = \gamma_1 \int_0^{x_2} \varphi(x_1, \chi) d\chi$$

$$\beta_2(x, \hat{\theta}_1) = \gamma_2 \left(k\frac{x_2^2}{2} + \epsilon L_g V_1(x_1) x_2\right) \quad (4.21)$$

$$+ \gamma_2 \int_0^{x_2} \varphi(x_1, \chi)^\top (\hat{\theta}_1 + \beta_1(x_1, \chi)) d\chi,$$

and $k > 0$, $\epsilon > 0$, $\gamma_1 > 0$, $\gamma_2 > 0$ *are constants. Then the update law (4.19) is well-defined, the closed-loop system (4.18), (4.19), (4.20) has a globally stable equilibrium at* $(x, \hat{\theta}) = (0, \theta)$, *and condition (4.6) holds.*

Proof. To begin with, define the off-the-manifold co-ordinates[7] $z_1 = \hat{\theta}_1 - \theta_1 + \beta_1(x)$, $z_2 = \hat{\theta}_2 - \theta_2^{-1} + \beta_2(x, \hat{\theta}_1)$ and consider the control law given in (4.20), which is in the spirit of (4.8) but taking into account the presence of θ_2. Let $z = [z_1^\top, z_2]^\top$ and $\hat{\theta} = [\hat{\theta}_1^\top, \hat{\theta}_2]^\top$, and note that the off-the-manifold dynamics are given by

$$\dot{z} = \left(I + \frac{\partial \beta}{\partial \hat{\theta}}\right)\dot{\hat{\theta}} + \frac{\partial \beta}{\partial x_1}(f(x_1) + g(x_1)x_2)$$
$$+ \frac{\partial \beta}{\partial x_2}\left(\theta_2 u + \varphi(x)^\top(\hat{\theta}_1 + \beta_1(x) - z_1)\right).$$

Using the definition of z_2 and substituting (4.20) into the above equation yields

$$\dot{z} = \left(I + \frac{\partial \beta}{\partial \hat{\theta}}\right)\dot{\hat{\theta}} + \frac{\partial \beta}{\partial x_1}(f(x_1) + g(x_1)x_2) + \frac{\partial \beta}{\partial x_2}\left(-kx_2 - \epsilon L_g V_1(x_1)\right.$$
$$\left. -\varphi(x)^\top z_1 + \theta_2 z_2 \left(-kx_2 - \epsilon L_g V_1(x_1) - \varphi(x)^\top (\hat{\theta}_1 + \beta_1(x))\right)\right).$$

Note that the matrix $I + \frac{\partial \beta}{\partial \hat{\theta}}$ is invertible, since the second term only contributes to non-diagonal elements. Substituting the update law (4.19) and using (4.21) yields

---
[7] Notice that the function $\beta_2(\cdot)$ depends not only on the state x but also on the estimate $\hat{\theta}_1$.

$$\dot{z} = \frac{\partial \beta}{\partial x_2}\Big(-\varphi(x)^\top z_1 + \theta_2 z_2 \Big(-kx_2 - \epsilon L_g V_1(x_1) - \varphi(x)^\top (\hat{\theta}_1 + \beta_1(x))\Big)\Big)$$
$$= -\Gamma \Phi(x, \hat{\theta}_1) \Phi(x, \hat{\theta}_1)^\top z, \tag{4.22}$$

where $\Gamma = \text{diag}(\gamma_1, \gamma_2 \theta_2^{-1})$ and

$$\Phi(x, \hat{\theta}_1) = \begin{bmatrix} \varphi(x) \\ \theta_2 \Big(kx_2 + \epsilon L_g V_1(x_1) + \varphi(x)^\top (\hat{\theta}_1 + \beta_1(x))\Big) \end{bmatrix}.$$

Consider now the function $V_2(z) = z^\top \Gamma^{-1} z$, whose time-derivative along the trajectories of (4.22) satisfies

$$\dot{V}_2(z) = -2 \Big(\Phi(x, \hat{\theta}_1)^\top z\Big)^2 \leq 0,$$

hence $z(t) \in \mathcal{L}_\infty$ and $\Phi(x(t), \hat{\theta}_1(t)) z(t) \in \mathcal{L}_2$. Note now that the system (4.18) can be rewritten as

$$\begin{aligned} \dot{x}_1 &= f(x_1) + g(x_1) x_2, \\ \dot{x}_2 &= -kx_2 - \epsilon L_g V_1(x_1) - \Phi(x, \hat{\theta}_1)^\top z, \end{aligned} \tag{4.23}$$

i.e., as an asymptotically stable system perturbed by an $\mathcal{L}_2$ signal. As a result, by using the Lyapunov function $W(x, z) = 2\epsilon V_1(x_1) + x_2^2 + k^{-1} V_2(z)$ and invoking similar arguments to those in the proof of Proposition 4.1, it follows that the equilibrium $(x, \hat{\theta}) = (0, \theta)$ is globally stable and (4.6) holds. $\square$

4.2.4 Unmatched Uncertainties

We now extend the result in Proposition 4.3 to a more general class of systems that do not necessarily satisfy the matching condition. As will become apparent in Example 4.2, the motivation is to allow for actuator dynamics to be appended to the system (4.18), thus allowing to deal with problems—often encountered in practice—where the transient characteristics of the actuators may be comparable to those of the control signal and hence cannot be neglected.

Consider the system

$$\begin{aligned} \dot{x}_1 &= f(x_1) + g(x_1) x_2, \\ \dot{x}_2 &= \theta_2 \nu_1 + \varphi(x, d)^\top \theta_1, \\ \dot{\nu}_1 &= \nu_2, \\ &\vdots \\ \dot{\nu}_{m-1} &= \nu_m, \\ \dot{\nu}_m &= \nu_{m+1} \triangleq h(\nu, d) + u, \end{aligned} \tag{4.24}$$

where $x = [x_1^\top, x_2]^\top \in \mathbb{R}^n \times \mathbb{R}$ and $\nu = [\nu_1, \ldots, \nu_m]^\top \in \mathbb{R}^m$ are the states, $u \in \mathbb{R}$ is the control input, $f(\cdot), g(\cdot)$ and $\varphi(\cdot)$ are smooth mappings, with $f(\cdot)$

satisfying Assumption 4.1, $\theta_1 \in \mathbb{R}^p$ and $\theta_2 \in \mathbb{R}$ are unknown parameters with $\theta_2 > 0$, and $d(t) \in \mathbb{R}^q$ is a known $\mathcal{C}^m$ signal with known derivatives.

Note that, if ν_1 were the control input, the design would be similar to the one in Proposition 4.3 with ν_1 equal to (4.20). Therefore, the objective is to find a control law u such that ν_1 asymptotically tracks the *reference* (4.20) and all signals remain bounded. This is achieved by extending the adaptive scheme of Section 4.2.3 and combining it with a backstepping construction.

Proposition 4.4. *Consider the system (4.24) and the adaptive state feedback control law*

$$\dot{\hat{\theta}} = -\left(I + \frac{\partial \beta}{\partial \hat{\theta}}\right)^{-1} \left(\frac{\partial \beta}{\partial x_1}\left(f(x_1) + g(x_1)x_2\right) + \frac{\partial \beta}{\partial d}\dot{d} + \frac{\partial \beta}{\partial \nu_1}\nu_2 \right.$$
$$\left. + \frac{\partial \beta}{\partial x_2}\left(-kx_2 - \epsilon L_g V_1(x_1) + (\hat{\theta}_3 + \beta_3)(\nu_1 - \nu_1^*)\right)\right), \quad (4.25)$$

$$\nu_1^* = -(\hat{\theta}_2 + \beta_2)\left(kx_2 + \epsilon L_g V_1(x_1) + \varphi(x,d)^\top(\hat{\theta}_1 + \beta_1)\right), \quad (4.26)$$

$$\nu_{i+1}^* = -\sigma_i + \frac{\partial \nu_i^*}{\partial \hat{\theta}}\dot{\hat{\theta}} + \frac{\partial \nu_i^*}{\partial x_1}(f(x_1) + g(x_1)x_2)$$
$$+ \frac{\partial \nu_i^*}{\partial x_2}\left(-kx_2 - \epsilon L_g V_1(x_1) + (\hat{\theta}_3 + \beta_3)(\nu_1 - \nu_1^*)\right)$$
$$+ \sum_{j=1}^{i-1}\left(\frac{\partial \nu_i^*}{\partial \nu_j}\nu_{j+1} + \frac{\partial \nu_i^*}{\partial d^{(j-1)}}d^{(j)}\right), \quad i=1,\ldots,m, \quad (4.27)$$

$$u = \nu_{m+1}^* - h(\nu, d), \quad (4.28)$$

where $\beta = [\beta_1(x,d)^\top, \beta_2(x,\hat{\theta}_1,d), \beta_3(x,\hat{\theta},\nu_1,d)]^\top$, *with*

$$\beta_1 = \gamma_1 \int_0^{x_2} \varphi(x_1,\chi,d)d\chi,$$

$$\beta_2 = \gamma_2\left(k\frac{x_2^2}{2} + \epsilon L_g V_1(x_1)x_2\right) + \gamma_2 \int_0^{x_2} \varphi(x_1,\chi,d)^\top(\hat{\theta}_1 + \beta_1(x_1,\chi,d))d\chi,$$

$$\beta_3 = \gamma_3 \nu_1 x_2 - \gamma_3 \int_0^{x_2} \nu_1^*(x_1,\chi,\hat{\theta},d)d\chi,$$

$k > 0$, $\epsilon > 0$, $\gamma_1 > 0$, $\gamma_2 > 0$, $\gamma_3 > 0$ *are constants, and*

$$\sigma_1 = \left(c_1 + \frac{\epsilon}{2}\left(\frac{\partial \nu_1^*}{\partial x_2}\right)^2\right)(\nu_1 - \nu_1^*) + (\hat{\theta}_3 + \beta_3)x_2,$$

$$\sigma_i = \left(c_i + \frac{\epsilon}{2}\left(\frac{\partial \nu_i^*}{\partial x_2}\right)^2\right)(\nu_i - \nu_i^*) + (\nu_{i-1} - \nu_{i-1}^*),$$

for $i = 2,\ldots,m$, *where* $c_i > 0$ *and* $\varepsilon > 0$ *are constants. Then the update law (4.25) is well-defined, the closed-loop system (4.24), (4.25), (4.28) has a globally stable equilibrium at* $(x,\nu,\hat{\theta}) = (0,0,\theta)$, *and condition (4.6) holds.*

4.2 Systems in Feedback Form   65

Proof. To begin with, define the off-the-manifold co-ordinates $z_1 = \hat{\theta}_1 - \theta_1 + \beta_1(x,d)$, $z_2 = \hat{\theta}_2 - \theta_2^{-1} + \beta_2(x, \hat{\theta}_1, d)$ and $z_3 = \hat{\theta}_3 - \theta_2 + \beta_3(x, \hat{\theta}, \nu_1, d)$. Note that, in contrast with the result in Section 4.2.3, here we do employ overparameterisation. Consider now the *virtual* control law (4.26), which is motivated by (4.20), and define the variable $\tilde{\nu}_1 = \nu_1 - \nu_1^*(x, \hat{\theta})$. The off-the-manifold dynamics are given by

$$\dot{z} = \left(I + \frac{\partial \beta}{\partial \hat{\theta}}\right)\dot{\hat{\theta}} + \frac{\partial \beta}{\partial x_1}\left(f(x_1) + g(x_1)x_2\right) + \frac{\partial \beta}{\partial d}\dot{d}$$
$$+ \frac{\partial \beta}{\partial x_2}\left(\theta_2\left(\nu_1^* + \tilde{\nu}_1\right) + \varphi(x,d)^\top\left(\hat{\theta}_1 + \beta_1 - z_1\right)\right) + \frac{\partial \beta}{\partial \nu_1}\nu_2.$$

Substituting (4.26) into the above equation yields

$$\dot{z} = \left(I + \frac{\partial \beta}{\partial \hat{\theta}}\right)\dot{\hat{\theta}} + \frac{\partial \beta}{\partial x_1}\left(f(x_1) + g(x_1)x_2\right) + \frac{\partial \beta}{\partial \nu_1}\nu_2$$
$$+ \frac{\partial \beta}{\partial x_2}\Big(-kx_2 - \epsilon L_g V_1(x_1) - \varphi(x,d)^\top z_1 + \theta_2 z_2$$
$$\times \left(-kx_2 - \epsilon L_g V_1(x_1) - \varphi(x,d)^\top\left(\hat{\theta}_1 + \beta_1\right)\right) + \theta_2 \tilde{\nu}_1\Big).$$

Consider now the update law (4.25) and note that the resulting error dynamics are given by

$$\dot{z} = -\frac{\partial \beta}{\partial x_2}\left(\varphi(x,d)^\top z_1 + \theta_2\left(kx_2 + \epsilon L_g V_1(x_1) + \varphi(x,d)^\top\left(\hat{\theta}_1 + \beta_1\right)\right)z_2 + \tilde{\nu}_1 z_3\right)$$
$$= -\Gamma \Phi \Phi^\top z, \tag{4.29}$$

where $\Gamma = \mathrm{diag}(\gamma_1, \gamma_2 \theta_2^{-1}, \gamma_3)$ and

$$\Phi(x, \hat{\theta}, \nu_1, d) = \begin{bmatrix} \varphi(x,d) \\ \theta_2\left(kx_2 + \epsilon L_g V_1(x_1) + \varphi(x,d)^\top\left(\hat{\theta}_1 + \beta_1\right)\right) \\ \nu_1 - \nu_1^*(x, \hat{\theta}, d) \end{bmatrix}.$$

Using the foregoing notation, the first two equations in (4.24) can be rewritten as

$$\dot{x}_1 = f(x_1) + g(x_1)x_2,$$
$$\dot{x}_2 = -kx_2 - \epsilon L_g V_1(x_1) - \Phi^\top z + (\hat{\theta}_3 + \beta_3)\tilde{\nu}_1. \tag{4.30}$$

Comparing with (4.12) and (4.23) and recalling the properties of (4.29) we conclude that, when $\tilde{\nu}_1 = 0$, the system (4.30) has a globally stable equilibrium at zero and (4.6) holds. It remains to prove that the control law (4.28) is such that all signals remain bounded and the last term in the second equation of (4.30) converges to zero. This can be shown recursively using the following backstepping procedure.

Step 1. The dynamics of $\tilde{\nu}_1$ are given by

$$\dot{\nu}_1 = \nu_2 - \frac{\partial \nu_1^*}{\partial \hat{\theta}}\dot{\hat{\theta}} - \frac{\partial \nu_1^*}{\partial x_1}(f(x_1) + g(x_1)x_2) - \frac{\partial \nu_1^*}{\partial d}\dot{d}$$
$$- \frac{\partial \nu_1^*}{\partial x_2}\Big(-kx_2 - \epsilon L_g V_1(x_1) - \Phi^\top z + (\hat{\theta}_3 + \beta_3)\tilde{\nu}_1\Big).$$

Consider ν_2 as a virtual control input and define the error $\tilde{\nu}_2 = \nu_2 - \nu_2^*(x, \hat{\theta}, \nu_1)$, where ν_2^* is given by (4.27). Note that the system that describes the dynamics of $\tilde{\nu}_1$ can be rewritten as

$$\dot{\tilde{\nu}}_1 = -\sigma_1 + \tilde{\nu}_2 + \frac{\partial \nu_1^*}{\partial x_2}\Phi^\top z. \qquad (4.31)$$

Step 2. The dynamics of $\tilde{\nu}_2$ are given by

$$\dot{\nu}_2 = \nu_3 - \frac{\partial \nu_2^*}{\partial \hat{\theta}}\dot{\hat{\theta}} - \frac{\partial \nu_2^*}{\partial x_1}(f(x_1) + g(x_1)x_2) - \frac{\partial \nu_2^*}{\partial d}\dot{d} - \frac{\partial \nu_2^*}{\partial \dot{d}}\ddot{d} - \frac{\partial \nu_2^*}{\partial \nu_1}\nu_2$$
$$- \frac{\partial \nu_2^*}{\partial x_2}\Big(-kx_2 - \epsilon L_g V_1(x_1) - \Phi^\top z + (\hat{\theta}_3 + \beta_3(x, \hat{\theta}, \nu_1))\tilde{\nu}_1\Big).$$

Consider ν_3 as a virtual control input and define the error $\tilde{\nu}_3 = \nu_3 - \nu_3^*(x, \hat{\theta}, \nu_1, \nu_2)$, where ν_3^* is given by (4.27). The dynamics of $\tilde{\nu}_2$ can be rewritten as

$$\dot{\tilde{\nu}}_2 = -\sigma_2 + \tilde{\nu}_3 + \frac{\partial \nu_2^*}{\partial x_2}\Phi^\top z. \qquad (4.32)$$

Step m. Finally, the dynamics of $\tilde{\nu}_m$ are given by

$$\dot{\nu}_m = h(\nu, d) + u - \frac{\partial \nu_m^*}{\partial \hat{\theta}}\dot{\hat{\theta}} - \frac{\partial \nu_m^*}{\partial x_1}(f(x_1) + g(x_1)x_2) - \sum_{j=1}^{m-1}\frac{\partial \nu_m^*}{\partial \nu_j}\nu_{j+1}$$
$$- \sum_{j=1}^{m-1}\frac{\partial \nu_m^*}{\partial d^{(j-1)}}d^{(j)} - \frac{\partial \nu_m^*}{\partial x_2}\Big(-kx_2 - \epsilon L_g V_1(x_1) - \Phi^\top z + (\hat{\theta}_3 + \beta_3)\tilde{\nu}_1\Big).$$

Substituting the control law (4.28) yields

$$\dot{\tilde{\nu}}_m = -\sigma_m + \frac{\partial \nu_m^*}{\partial x_2}\Phi^\top z. \qquad (4.33)$$

Consider now the function $W(x, \tilde{\nu}) = 2\epsilon V_1(x_1) + x_2^2 + |\tilde{\nu}|^2$, whose time-derivative along the trajectories of (4.24) is

$$\dot{W}(x, \tilde{\nu}) = 2\epsilon L_f V_1(x_1) - 2kx_2^2 - 2x_2\Phi^\top z + 2x_2(\hat{\theta}_3 + \beta_3)\tilde{\nu}_1 - 2\sigma_1\tilde{\nu}_1$$
$$+ 2\tilde{\nu}_1\tilde{\nu}_2 + 2\tilde{\nu}_1\frac{\partial \nu_1^*}{\partial x_2}\Phi^\top z - 2\sigma_2\tilde{\nu}_2 + 2\tilde{\nu}_2\tilde{\nu}_3 + 2\tilde{\nu}_2\frac{\partial \nu_2^*}{\partial x_2}\Phi^\top z - \cdots$$
$$- 2\sigma_m\tilde{\nu}_m + 2\tilde{\nu}_m\frac{\partial \nu_m^*}{\partial x_2}\Phi^\top z$$

$$\leq -2\epsilon\kappa(x_1) - kx_2^2 + \frac{1}{k}\left(\Phi^\top z\right)^2 + 2x_2\big(\hat{\theta}_3 + \beta_3\big)\tilde{\nu}_1 - 2\sigma_1\tilde{\nu}_1$$

$$+2\tilde{\nu}_1\tilde{\nu}_2 + \varepsilon\left(\frac{\partial \nu_1^*}{\partial x_2}\right)^2\tilde{\nu}_1^2 + \frac{1}{\varepsilon}\left(\Phi^\top z\right)^2 - 2\sigma_2\tilde{\nu}_2 + 2\tilde{\nu}_2\tilde{\nu}_3 + \varepsilon\left(\frac{\partial \nu_2^*}{\partial x_2}\right)^2\tilde{\nu}_2^2$$

$$+\frac{1}{\varepsilon}\left(\Phi^\top z\right)^2 - \cdots - 2\sigma_m\tilde{\nu}_m + \varepsilon\left(\frac{\partial \nu_m^*}{\partial x_2}\right)^2\tilde{\nu}_m^2 + \frac{1}{\varepsilon}\left(\Phi^\top z\right)^2$$

$$= -2\epsilon\kappa(x_1) - kx_2^2 - 2\sum_{j=1}^m c_j\tilde{\nu}_j^2 + \left(\frac{1}{k} + \frac{m}{\varepsilon}\right)\left(\Phi^\top z\right)^2.$$

The proof is completed by considering the Lyapunov function $W(x,\tilde{\nu}) + \left(k^{-1} + m\varepsilon^{-1}\right)z^\top \Gamma^{-1} z$ and invoking similar arguments to those in the proof of Proposition 4.1. □

The purpose of the functions $\sigma_i(\cdot)$ introduced at each step is not only to stabilise the zero equilibrium of each subsystem but also to render the system (4.31), (4.32), (4.33) $\mathcal{L}_2$ stable with respect to the perturbation $\Phi^\top z$. In contrast with the unmatched cases dealt with in Propositions 4.1, 4.2 and 4.3, the latter cannot be achieved simply with a linear feedback $c_i\tilde{\nu}_i$. Therefore, $\sigma_i(\cdot)$ must also contain a *nonlinear damping* term. Note, however, that the constant ε multiplying this term can be taken arbitrarily small.

4.3 Lower Triangular Systems

In this section it is shown that the I&I approach provides, in some cases (see Assumption 4.2), an alternative to the classical adaptive backstepping design.

Consider a class of systems that can be described in suitable co-ordinates by equations of the form[8]

$$\begin{aligned}
\dot{x}_1 &= x_2 + \varphi_1(x_1)^\top \theta, \\
\dot{x}_2 &= x_3 + \varphi_2(x_1, x_2)^\top \theta, \\
&\vdots \\
\dot{x}_i &= x_{i+1} + \varphi_i(x_1, \ldots, x_i)^\top \theta, \\
&\vdots \\
\dot{x}_n &= u + \varphi_n(x_1, \ldots, x_n)^\top \theta,
\end{aligned} \quad (4.34)$$

with states $x_i \in \mathbb{R}$, $i = 1,\ldots,n$, where $u \in \mathbb{R}$ is the control input, $\varphi_i(\cdot)$ are $\mathcal{C}^{n-i}$ mappings, and $\theta \in \mathbb{R}^p$ is an unknown constant vector. The control problem is to find an adaptive state feedback control law of the form (4.5)

---

[8]The system (4.34) is in the so-called parametric strict feedback form, extensively studied in [123] and related references.

such that all trajectories of the closed-loop system (4.34), (4.5) are bounded and
$$\lim_{t\to\infty}(x_1(t)-x_1^*)=0, \qquad (4.35)$$
where $x_1^* = x_1^*(t)$ is an arbitrary $\mathcal{C}^n$ reference signal.

4.3.1 Estimator Design

We first construct an estimator (of order np) for the unknown parameter vector θ. As will become clear, the overparameterisation is needed to achieve the desired modularity while preserving the triangular structure of the system.

To begin with, define the estimation errors
$$z_i = \hat{\theta}_i - \theta + \beta_i(x_1,\ldots,x_i), \qquad (4.36)$$
for $i = 1,\ldots,n$, where $\hat{\theta}_i$ are the estimator states and $\beta_i(\cdot)$ are $\mathcal{C}^{n-i}$ functions yet to be specified. Using the above definitions the dynamics of z_i are given by
$$\dot{z}_i = \dot{\hat{\theta}}_i + \sum_{k=1}^{i}\frac{\partial\beta_i}{\partial x_k}\left(x_{k+1}+\varphi_k(x_1,\ldots,x_k)^\top\theta\right)$$
$$= \dot{\hat{\theta}}_i + \sum_{k=1}^{i}\frac{\partial\beta_i}{\partial x_k}\left(x_{k+1}+\varphi_k(x_1,\ldots,x_k)^\top\left(\hat{\theta}_i+\beta_i(x_1,\ldots,x_i)-z_i\right)\right),$$
where $x_{n+1} \triangleq u$. Selecting the update laws $\dot{\hat{\theta}}_i$ to cancel the known quantities in the dynamics of z_i, i.e.,
$$\dot{\hat{\theta}}_i = -\sum_{k=1}^{i}\frac{\partial\beta_i}{\partial x_k}\left(x_{k+1}+\varphi_k(x_1,\ldots,x_k)^\top\left(\hat{\theta}_i+\beta_i(x_1,\ldots,x_i)\right)\right), \qquad (4.37)$$
yields the error dynamics
$$\dot{z}_i = -\left[\sum_{k=1}^{i}\frac{\partial\beta_i}{\partial x_k}\varphi_k(x_1,\ldots,x_k)^\top\right]z_i. \qquad (4.38)$$

Note that the system (4.38), for $i = 1,\ldots,n$, can be regarded as a linear time-varying system with a block diagonal dynamic matrix whose diagonal blocks have to be rendered negative-semidefinite. To this end, following the idea of Proposition 4.1, we select the functions $\beta_i(\cdot)$ as
$$\beta_i(x_1,\ldots,x_i) = \gamma_i\int_0^{x_i}\varphi_i(x_1,\ldots,x_{i-1},\chi)\mathrm{d}\chi + \varepsilon_i(x_i), \qquad (4.39)$$
where $\gamma_i > 0$ are constants and $\varepsilon_i(x_i)$ are $\mathcal{C}^{n-i}$ functions with $\varepsilon_1(x_1) = 0$.

Consider now the following assumption.

4.3 Lower Triangular Systems

Assumption 4.2. There exist functions $\varepsilon_i(x_i)$ satisfying the partial differential (matrix) inequality

$$F_i(x_1,\ldots,x_i)^\top + F_i(x_1,\ldots,x_i) \geq 0, \tag{4.40}$$

for $i = 2,\ldots,n$, where

$$F_i(x_1,\ldots,x_i) = \gamma_i \sum_{k=1}^{i-1} \frac{\partial}{\partial x_k}\left(\int_0^{x_i} \varphi_i(x_1,\ldots,x_{i-1},\chi)\mathrm{d}\chi\right)\varphi_k(x_1,\ldots,x_k)^\top$$
$$+ \frac{\partial \varepsilon_i}{\partial x_i}\varphi_i(x_1,\ldots,x_i)^\top.$$

Note that, in the special case when $\varphi_i(\cdot)$ is a function of x_i only, the partial differential (matrix) inequality (4.40) admits the trivial solution $\varepsilon_i(x_i) = 0$ for $i = 2,\ldots,n$. The same simplification occurs when only one of the functions $\varphi_i(\cdot)$ in (4.34) is nonzero (see Section 4.2.4). In general, the solvability of (4.40) depends strongly on the structure of the regressors $\varphi_i(\cdot)$.

Assumption 4.2 allows to establish the following lemma, which is instrumental for the control design.

Lemma 4.1. *Consider the system (4.38), where the functions $\beta_i(\cdot)$ are given by (4.39), and suppose that Assumption 4.2 holds. Then the system (4.38) has a uniformly globally stable equilibrium at the origin, $z_i(t) \in \mathcal{L}_\infty$ and $\varphi_i(x_1(t),\ldots,x_i(t))^\top z_i(t) \in \mathcal{L}_2$, for all $i = 1,\ldots,n$ and for all $x_1(t),\ldots,x_i(t)$. If in addition $\varphi_i(\cdot)$ and its time-derivative are bounded, then $\varphi_i(x_1,\ldots,x_i)^\top z_i$ converges to zero.*

Proof. Consider the Lyapunov function $V(z) = \sum_{i=1}^n z_i^\top z_i$ whose time-derivative along the trajectories of (4.38) satisfies

$$\dot{V}(z) = -\sum_{i=1}^n z_i^\top \left[2\gamma_i\varphi_i(x_1,\ldots,x_i)\varphi_i(x_1,\ldots,x_i)^\top + F_i(x_1,\ldots,x_i)\right.$$
$$\left. + F_i(x_1,\ldots,x_i)^\top\right]z_i$$
$$\leq -\sum_{i=1}^n 2\gamma_i\left(\varphi_i(x_1,\ldots,x_i)^\top z_i\right)^2,$$

where we have used Assumption 4.2 to obtain the last inequality. As a result, the system (4.38) has a uniformly globally stable equilibrium at the origin, $z_i(t) \in \mathcal{L}_\infty$ and $\varphi_i(x_1(t),\ldots,x_i(t))^\top z_i(t) \in \mathcal{L}_2$, for all $i = 1,\ldots,n$. Moreover, the above holds independently of the behaviour of the states $x_1,\ldots,x_i$. Finally, convergence of $\varphi_i(x_1,\ldots,x_i)^\top z_i$ to zero follows directly from Corollary A.1. □

By definition (4.36), the result in Lemma 4.1 implies that an asymptotically converging estimate of each term $\varphi_i(x_1,\ldots,x_i)^\top \theta$ in (4.34) is given by

70   4 I&I Adaptive Control: Systems in Special Forms

$$\varphi_i(x_1,\ldots,x_i)^\top(\hat{\theta}_i+\beta_i(x_1,\ldots,x_i)).$$

Note that we do not estimate the parameter θ, but only the "perturbation function" $\varphi_i(x_1,\ldots,x_i)^\top\theta$.

Example 4.1. Consider the two-dimensional system

$$\begin{aligned}\dot{x}_1 &= x_2+\varphi_1(x_1)^\top\theta,\\ \dot{x}_2 &= u+\varphi_2(x_1,x_2)^\top\theta,\end{aligned} \qquad (4.41)$$

and note that the matrix $F_2(\cdot)$ in (4.40) is given by

$$F_2(x_1,x_2)=\gamma_2\frac{\partial}{\partial x_1}\left(\int_0^{x_2}\varphi_2(x_1,\chi)\mathrm{d}\chi\right)\varphi_1(x_1)^\top+\frac{\partial\varepsilon_2}{\partial x_2}\varphi_2(x_1,x_2)^\top. \qquad (4.42)$$

Let $\varphi_1(x_1)=[x_1,\ 0]^\top$ and $\varphi_2(x_1,x_2)=[0,\ x_1^r x_2^q]^\top$, where $r,q\geq 1$. Substituting into (4.42) yields

$$\begin{aligned}F_2(x_1,x_2)&=\gamma_2\begin{bmatrix}0\\ \frac{rx_1^{r-1}x_2^{q+1}}{q+1}\end{bmatrix}\begin{bmatrix}x_1 & 0\end{bmatrix}+\frac{\partial\varepsilon_2}{\partial x_2}\begin{bmatrix}0 & x_1^r x_2^q\end{bmatrix}\\ &=\gamma_2\begin{bmatrix}0 & 0\\ \frac{rx_1^r x_2^{q+1}}{q+1} & 0\end{bmatrix}+\frac{\partial\varepsilon_2}{\partial x_2}\begin{bmatrix}0 & x_1^r x_2^q\end{bmatrix},\end{aligned}$$

which suggests selecting

$$\varepsilon_2(x_2)=-\gamma_2\begin{bmatrix}\frac{rx_2^2}{2(q+1)}\\ 0\end{bmatrix},$$

so that the matrix (4.42) becomes skew-symmetric, hence condition (4.40) holds. Note that the regressors $\varphi_1(x_1)$ and $\varphi_2(x_1,x_2)$ are such that the parameters entering in each equation of the system (4.41) are independent. If in addition $r=0$, then $\varepsilon_2(x_2)=0$ and the overparameterisation is *eliminated* since only the ith element of each function $\beta_i(\cdot)$ in (4.39) is nonzero. Letting instead $\varphi_1(x_1)=x_1$ and $\varphi_2(x_1,x_2)=x_1^r x_2^q$ in (4.42) yields the matrix

$$F(x_1,x_2)=\gamma_2\frac{rx_1^r x_2^{q+1}}{q+1}+\frac{\partial\varepsilon_2}{\partial x_2}x_1^r x_2^q,$$

which satisfies condition (4.40) for $\varepsilon_2(x_2)=-\gamma_2\frac{rx_2^2}{2(q+1)}$.   ∎

4.3.2 Controller Design

In this section we propose a control law that renders the closed-loop system $\mathcal{L}_2$ stable from the "perturbation" inputs $\varphi_i(x_1,\ldots,x_i)^\top z_i$ to the output $x_1-x_1^\star$ and keeps all signals bounded. As pointed out in Lemma 4.1, these two properties imply that $\varphi_i(x_1,\ldots,x_i)^\top z_i$ converge to zero, hence (4.35) holds. The result is summarised in the following statement.

4.3 Lower Triangular Systems

Proposition 4.5. *Consider the system (4.34), (4.37), where the functions $\beta_i(\cdot)$ are given by (4.39) with $\varepsilon_i(\cdot)$ such that Assumption 4.2 holds, and the control law*

$$
\begin{aligned}
x_{i+1}^* &= -\sigma_i - \varphi_i(x_1,\ldots,x_i)^\top \left(\hat{\theta}_i + \beta_i(x_1,\ldots,x_i)\right) \\
&\quad + \sum_{k=1}^{i-1} \frac{\partial x_i^*}{\partial x_k}\left[x_{k+1} + \varphi_k(x_1,\ldots,x_k)^\top\left(\hat{\theta}_k + \beta_k(x_1,\ldots,x_k)\right)\right] \\
&\quad + \sum_{k=1}^{i-1} \frac{\partial x_i^*}{\partial \hat{\theta}_k}\dot{\hat{\theta}}_k + x_1^{*\,(i)}, \qquad i = 1,\ldots,n, \\
u &= x_{n+1}^*, \tag{4.43}
\end{aligned}
$$

with

$$
\sigma_1 = \left(c_1 + \frac{\epsilon}{2}\right)(x_1 - x_1^*),
$$

$$
\sigma_i = \left(c_i + \frac{\epsilon}{2}\right)(x_i - x_i^*) + \frac{\epsilon}{2}\sum_{k=1}^{i-1}\left(\frac{\partial x_i^*}{\partial x_k}\right)^2(x_i - x_i^*) + (x_{i-1} - x_{i-1}^*),
$$

for $i = 2,\ldots,n$, where $c_i > 0$ and $\epsilon > 0$ are constants. Then all trajectories of the closed-loop system (4.34), (4.37), (4.43) are bounded and (4.35) holds.

Proof. To begin with, note that for the error variables defined in (4.36) the result in Lemma 4.1 holds. We now prove that the control law defined recursively in (4.43) is such that all signals remain bounded and (4.35) holds. Consider the change of co-ordinates

$$\tilde{x}_i = x_i - x_i^*,$$

for $i = 1,\ldots,n$, and note that the closed-loop system can be partly described in these co-ordinates by the equations

$$
\begin{aligned}
\dot{\tilde{x}}_1 &= -\sigma_1 + \tilde{x}_2 - \varphi_1(x_1)^\top z_1 \\
\dot{\tilde{x}}_2 &= -\sigma_2 + \tilde{x}_3 + \frac{\partial x_2^*}{\partial x_1}\varphi_1(x_1)^\top z_1 - \varphi_2(x_1,x_2)^\top z_2 \\
&\;\;\vdots \\
\dot{\tilde{x}}_n &= -\sigma_n + \sum_{k=1}^{n-1}\frac{\partial x_n^*}{\partial x_k}\varphi_k(x_1,\ldots,x_k)^\top z_k - \varphi_n(x_1,\ldots,x_n)^\top z_n.
\end{aligned} \tag{4.44}
$$

It remains to show that the functions $\sigma_i(\cdot)$ in Proposition 4.5 are such that the system (4.38), (4.44) has a globally stable equilibrium at the origin and (4.35) holds. To this end, consider the function $W(\tilde{x}) = \sum_{k=1}^{n}\tilde{x}_k^2$, whose time-derivative along the trajectories of (4.44) satisfies

$$\dot{W} = -2\tilde{x}_1\sigma_1 + 2\tilde{x}_1\tilde{x}_2 - 2\tilde{x}_1\varphi_1(x_1)^\top z_1 - 2\tilde{x}_2\sigma_2 + 2\tilde{x}_2\tilde{x}_3$$
$$+2\tilde{x}_2\frac{\partial x_2^*}{\partial x_1}\varphi_1(x_1)^\top z_1 - 2\tilde{x}_2\varphi_2(x_1,x_2)^\top z_2 + \cdots - 2\tilde{x}_n\sigma_n$$
$$+2\tilde{x}_n\sum_{k=1}^{n-1}\frac{\partial x_n^*}{\partial x_k}\varphi_k(x_1,\ldots,x_k)^\top z_k - 2\tilde{x}_n\varphi_n(x_1,\ldots,x_n)^\top z_n$$
$$\leq -2\tilde{x}_1\sigma_1 + 2\tilde{x}_1\tilde{x}_2 + \epsilon\tilde{x}_1^2 + \frac{1}{\epsilon}\left(\varphi_1(x_1)^\top z_1\right)^2$$
$$-2\tilde{x}_2\sigma_2 + 2\tilde{x}_2\tilde{x}_3 + \epsilon\left(\frac{\partial x_2^*}{\partial x_1}\right)^2\tilde{x}_2^2 + \frac{1}{\epsilon}\left(\varphi_1(x_1)^\top z_1\right)^2$$
$$+\epsilon\tilde{x}_2^2 + \frac{1}{\epsilon}\left(\varphi_2(x_1,x_2)^\top z_2\right)^2 + \cdots - 2\tilde{x}_n\sigma_n + \epsilon\sum_{k=1}^{n-1}\left(\frac{\partial x_n^*}{\partial x_k}\right)^2\tilde{x}_n^2$$
$$+\frac{1}{\epsilon}\sum_{k=1}^{n-1}\left(\varphi_k(x_1,\ldots,x_k)^\top z_k\right)^2 + \epsilon\tilde{x}_n^2 + \frac{1}{\epsilon}\left(\varphi_n(x_1,\ldots,x_n)^\top z_n\right)^2.$$

Selecting the functions $\sigma_i(\cdot)$ as in Proposition 4.5 yields

$$\dot{W}(\tilde{x}) \leq -2\sum_{i=1}^n c_i\tilde{x}_i^2 + \frac{1}{\epsilon}\sum_{i=1}^n (n-i+1)\left(\varphi_i(x_1,\ldots,x_i)^\top z_i\right)^2.$$

The proof is completed by combining the function $W(\tilde{x})$ given above with the one in the proof of Lemma 4.1 and invoking similar arguments to those in the proof of Proposition 4.1. □

Example 4.2 (Aircraft wing rock). In this example the I&I methodology is applied to the problem of wing rock elimination in high-performance aircrafts. Wing rock is a limit cycle oscillation which appears in the rolling motion of slender delta wings at high angles of attack[9]. The motion of the wing can be described by the equations

$$\begin{aligned}\dot{x}_1 &= x_2,\\ \dot{x}_2 &= x_3 + \varphi(x_1,x_2)^\top\theta,\\ \dot{x}_3 &= \frac{1}{\tau}u - \frac{1}{\tau}x_3,\end{aligned} \quad (4.45)$$

where the states x_1, x_2 and x_3 represent the roll angle, roll rate and aileron deflection angle, respectively, τ is the aileron time constant, u is the control input, $\theta \in \mathbb{R}^5$ is an unknown constant vector and

$$\varphi(x_1,x_2)^\top = \begin{bmatrix} 1, & x_1, & x_2, & |x_1|x_2, & |x_2|x_2 \end{bmatrix}.$$

---

[9] For details on this problem see *e.g.*, [65, 69] and the references in [123, Section 4.6]. This example has been adopted from [146] and [123, Section 4.6], where a controller based on the adaptive backstepping method has been proposed.

The control objective is to regulate x_1 to zero with all signals bounded.

To begin with, consider the estimation error $z = \hat{\theta} - \theta + \beta(x_1, x_2)$ and note that the function $\beta(\cdot)$ can be selected according to Proposition 4.4, namely

$$\beta(x_1, x_2) = k \left[x_2,\ x_1 x_2,\ \frac{1}{2} x_2^2,\ \frac{1}{2}|x_1| x_2^2,\ \frac{1}{3}|x_2| x_2^2 \right],$$

yielding the error dynamics

$$\dot{z} = -k \varphi(x_1, x_2) \varphi(x_1, x_2)^\top z. \tag{4.46}$$

Following the construction in the proof of Proposition 4.4 and applying the control law

$$\frac{1}{\tau} u = \frac{1}{\tau} x_3 - \sigma_3(x_1, x_2, x_3, \hat{\theta}) + \frac{\partial x_3^*}{\partial \hat{\theta}} \dot{\hat{\theta}} + \frac{\partial x_3^*}{\partial x_1} x_2$$
$$+ \frac{\partial x_3^*}{\partial x_2} \left[x_3 + \varphi(x_1, x_2)^\top (\hat{\theta} + \beta(x_1, x_2)) \right], \tag{4.47}$$

the closed-loop system can be written as

$$\begin{aligned}
\dot{x}_1 &= -\sigma_1(x_1) + \tilde{x}_2, \\
\dot{\tilde{x}}_2 &= -\sigma_2(x_1, x_2, \hat{\theta}) + \tilde{x}_3 - \varphi(x_1, x_2)^\top z, \\
\dot{\tilde{x}}_3 &= -\sigma_3(x_1, x_2, x_3, \hat{\theta}) + \frac{\partial x_3^*}{\partial x_2} \varphi(x_1, x_2)^\top z.
\end{aligned} \tag{4.48}$$

Consider now the function

$$W(x_1, \tilde{x}_2, \tilde{x}_3) = \frac{1}{2} \left(x_1^2 + \tilde{x}_2^2 + \tilde{x}_3^2 \right),$$

whose time-derivative along the trajectories of (4.48) satisfies

$$\begin{aligned}
\dot{W} &= -x_1 \sigma_1(x_1) + x_1 \tilde{x}_2 - \tilde{x}_2 \sigma_2(x_1, x_2, \hat{\theta}) + \tilde{x}_2 \tilde{x}_3 - \tilde{x}_2 \varphi(x_1, x_2)^\top z \\
&\quad - \tilde{x}_3 \sigma_3(x_1, x_2, x_3, \hat{\theta}) + \tilde{x}_3 \frac{\partial x_3^*}{\partial x_2} \varphi(x_1, x_2)^\top z \\
&\leq -x_1 \sigma_1(x_1) + x_1 \tilde{x}_2 - \tilde{x}_2 \sigma_2(x_1, x_2, \hat{\theta}) + \tilde{x}_2 \tilde{x}_3 \\
&\quad + \frac{\epsilon}{2} \left(\varphi(x_1, x_2)^\top z \right)^2 + \frac{1}{2\epsilon} \tilde{x}_2^2 - \tilde{x}_3 \sigma_3(x_1, x_2, x_3, \hat{\theta}) \\
&\quad + \frac{\epsilon}{2} \left(\varphi(x_1, x_2)^\top z \right)^2 + \frac{1}{2\epsilon} \left(\frac{\partial x_3^*}{\partial x_2} \right)^2 \tilde{x}_3^2.
\end{aligned}$$

Selecting the functions $\sigma_i(\cdot)$ as in Proposition 4.4, namely

$$\sigma_1 = c_1 x_1, \quad \sigma_2 = x_1 + \left(c_2 + \frac{1}{2\epsilon} \right) \tilde{x}_2, \quad \sigma_3 = \tilde{x}_2 + \left(c_3 + \frac{1}{2\epsilon} \left(\frac{\partial x_3^*}{\partial x_2} \right)^2 \right) \tilde{x}_3,$$

where c_1, c_2, c_3 are positive constants, yields

$$\dot{W} \leq -c_1 x_1^2 - c_2 \tilde{x}_2^2 - c_3 \tilde{x}_3^2 + \epsilon \left(\varphi(x_1, x_2)^\top z\right)^2.$$

Hence, from (4.46), the closed-loop system has a globally stable equilibrium with the Lyapunov function $W(x_1, \tilde{x}_2, \tilde{x}_3) + \frac{\epsilon}{2k} z^\top z$ and, moreover, $\lim_{t \to \infty} x_1(t) = 0$.

The system (4.45) in closed loop with the controller (4.47) has been simulated using the parameters $\tau = 1/15$ and $\theta = [0, -26.67, 0.76485, -2.9225, 0]$, while the design parameters are set to $c_1 = c_2 = c_3 = 5$, $k = 100$ and $\epsilon = 5000$. For comparison purposes, we have also implemented the full-information controller, which is obtained by assuming the parameters are known and applying standard feedback linearisation.

Figure 4.1 shows the trajectory of the controlled system for the initial conditions $x_1(0) = 0.4$, $x_2(0) = x_3(0) = 0$, $\hat{\theta}(0) = 0$, and for different values of the adaptive gain γ. Observe that, as the parameter γ increases, the adaptive scheme recovers the performance of the full-information controller. Note that, due to the form of the error dynamics (4.46), which is imposed by the selection of the function $\beta(\cdot)$, the norm of the error z is decreasing with time at a rate that is directly related to the gain γ.

In contrast, the adaptive backstepping controller exhibits an "underdamped" behaviour for all values of the adaptive gain γ. In particular, for increasing values of γ the behaviour close to the origin improves but further from the origin it is increasingly oscillatory leading to unacceptably large values for the state x_3, which corresponds to the actuator output. This undesirable effect is due to the strong coupling between the dynamics of the plant states and those of the estimation error. ∎

4.3.3 Estimator Design using Dynamic Scaling

In this section we show how Assumption 4.2 (which, as we have seen, is restrictive in the general case) can be removed by adding a *dynamic* scaling factor in the estimator dynamics[10].

Consider again the system (4.34) which can be written in the form

$$\dot{x}_i = x_{i+1} + \varphi_i(x_1, \ldots, x_i)^\top \theta_i, \qquad (4.49)$$

with states $x_i \in \mathbb{R}$, $i = 1, \ldots, n$ and control input $u \triangleq x_{n+1}$, where $\varphi_i(\cdot)$ are C^{n-i} mappings, and $\theta_i \in \mathbb{R}^{p_i}$ are unknown constant vectors. Let

$$z_i = \frac{\hat{\theta}_i - \theta_i + \beta_i(\hat{x}_1, \ldots, \hat{x}_{i-1}, x_i)}{r_i}, \qquad (4.50)$$

---

[10] Dynamic scaling has been widely used in the framework of high-gain observers, see *e.g.*, [176] and [121].

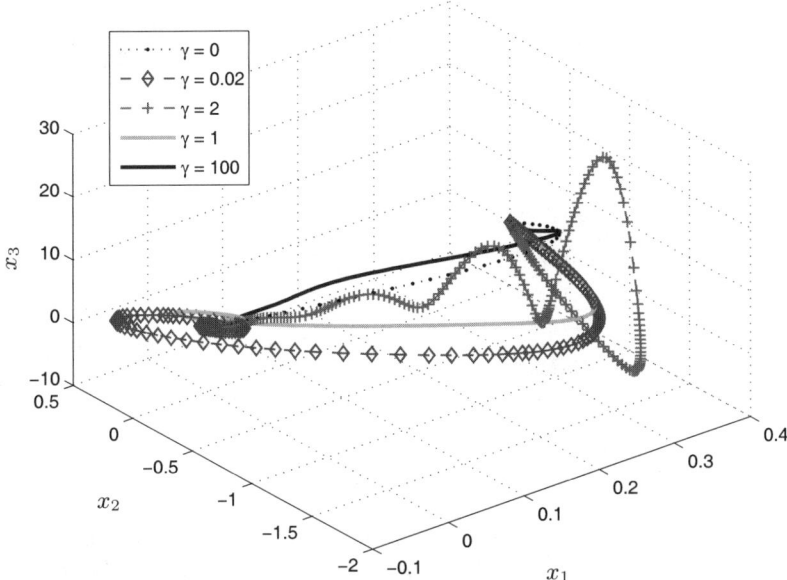

Fig. 4.1. Trajectory of the aircraft wing rock system for the initial conditions $x_1(0) = 0.4$, $x_2(0) = x_3(0) = 0$. Dotted line: full-information controller ($\gamma = 0$). Dashed lines: adaptive backstepping controller. Solid lines: I&I controller.

for $i = 1, \ldots, n$, where $\hat{\theta}_i$ are the estimator states, r_i are scaling factors, $\beta_i(\cdot)$ are $\mathcal{C}^{n-i}$ functions yet to be specified, and the auxiliary states $\hat{x}_i$ are obtained from the filter

$$\dot{\hat{x}}_i = x_{i+1} + \varphi_i(x_1, \ldots, x_i)^\top \left(\hat{\theta}_i + \beta_i\right) - k_i(\hat{x}_i - x_i), \tag{4.51}$$

for $i = 1, \ldots, n-1$, where $k_i = k_i(\cdot)$ are positive functions. Using the above definitions and the update laws

$$\dot{\hat{\theta}}_i = -\sum_{j=1}^{i-1} \frac{\partial \beta_i}{\partial \hat{x}_j} \dot{\hat{x}}_j - \frac{\partial \beta_i}{\partial x_i} \left(x_{i+1} + \varphi_i(x_1, \ldots, x_i)^\top \left(\hat{\theta}_i + \beta_i\right)\right) \tag{4.52}$$

yields the error dynamics

$$\dot{z}_i = -\frac{\partial \beta_i}{\partial x_i} \varphi_i(x_1, \ldots, x_i)^\top z_i - \frac{\dot{r}_i}{r_i} z_i. \tag{4.53}$$

Note that, similarly to (4.38), the system (4.53), for $i = 1, \ldots, n$, can be regarded as a linear time-varying system with a block diagonal dynamic matrix. In order to render the diagonal blocks negative-semidefinite, we select the functions $\beta_i(\cdot)$ as

$$\beta_i(\hat{x}_1,\ldots,\hat{x}_{i-1},x_i) = \gamma_i \int_0^{x_i} \varphi_i(\hat{x}_1,\ldots,\hat{x}_{i-1},\chi)\mathrm{d}\chi, \qquad (4.54)$$

where γ_i are positive constants.

Let $e_i = \hat{x}_i - x_i$ and note that, since $\varphi_i(\cdot)$ is continuously differentiable, we can write

$$\varphi_i(\hat{x}_1,\ldots,\hat{x}_{i-1},x_i) = \varphi_i(x_1,\ldots,x_i) - \sum_{j=1}^{i-1} e_j \delta_{ij}(x_1,\ldots,x_i,e_1,\ldots,e_j),$$

for some functions $\delta_{ij}(\cdot)$. Using the above equation and substituting (4.54) into (4.53) yields the error dynamics[11]

$$\dot{z}_i = -\gamma_i \varphi_i \varphi_i^\top z_i + \gamma_i \sum_{j=1}^{i-1} e_j \delta_{ij} \varphi_i^\top z_i - \frac{\dot{r}_i}{r_i} z_i. \qquad (4.55)$$

The above system has an equilibrium at zero and this can be rendered uniformly globally stable by selecting the dynamics of the scaling factors r_i as

$$\dot{r}_i = c_i r_i \sum_{j=1}^{i-1} e_j^2 |\delta_{ij}|^2, \qquad r_i(0) = 1, \qquad (4.56)$$

with

$$c_i \geq \gamma_i \frac{i-1}{2}.$$

From (4.56) we have that $r_i(t) \geq 1$, for all $t \geq 0$. Obviously, we can select $c_1 = 0$ and, since the filter is of order $n-1$, the total order of the estimator becomes $\sum_{i=1}^n p_i + 2(n-1)$.

Finally, note that from (4.49) and (4.51) the dynamics of $e_i = \hat{x}_i - x_i$ are given by

$$\dot{e}_i = -k_i e_i + r_i \varphi_i^\top z_i, \qquad (4.57)$$

hence selecting the functions k_i to satisfy

$$k_i \geq \lambda_i r_i^2 + \epsilon \sum_{j=0}^{n-i-1} c_{n-j} r_{n-j}^2 |\delta_{(n-j)i}|^2, \qquad (4.58)$$

where $\lambda_i > 0$ and $\epsilon > 0$ are constants, ensures that r is bounded and the system (4.55), (4.57) has a uniformly globally stable equilibrium at $(z,e) = (0,0)$.

Note that in the special case when $\varphi_i(\cdot)$ is a function of x_i only, the auxiliary states (4.51) are not used in the adaptive law and $\delta_{ij}(\cdot) = 0$ which

---

[11] In what follows, for compactness we drop the function arguments.

4.3 Lower Triangular Systems

implies that $\dot{r}_i = 0$, hence we can simply fix the scaling factors r_i to be equal to one. The same simplification occurs when only one of the $\varphi_i(\cdot)$ functions is nonzero.

The following lemma establishes the properties of the proposed estimator.

Lemma 4.2. *The system (4.55), (4.56), (4.57) has a uniformly globally stable equilibrium manifold defined by $\{(z, r, e) \mid z = e = 0\}$. Moreover, z_i, r_i and e_i are bounded, $e_i(t) \in \mathcal{L}_2$, and $\varphi_i(x_1(t), \ldots, x_i(t))^\top z_i(t) \in \mathcal{L}_2$, for all $i = 1, \ldots, n$, and for all $x_1(t), \ldots, x_i(t)$. If, in addition, $\varphi_i(\cdot)$ and its time-derivative are bounded, then the signals $\varphi_i(x_1(t), \ldots, x_i(t))^\top z_i(t)$ converge to zero.*

Proof. Consider the positive-definite and proper function

$$V_i(z_i) = \frac{1}{2\gamma_i}|z_i|^2,$$

whose time-derivative along the trajectories of (4.55) satisfies

$$\dot{V}_i \leq -\left(\varphi_i^\top z_i\right)^2 + \sum_{j=1}^{i-1}\left[\frac{1}{2(i-1)}\left(\varphi_i^\top z_i\right)^2 + \frac{i-1}{2}e_j^2\left(\delta_{ij}^\top z_i\right)^2\right] - \frac{\dot{r}_i}{\gamma_i r_i}|z_i|^2,$$

where for compactness we have dropped the function arguments. Applying Cauchy's inequality $|\delta_{ij}^\top z_i| \leq |\delta_{ij}||z_i|$, and substituting $\dot{r}_i$ from (4.56), yields

$$\dot{V}_i \leq -\frac{1}{2}\left(\varphi_i^\top z_i\right)^2,$$

which implies that the system (4.55) has a uniformly globally stable equilibrium at the origin, $z_i(t) \in \mathcal{L}_\infty$ and $\varphi_i(x_1(t), \ldots, x_i(t))^\top z_i(t) \in \mathcal{L}_2$, for all $i = 1, \ldots, n$. Moreover, the above holds true independently of the behaviour of the states $x_1, \ldots, x_i$ and $e_1, \ldots, e_{i-1}$.

Consider now the function

$$W_i(e_i, z_i) = \frac{1}{2}|e_i|^2 + \frac{1}{\lambda_i}V_i(z_i),$$

whose time-derivative along the trajectories of (4.55), (4.57) satisfies

$$\dot{W}_i \leq -k_i e_i^2 + \frac{1}{2\lambda_i}\left(\varphi_i^\top z_i\right)^2 + \frac{\lambda_i}{2}r_i^2 e_i^2 - \frac{1}{2\lambda_i}\left(\varphi_i^\top z_i\right)^2 \leq -\frac{\lambda_i}{2}e_i^2,$$

for any $\lambda_i > 0$, from where we conclude that the system (4.55), (4.57) has a uniformly globally stable equilibrium at $(z_i, e_i) = (0, 0)$ and $e_i(t) \in \mathcal{L}_2 \cap \mathcal{L}_\infty$.

It remains to show that the signals r_i are bounded. To this end, consider the combined Lyapunov function

$$U(e, z, r) = \sum_{i=1}^{n}\left[W_i(e_i, z_i) + \frac{\epsilon}{2}r_i^2\right],$$

whose time-derivative along the trajectories of (4.55), (4.56), (4.57) satisfies

$$\dot{U} \leq -\sum_{i=1}^{n}\left(k_i - \frac{\lambda_i}{2}r_i^2\right)e_i^2 + \epsilon\sum_{i=2}^{n}\left[c_i r_i^2 \sum_{j=1}^{i-1} e_j^2 |\delta_{ij}|^2\right].$$

Note now that the last term is equal to

$$\epsilon \sum_{i=2}^{n} \sum_{j=0}^{n-i-1} c_{n-j} r_{n-j}^2 |\delta_{(n-j)i}|^2 e_i^2,$$

hence selecting k_i from (4.58) ensures that

$$\dot{U} \leq -\sum_{i=1}^{n} \frac{\lambda_i}{2} e_i^2,$$

which proves that $r_i(t) \in \mathcal{L}_\infty$ and $\lim_{t \to \infty} e_i(t) = 0$. Finally, when $\varphi_i(\cdot)$ and its time-derivative are bounded, it follows from Barbalat's Lemma that $\varphi_i^\top z_i$ converge to zero. □

The above estimation design procedure can easily be extended to systems of the form

$$\dot{x} = f(x, u) + \Phi(x)\theta, \tag{4.59}$$

with state $x \in \mathbb{R}^n$ and input $u \in \mathbb{R}^m$, where $\theta \in \mathbb{R}^p$ is an unknown constant vector and each element of the vector $\Phi(x)\theta$ has the form $\varphi_i(x)^\top \theta_i$. Note that, in the general case, the filter (4.51) must be of order n, since the functions $\varphi_i(\cdot)$, $i = 1, \ldots, n-1$, may depend on x_n. Finally, it is possible to derive a counterpart of Proposition 4.5 with the dynamic scaling-based estimator replacing the estimator of Section 4.3.1.

4.4 Linear Systems

In this section we discuss how the I&I approach can be used to design adaptive controllers for linear systems. First, we show that the result in Proposition 4.4 can be applied to single-input single-output (SISO) linear time-invariant (LTI) systems with unknown parameters, under the standard assumptions that the plant is minimum-phase and the sign of the high-frequency gain is known.

We then show that it is possible to globally adaptively stabilise linear multivariable systems with reduced prior knowledge on the high-frequency gain. In particular we relax a restrictive (non-generic) symmetry condition usually required to solve this problem.

4.4.1 Linear SISO Systems

Consider a SISO LTI plant with input u and output y described by the transfer function

$$\frac{y(s)}{u(s)} = \frac{b_r s^{n-r} + \cdots + b_{n-1} s + b_n}{s^n + a_1 s^{n-1} + \cdots + a_{n-1} s + a_n} \triangleq \frac{N(s)}{D(s)}, \qquad (4.60)$$

where $0 < r \leq n$ and the polynomials $N(s)$ and $D(s)$ are coprime with unknown coefficients. It is assumed that the order n and the relative degree r of the plant are known. We design an adaptive controller for the system (4.60) using the result in Proposition 4.4. As in certainty-equivalent direct adaptive control, also known in the literature as model reference adaptive control (MRAC)[12], the main assumptions are that the system (4.60) is minimum-phase and that the sign of the first nonzero Markov parameter—the high-frequency gain—is known[13].

Motivated by the classical MRAC parameterisation and in order to obtain a form similar to (4.24), we define the asymptotically stable input/output filters

$$\begin{aligned} \dot{\zeta} &= A\zeta + Bu, \\ \dot{\xi} &= A\xi - By, \end{aligned} \qquad (4.61)$$

where $\zeta \in \mathbb{R}^{n-1}, \xi \in \mathbb{R}^{n-1}$,

$$A = \begin{bmatrix} 0 & 1 & \cdots & 0 \\ \vdots & & \ddots & \\ 0 & 0 & \cdots & 1 \\ -\lambda_1 & -\lambda_2 & \cdots & -\lambda_{n-1} \end{bmatrix}, \quad B = \begin{bmatrix} 0 \\ \vdots \\ 0 \\ 1 \end{bmatrix},$$

and the vector $\Lambda = [\lambda_1, \lambda_2, \ldots, \lambda_{n-1}]^\top \in \mathbb{R}^{n-1}$ is such that all eigenvalues of the matrix A have negative real part.

A state-space realisation of the transfer function (4.60) is given by the equations

$$\begin{aligned} \dot{x}_1 &= x_2 - a_1 x_1, \\ &\vdots \\ \dot{x}_r &= x_{r+1} - a_r x_1 + b_r u, \\ &\vdots \\ \dot{x}_n &= -a_n x_1 + b_n u, \\ y &= x_1. \end{aligned} \qquad (4.62)$$

Define the unknown parameters

---

[12] See, e.g., [77, 152, 131, 191].
[13] This assumption can be relaxed using exhaustive search procedures in parameter space. It is well known, however, that these schemes may exhibit practically inadmissible transient behaviours, see, e.g., [131].

$$\theta_2 = b_r, \qquad \theta_1 = [b_{r+1}, \ldots, b_n, a_1 - \lambda_{n-1}, \ldots, a_{n-1} - \lambda_1, a_n]^\top$$

and note that the system (4.62) can be transformed into the system

$$\dot{y} = \theta_2 \zeta_{n-r+1} + \varphi(\xi, y, \zeta_1, \ldots, \zeta_{n-r})^\top \theta_1 + e_1^\top \eta,$$
$$\dot{\eta} = A\eta,$$

where $\eta \in \mathbb{R}^{n-1}$, ζ_n and ξ_n are defined as $\zeta_n \triangleq \dot{\zeta}_{n-1} = -\Lambda^\top \zeta + u$ and $\xi_n \triangleq \dot{\xi}_{n-1} = -\Lambda^\top \xi - y$, respectively, and

$$\varphi(\xi, y, \zeta_1, \ldots, \zeta_{n-r}) = [\zeta_{n-r}, \ldots, \zeta_1, \xi_n, \ldots, \xi_1]^\top.$$

In the following we neglect the exponentially decaying state η, the presence of which, as will become clear, does not alter the stability properties of the closed-loop system.

Note 4.1. The foregoing parameterisation is conceptually similar to the standard parameterisation used in classical adaptive control (see, e.g., [77]), the only difference being that the filters used are of order $n - 1$ instead of order n. This is to ensure that the derivative of the output (rather than the output itself) depends linearly on the unknown parameters. ◁

The overall system can be rewritten as

$$\dot{\xi} = A\xi - By,$$
$$\dot{y} = \theta_2 \zeta_{n-r+1} + \varphi(\xi, y, \zeta_1, \ldots, \zeta_{n-r})^\top \theta_1,$$
$$\dot{\zeta}_{n-r+1} = \zeta_{n-r+2},$$
$$\vdots$$
$$\dot{\zeta}_{n-1} = -\Lambda^\top \zeta + u,$$

which is of the form (4.24) with $x_1 = \xi$, $x_2 = y$, $d = [\zeta_1, \ldots, \zeta_{n-r}]^\top$ and $\nu = [\zeta_{n-r+1}, \ldots, \zeta_{n-1}]^\top$, hence the result in Proposition 4.4 is applicable, while stability of the internal dynamics with input ζ_{n-r+1}, namely

$$\dot{\zeta}_1 = \zeta_2, \quad \cdots \quad \dot{\zeta}_{n-r} = \zeta_{n-r+1},$$

follows from the minimum-phase property. The properties of the resulting adaptive controller are illustrated in the following statement.

Corollary 4.1. *Consider the system (4.61), (4.62) and the adaptive state feedback control law (4.25)–(4.28), where x_1, x_2, d, and ν are replaced by ξ, y, $[\zeta_1, \ldots, \zeta_{n-r}]^\top$, and $[\zeta_{n-r+1}, \ldots, \zeta_{n-1}]^\top$, respectively, and $V_1 = \xi^\top P \xi$ with P a positive-definite matrix satisfying the Lyapunov equation $A^\top P + PA = -I$. Then the update law (4.25) is well-defined, all trajectories of the closed-loop system are bounded and $\lim_{t \to \infty} y(t) = 0$.*

4.4.2 Linear Multivariable Systems

The result in the previous section relies on the assumption that the sign of the high-frequency gain is known. For multivariable plants, where the high-frequency gain is a matrix, say K_p, MRAC typically assumes[14] the knowledge of a (nonsingular) matrix Γ such that

$$K_p \Gamma^\top = \Gamma K_p^\top > 0. \tag{4.63}$$

It is important to note that (4.63) implies a symmetry condition which is quite restrictive and non-generic since it involves *equality constraints*. In this section we show that the weaker assumption

$$K_p \Gamma^\top + \Gamma K_p^\top > 0 \tag{4.64}$$

is sufficient for global stabilisation (and global tracking) with an adaptive scheme stemming from the application of the I&I approach.

We consider first the simplest example that captures the central issue of prior knowledge on K_p. Namely, we consider the problem of stabilisation of the zero equilibrium of the system

$$\dot{y} = K_p u, \tag{4.65}$$

where $u, y \in \mathbb{R}^m$, and K_p is unknown and nonsingular. We then show that this basic result can be immediately extended to the problem of asymptotic tracking for a trajectory $y^* = y^*(t)$. Finally, we treat the case of arbitrary minimum-phase systems of known order and (vector) relative degree $\{1, \ldots, 1\}$.

Note 4.2. To understand the nature of the assumptions, and place the I&I controller in perspective, we review the MRAC solution for controlling the plant (4.65). Consider the regressor matrix

$$\Phi(y) = \begin{bmatrix} y^\top & 0 & \cdots & 0 \\ 0 & y^\top & \cdots & 0 \\ \vdots & \vdots & \ddots & \vdots \\ 0 & 0 & \cdots & y^\top \end{bmatrix} \in \mathbb{R}^{m \times m^2} \tag{4.66}$$

and introduce the parameterisation

$$-K_p^{-1} y = \Phi(y)\theta,$$

where $\theta \in \mathbb{R}^{m^2}$ contains the rows of $-K_p^{-1}$. The control law is defined as

$$u = \Phi(y)\hat{\theta},$$
$$\dot{\hat{\theta}} = -\Phi^\top(y)\Gamma^{-1} y,$$

---
[14] See, e.g., [77, Section 9.7.3].

where, for simplicity, we have taken unitary adaptation gain and selected the reference model dynamics as[15] $\frac{1}{s+1}I$. The error equations

$$\dot{y} = -y + K_p \Phi(y)\tilde{\theta},$$
$$\dot{\tilde{\theta}} = -\Phi^\top(y)\Gamma^{-1}y,$$
(4.67)

are obtained immediately from the equations above and the definition of the parameter estimation error $\tilde{\theta} = \hat{\theta} - \theta$, and stability is analysed with the Lyapunov function candidate

$$V_0(y, \tilde{\theta}) = \frac{1}{2}\left(y^\top P^{-1} y + |\tilde{\theta}|^2\right),$$

where $P = P^\top > 0$ is a matrix to be determined. Taking the derivative of $V_0(y, \tilde{\theta})$ along the trajectories of (4.67) gives

$$\dot{V}_0 = -y^\top P^{-1} y + y^\top \left(P^{-1} K_p - \Gamma^{-\top}\right)\Phi(y)\tilde{\theta}.$$

It is then clear that, if condition (4.63) holds, setting $P = K_p \Gamma^\top$ cancels the second right-hand side term. (Obviously, the symmetry condition is necessary for this construction, hence the weaker assumption (4.64) is not sufficient.) The proof that $\lim_{t\to\infty} y(t) = 0$ is completed by noting that $\dot{V}_0 = -y^\top P^{-1} y \leq 0$, hence $y(t) \in \mathcal{L}_2 \cap \mathcal{L}_\infty$ and $\tilde{\theta}(t) \in \mathcal{L}_\infty$, and invoking Corollary A.1. ◁

We now present an alternative solution by applying the adaptive I&I approach to system (4.65). Similarly to the classical MRAC solution consider the parameterisation

$$-K_p^{-1} y_f = \Phi_f(y_f)\theta,$$
(4.68)

where we have introduced the filtered output

$$\dot{y}_f = -y_f + y,$$

the filtered regressor

$$\Phi_f(y_f) = \begin{bmatrix} y_f^\top & 0 & \cdots & 0 \\ 0 & y_f^\top & \cdots & 0 \\ \vdots & \vdots & \ddots & \vdots \\ 0 & 0 & \cdots & y_f^\top \end{bmatrix} \in \mathbb{R}^{m \times m^2},$$
(4.69)

and the filtered input

$$\dot{u}_f = -u_f + u.$$
(4.70)

It is clear[16] from (4.65) and (4.70) that $\dot{y}_f = K_p u_f + \epsilon_t$, where ϵ_t denotes an exponentially decaying term. To simplify the notation we omit in the sequel

---

[15] As will become clear, we could assign the reference model dynamics $\dot{y} = \Lambda y$, with Λ a Hurwitz matrix, with a suitable redefinition of the regressor matrix $\Phi(y)$ and the Lyapunov function candidate.

[16] Setting $\eta = -K_p u_f - y_f + y$ and taking into account the dynamics of the filters yields the system $\dot{y}_f = K_p u_f + \eta$, $\dot{\eta} = -\eta$.

this term that does not affect the validity of the analysis[17]. The adaptive I&I design is carried out using the filtered representation of the plant, namely

$$\dot{y}_f = K_p u_f. \tag{4.71}$$

Define the adaptive control law

$$\begin{aligned} u_f &= \Phi_f(y_f)(\hat{\theta} + \beta_1(y_f)), \\ \dot{\hat{\theta}} &= \beta_2(y_f), \end{aligned} \tag{4.72}$$

where $\beta_1(y_f)$ and $\beta_2(y_f)$ are vector-valued functions to be determined, and the off-the-manifold co-ordinates

$$z = \hat{\theta} - \theta + \beta_1(y_f), \tag{4.73}$$

which should be driven to zero. Replacing the definition of z in (4.71) and using (4.68) yields the error equation

$$\dot{y}_f = -y_f + K_p \Phi_f(y_f) z. \tag{4.74}$$

Comparing (4.74) with the first error equation of MRAC in (4.67) we observe that the off-the-manifold co-ordinate z plays the same role as the parameter estimation error $\tilde{\theta}$. The novelty of adaptive I&I resides in the way to generate the dynamics of z, which proceeds as follows. First, from (4.73) we evaluate

$$\begin{aligned} \dot{z} &= \dot{\hat{\theta}} + \frac{\partial \beta_1}{\partial y_f}(y_f)\dot{y}_f \\ &= \beta_2(y_f) + \frac{\partial \beta_1}{\partial y_f}(y_f)\left(-y_f + K_p \Phi_f(y_f) z\right), \end{aligned}$$

where we have replaced (4.72) and (4.74) to obtain the second line. Observation of the latter, and recalling that z is not measurable, suggests the choice

$$\beta_1(y_f) = -\Phi_f^\top(y_f)\Gamma^{-1} y_f$$

yielding

$$\begin{aligned} \dot{z} &= \beta_2(y_f) - \Phi_f^\top(y_f)\Gamma^{-1}\left(-y_f + K_p \Phi_f(y_f) z\right) - \dot{\Phi}_f^\top(y_f)\Gamma^{-1} y_f \\ &= \beta_2(y_f) - \Phi_f^\top(y_f)\Gamma^{-1} K_p \Phi_f(y_f) z + \left[2\Phi_f(y_f) - \Phi(y)\right]^\top \Gamma^{-1} y_f, \end{aligned}$$

where we have used $\dot{\Phi}_f(y_f) = -\Phi_f(y_f) + \Phi(y)$. Setting

$$\beta_2(y_f) = -\left[2\Phi_f(y_f) - \Phi(y)\right]^\top \Gamma^{-1} y_f$$

yields the error equation

---
[17] See also [77, Section 6.5.4].

$$\dot{z} = -\Phi_f^\top(y_f)\Gamma^{-1}K_p\Phi_f(y_f)z. \qquad (4.75)$$

We now prove that the origin of the system (4.74), (4.75) is globally stable and $\lim_{t\to\infty} y_f(t) = 0$. To this end, consider the quadratic Lyapunov function candidate

$$V(y_f, z) = \frac{1}{2}|y_f|^2 + \frac{\alpha}{2}|z|^2, \qquad (4.76)$$

with $\alpha > 0$. The derivative of V along the trajectories of (4.74), (4.75) gives

$$\dot{V} = -|y_f|^2 + y_f^\top K_p z_\varphi - \alpha z_\varphi^\top M z_\varphi,$$

where $z_\varphi = \Phi_f(y_f)z$ and

$$M = \frac{1}{2}\left(\Gamma^{-1}K_p + K_p^\top \Gamma^{-\top}\right) = M^\top,$$

which, in view of (4.64), is positive-definite. Setting $\alpha > \lambda_{max}\{K_p M^{-1} K_p^\top\}$ yields

$$\begin{bmatrix} I & -\frac{1}{2}K_p \\ -\frac{1}{2}K_p^\top & \alpha M \end{bmatrix} > 0,$$

hence for all sufficiently large α there exists $\delta > 0$ such that

$$\dot{V} \leq -\delta\left(|y_f|^2 + |z_\varphi|^2\right).$$

This implies stability of the zero equilibrium and

$$\lim_{t\to\infty} y_f(t) = 0, \qquad \lim_{t\to\infty} z_\varphi(t) = 0,$$

hence, from (4.74), $\lim_{t\to\infty} \dot{y}_f(t) = 0$. Finally, from $\dot{y}_f = -y_f + y$, we conclude that $\lim_{t\to\infty} y(t) = 0$ as desired.

To complete the design we must recover the actual control u from the expression of u_f derived above, namely

$$u_f = \Phi_f(y_f)\left(\hat{\theta} - \Phi_f^\top(y_f)\Gamma^{-1} y_f\right).$$

Replacing this expression in $u = \dot{u}_f + u_f$ and using the fact that

$$\frac{d}{dt}\left(\hat{\theta} - \Phi_f^\top(y_f)\Gamma^{-1} y_f\right) = -\Phi_f^\top(y_f)\Gamma^{-1} y$$

yield

$$u = \Phi(y)\left(\hat{\theta} - \Phi_f^\top(y_f)\Gamma^{-1} y_f\right) - \Phi_f(y_f)\Phi_f^\top(y_f)\Gamma^{-1} y.$$

We have thus established the following result.

Proposition 4.6. *Consider the system (4.65), where $u \in \mathbb{R}^m$, $y \in \mathbb{R}^m$ and $\det(K_p) \neq 0$. Assume known a (nonsingular) matrix Γ such that (4.64) holds. Then the I&I adaptive controller*

4.4 Linear Systems

$$u = \Phi(y)(\hat{\theta} - \Phi_f^\top(y_f)\Gamma^{-1}y_f) - \Phi_f(y_f)\Phi_f^\top(y_f)\Gamma^{-1}y, \quad (4.77)$$
$$\dot{\hat{\theta}} = -\left[2\Phi_f(y_f) - \Phi(y)\right]^\top \Gamma^{-1}y_f, \quad (4.78)$$
$$\dot{y}_f = -y_f + y, \quad (4.79)$$

where $\Phi(y)$ and $\Phi_f(y_f)$ are given by (4.66) and (4.69), respectively, ensures $\lim_{t\to\infty} y(t) = 0$ with all signals bounded for all initial conditions $y(0) \in \mathbb{R}^m$, $y_f(0) \in \mathbb{R}^m$ and $\hat{\theta}(0) \in \mathbb{R}^{m^2}$.

The above result can be extended to the case of tracking an arbitrary, differentiable, reference signal, as the following statement shows.

Proposition 4.7. *Consider the system (4.65) and a bounded reference trajectory $y^* = y^*(t)$ with bounded derivative $\dot{y}^*$. Assume known a (nonsingular) matrix Γ verifying (4.64). Then the I&I adaptive controller*

$$u = \Psi(\hat{\theta} - \Psi_f^\top \Gamma^{-1}e_f) - \Psi_f\Psi_f^\top \Gamma^{-1}e,$$
$$\dot{\hat{\theta}} = -(2\Psi_f - \Psi)^\top \Gamma^{-1}e_f,$$
$$\dot{e}_f = -e_f + e,$$
$$\dot{y}_f^* = -y_f^* + y^*,$$
$$e = y - y^*,$$

where

$$\Psi = \begin{bmatrix} (e - \dot{y}^*)^\top & 0 & \cdots & 0 \\ 0 & (e - \dot{y}^*)^\top & \cdots & 0 \\ \vdots & \vdots & \ddots & \vdots \\ 0 & 0 & \cdots & (e - \dot{y}^*)^\top \end{bmatrix} \in \mathbb{R}^{m \times m^2}$$

and

$$\Psi_f = \begin{bmatrix} (e_f - \dot{y}_f^*)^\top & 0 & \cdots & 0 \\ 0 & (e_f - \dot{y}_f^*)^\top & \cdots & 0 \\ \vdots & \vdots & \ddots & \vdots \\ 0 & 0 & \cdots & (e_f - \dot{y}_f^*)^\top \end{bmatrix} \in \mathbb{R}^{m \times m^2},$$

ensures $\lim_{t\to\infty} e(t) = 0$ with all signals bounded for all initial conditions $y(0) \in \mathbb{R}^m$, $e_f(0) \in \mathbb{R}^m$, $y_f^*(0) \in \mathbb{R}^m$ and $\hat{\theta}(0) \in \mathbb{R}^{m^2}$, and all bounded trajectories y^* with bounded derivative $\dot{y}^*$. Moreover, if $\dot{y}^*$ is persistently exciting, i.e., there exist T and $\delta > 0$ such that $\int_t^{t+T} \dot{y}^*(\tau)\dot{y}^*(\tau)^\top \mathrm{d}\tau \geq \delta I$, for all t, then convergence is exponential.

Proof. The proof is established with the parameterisation

$$\Psi_f \theta = -K_p^{-1}\left(e_f - \ddot{y}_f^*\right).$$

The I&I control takes the form $u_f = \Psi_f(\hat{\theta} + \beta_1(e_f))$, with the same definition of z given in (4.73), which leads to the error equation

$$\dot{e}_f = -e_f + K_p \Psi_f z.$$

The term $\beta_1(\cdot)$ and the parameter update law, which result in the controller given in Proposition 4.7, are obtained following the construction of Proposition 4.6. Finally, the exponential convergence claim follows from standard results[18]. □

Finally, I&I allows also relaxation of the symmetry condition (4.63) for the class of vector relative degree $\{1,\ldots,1\}$ systems considered in MRAC[19].

Proposition 4.8. *Consider a square multivariable LTI system described by the transfer matrix $G(s)$ and suppose the following hold.*

- *$G(s)$ is strictly proper, full-rank and minimum-phase, and an upper bound $\bar{\nu}$ on its observability index ν is known.*
- *$G(s)$ has vector relative degree $\{1,\ldots,1\}$, and the interactor matrix is diagonal. In particular, assume the interactor matrix to be $\xi_m(s) = \mathrm{diag}\{s+1,\ldots,s+1\}$.*
- *There exists a known (nonsingular) matrix Γ verifying (4.64), where $\lim_{s\to\infty} \xi_m(s)G(s) = K_p$.*
- *$y^* \in \mathbb{R}^m$ is a reference trajectory generated as $y^* = \xi_m^{-1} r$, where $r = r(t) \in \mathbb{R}^m$ is a bounded signal.*

Then the I&I adaptive controller

$$\begin{aligned}
u &= W(\hat{\theta} - W_f^\top \Gamma^{-1} e_f) - W_f W_f^\top \Gamma^{-1} e, \\
\dot{\hat{\theta}} &= -(2W_f - W)^\top \Gamma^{-1} e_f, \\
\dot{e}_f &= -e_f + e, \\
\dot{w}_f &= -w_f + w, \\
e &= y - y^*,
\end{aligned}$$

where

$$W = \begin{bmatrix} w^\top & 0 & \cdots & 0 \\ 0 & w^\top & \cdots & 0 \\ \vdots & \vdots & \ddots & \vdots \\ 0 & 0 & \cdots & w^\top \end{bmatrix} \in \mathbb{R}^{m \times 2m^2(\bar{\nu}+1)},$$

---

[18] See [191, Theorem 1.5.2].
[19] For further details on the system assumptions we refer the interested reader to the textbooks [191, 152, 77] or the more recent paper [161].

$$W_f = \begin{bmatrix} w_f^T & 0 & \cdots & 0 \\ 0 & w_f^T & \cdots & 0 \\ \vdots & \vdots & \ddots & \vdots \\ 0 & 0 & \cdots & w_f^T \end{bmatrix} \in \mathbb{R}^{m \times 2m^2(\bar{\nu}+1)},$$

$$w = \begin{bmatrix} w_u \\ w_y \\ y \\ r \end{bmatrix} \in \mathbb{R}^{2m(\bar{\nu}+1)}, \quad w_u = \frac{1}{\lambda(s)} \begin{bmatrix} u^{(\bar{\nu}-1)} \\ \vdots \\ \dot{u} \\ u \end{bmatrix} \in \mathbb{R}^{m\bar{\nu}},$$

$$w_y = \frac{1}{\lambda(s)} \begin{bmatrix} y^{(\bar{\nu}-1)} \\ \vdots \\ \dot{y} \\ y \end{bmatrix} \in \mathbb{R}^{m\bar{\nu}},$$

and $\lambda(s)$ is a monic Hurwitz polynomial of degree $\bar{\nu}$, ensures $\lim_{t \to \infty} e(t) = 0$ with all signals bounded for all initial conditions $y(0) \in \mathbb{R}^m$, $e_f(0) \in \mathbb{R}^m$, $w_f(0) \in \mathbb{R}^{2m(\bar{\nu}+1)}$, $\hat{\theta}(0) \in \mathbb{R}^{2m^2(\bar{\nu}+1)}$, and all bounded reference signals r.

Proof. The proof is established using the transfer matrix identity[20] which, given $\lambda(s), G(s)$ and $\xi_m(s)$, ensures the existence of monic polynomial matrices $C(s), D(s)$, of degrees $\bar{\nu} - 1$, such that

$$I - \frac{1}{\lambda(s)} C(s) - \frac{1}{\lambda(s)} D(s) G(s) = K_p^{-1} \xi_m(s) G(s).$$

After some suitable term re-ordering and filtering, this identity leads to the parameterisation

$$W_f \theta = u_f - K_p^{-1} \xi_m(s) e_f.$$

The I&I control takes again the form $u_f = W_f(\hat{\theta} + \beta_1(W_f, e_f))$, where $\beta_1(\cdot)$ depends now explicitly on e_f and W_f. Mimicking the derivations in the proof of Proposition 4.7 leads to the error equation

$$\xi_m e_f = K_p W_f z,$$

which, given the definition of the interactor matrix, is exactly the error equation of Proposition 4.7. The proof is completed following the construction of Proposition 4.6. □

Note 4.3. It is well known that passivity is the key property required for stabilisation in MRAC, see [131] for a discussion. For the system (4.65) the passivity condition is expressed in terms of positive realness of the transfer matrix

---

[20] See equation (6.3.10) of [191].

88   4 I&I Adaptive Control: Systems in Special Forms

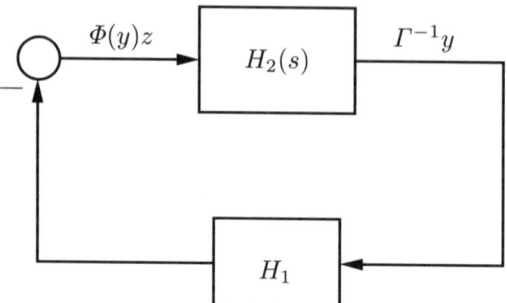

Fig. 4.2. Error equations of MRAC.

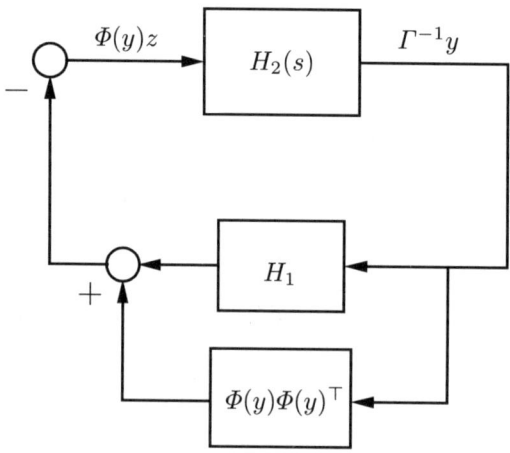

Fig. 4.3. Error equations of MRAC with PI adaptation.

$$H_2(s) = \frac{1}{s+1}\Gamma^{-1}K_p,$$

which defines the map $\Phi(y)\tilde{\theta} \mapsto \Gamma^{-1}y$. This operator is then placed in negative feedback with the operator $H_1 : \Gamma^{-1}y \mapsto -\Phi(y)\tilde{\theta}$ defined by the gradient estimator, which is known to be passive, see Figure 4.2. If $H_2(s)$ is strictly positive real, stability of the overall system follows from standard passivity theory [224]. Note that the symmetry condition (4.63) is *necessary* for strict positive realness of $H_2(s)$. Hence, if this condition is violated, stabilisation cannot be established invoking passivity arguments.

Since, introducing a small time constant in the filter (4.79) we can make the filtered output y_f track y arbitrarily fast, it is interesting to study the behaviour of the controller if we assume $y_f \approx y$ and $\Phi_f(y_f) \approx \Phi(y)$. Under this approximation, equations (4.77) and (4.78) of the I&I adaptive controller reduce to

$$u = \Phi(y)\hat{\theta} - \Phi(y)\Phi^\top(y)\Gamma^{-1}y,$$

$$\dot{\hat{\theta}} = -\Phi^\top(y)\Gamma^{-1}y,$$

which coincides with the equations of standard MRAC, but with an adaptation law consisting of integral and proportional terms. A schematic diagram of the resulting error system is shown in Figure 4.3, where we clearly see that the proportional term introduces a by-pass term to the gradient estimator. This term strengthens the passivity of H_1, and in some instances may overcome the lack of passivity of $H_2(s)$, see, e.g., [165]. ◁

5

Nonlinear Observer Design

In this chapter a general framework for constructing globally convergent (reduced-order) observers for classes of nonlinear systems is presented. Instrumental to this development is to formulate the observer design problem as a problem of rendering invariant and attractive an appropriately selected manifold in the extended state-space of the plant and the observer.

There are two current trends in nonlinear observer design[1]. The first one is to require that the nonlinearities appearing in the system equations are either *linear* functions of the unmeasured states, which leads to the so-called extended Luenberger observer, or *monotonic* functions of a linear combination of the states, in which case passivity can be exploited[2]. In the second one the observer consists of a linear filter and a nonlinear output map which is obtained by solving a set of partial differential equations[3]. This approach, similarly to the one in this book, requires the existence of a manifold which should be rendered invariant and attractive. However, while in this approach attractivity is guaranteed by asymptotic stability of the linear filter (which is partly fixed) and invariance is enforced by solving the PDE, in the I&I approach the manifold, which (as in the adaptive case) is parameterised by a function β, is rendered invariant by a *nonlinear* filter and attractive by a proper selection of the function β.

In this chapter we show how this extra degree of freedom can be used to deal with general nonlinear systems and, in particular, with a class of systems with nonlinearities which are not necessarily monotonic in the unmeasured

---

[1] A third one, not considered in this book, is to assume growth conditions on the nonlinearities and use high gain to dominate them, see [111] for a survey and [63], [127], [23] for some examples.
[2] See, *e.g.*, [12].
[3] For more details on this approach, see [7, 108, 116, 119, 120].

states. It is worth noting that the I&I observer encompasses the (extended) Luenberger observer, as well as the ones based on monotonicity[4].

The method is used to obtain new solutions to two practical problems. The first is the problem of estimating the three-dimensional motion of an object using two-dimensional images obtained from a single camera. The second is the problem of estimating the velocities of a two-degrees-of-freedom mechanical system by measuring the positions.

Note 5.1. The problem of constructing observers for nonlinear systems has received a great deal of attention due to its importance in practical applications, where some of the states may not be available for measurement. In the case of linear systems a complete theory on *asymptotic* (reduced-order) observers can be found in [130], while an observer with *finite-time* convergence has been developed in [51].

The classical approach to nonlinear observer design consists in finding a transformation that linearises the plant up to an output injection term and then applying standard linear observer design techniques. The existence of such a transformation, however, relies on a set of stringent assumptions [117, 118] which are hard to verify in practice, see also [24] for an extension using system immersion. Lyapunov-like conditions for the existence of a nonlinear observer with asymptotically stable error dynamics have been given in [222]. An observer for uniformly observable nonlinear systems in canonical form has been developed in [61, 63, 62], based on a global Lipschitz condition and a gain assignment technique. Some extensions of this result, which avoid the transformation to canonical form and allow for more flexibility in the selection of the observer gain, have been proposed in [45]. More recently, in [108] conditions for the existence of a linear observer with a nonlinear output map have been given in terms of the local solution of a partial differential equation (PDE), thus extending Luenberger's early ideas [129, 130] to the nonlinear case. Extensions to this fundamental result have been developed in [119, 120, 7].

A globally convergent reduced-order observer for systems in canonical form has been proposed in [23] using the notion of output-to-state stability. A high-gain design for a similar class of systems that presumes a bound on the system trajectories has been proposed in [127]. See also [111] for a survey on high-gain observers. ◁

5.1 Introduction

We consider nonlinear, time-varying systems described by equations of the form

$$\begin{aligned} \dot{\eta} &= f_1(\eta, y, t), \\ \dot{y} &= f_2(\eta, y, t), \end{aligned} \tag{5.1}$$

where $\eta \in \mathbb{R}^n$ is the unmeasured part of the state and $y \in \mathbb{R}^m$ is the measurable part of the state. It is assumed that the vector fields $f_1(\cdot)$ and $f_2(\cdot)$ are forward complete, *i.e.*, trajectories starting at time t_0 are defined for all

---

[4]Even in the special case of monotonic nonlinearities with constant coefficients, the proposed result is more general than the one in [12] (see the example of Section 5.3).

times $t \geq t_0$. (This assumption can be removed under certain conditions, see Corollary 5.2.)

Definition 5.1. [5] *The dynamical system*

$$\dot{\xi} = \alpha(\xi, y, t), \tag{5.2}$$

with $\xi \in \mathbb{R}^p$, $p \geq n$, is called an observer for the system (5.1), if there exist mappings $\beta : \mathbb{R}^p \times \mathbb{R}^m \times \mathbb{R} \to \mathbb{R}^p$ and $\phi : \mathbb{R}^n \times \mathbb{R}^m \times \mathbb{R} \to \mathbb{R}^p$ that are left-invertible (with respect to their first argument)[6] *and such that the manifold*

$$\mathcal{M} = \{\, (\eta, y, \xi, t) \in \mathbb{R}^n \times \mathbb{R}^m \times \mathbb{R}^p \times \mathbb{R} \,:\, \beta(\xi, y, t) = \phi(\eta, y, t) \,\} \tag{5.3}$$

has the following properties.

(i) All trajectories of the extended system (5.1), (5.2) that start on the manifold $\mathcal{M}$ remain there for all future times, i.e., $\mathcal{M}$ is positively invariant.

(ii) All trajectories of the extended system (5.1), (5.2) that start in a neighbourhood of $\mathcal{M}$ asymptotically converge to $\mathcal{M}$.

The above definition implies that an asymptotically converging estimate of the state η is given by

$$\hat{\eta} = \phi^{\mathrm{L}}(\beta(\xi, y, t), y, t),$$

where ϕ^{L} denotes a left-inverse of ϕ. Note that the state estimation error $\hat{\eta} - \eta$ is zero on the manifold $\mathcal{M}$. Moreover, if the property (ii) holds for any $(\eta(t_0), y(t_0), \xi(t_0), t_0) \in \mathbb{R}^n \times \mathbb{R}^m \times \mathbb{R}^p \times \mathbb{R}$ then (5.2) is a *global* observer for the system (5.1).

5.2 Reduced-order Observers

In this section we present a general tool for constructing nonlinear (reduced-order) observers of the form given in Definition 5.1.

Theorem 5.1. *Consider the system (5.1), (5.2) and suppose that there exist $\mathcal{C}^1$ mappings $\beta(\xi, y, t) : \mathbb{R}^p \times \mathbb{R}^m \times \mathbb{R} \to \mathbb{R}^p$ and $\phi(\eta, y, t) : \mathbb{R}^n \times \mathbb{R}^m \times \mathbb{R} \to \mathbb{R}^p$, with a left-inverse $\phi^{\mathrm{L}} : \mathbb{R}^p \times \mathbb{R}^m \times \mathbb{R} \to \mathbb{R}^n$, such that the following hold.*

(A1) For all y, ξ and t, $\beta(\xi, y, t)$ is left-invertible with respect to ξ and

$$\det\left(\frac{\partial \beta}{\partial \xi}\right) \neq 0.$$

---

[5]This definition is in the spirit of the one given in [222].

[6]A mapping $\psi(x, y, t) : \mathbb{R}^l \times \mathbb{R}^m \times \mathbb{R} \to \mathbb{R}^p$ is left-invertible (with respect to x) if there exists a mapping $\psi^{\mathrm{L}} : \mathbb{R}^p \times \mathbb{R}^m \times \mathbb{R} \to \mathbb{R}^l$ such that $\psi^{\mathrm{L}}(\psi(x, y, t), y, t) = x$, for all $x \in \mathbb{R}^l$ (and for all y and t).

(A2) The system

$$\dot{z} = -\frac{\partial \beta}{\partial y}(f_2(\hat{\eta},y,t) - f_2(\eta,y,t)) + \left.\frac{\partial \phi}{\partial y}\right|_{\eta=\hat{\eta}} f_2(\hat{\eta},y,t) - \frac{\partial \phi}{\partial y} f_2(\eta,y,t)$$
$$+ \left.\frac{\partial \phi}{\partial \eta}\right|_{\eta=\hat{\eta}} f_1(\hat{\eta},y,t) - \frac{\partial \phi}{\partial \eta} f_1(\eta,y,t) + \left.\frac{\partial \phi}{\partial t}\right|_{\eta=\hat{\eta}} - \frac{\partial \phi}{\partial t}, \qquad (5.4)$$

with $\hat{\eta} = \phi^L(\phi(\eta,y,t) + z)$, has a (globally) asymptotically stable equilibrium at $z = 0$, uniformly in η, y and t.

Then the system (5.2) with

$$\alpha(\xi,y,t) = -\left(\frac{\partial \beta}{\partial \xi}\right)^{-1}\left(\frac{\partial \beta}{\partial y}f_2(\hat{\eta},y,t) + \frac{\partial \beta}{\partial t} - \left.\frac{\partial \phi}{\partial y}\right|_{\eta=\hat{\eta}} f_2(\hat{\eta},y,t)\right.$$
$$\left. - \left.\frac{\partial \phi}{\partial \eta}\right|_{\eta=\hat{\eta}} f_1(\hat{\eta},y,t) - \left.\frac{\partial \phi}{\partial t}\right|_{\eta=\hat{\eta}}\right), \qquad (5.5)$$

where $\hat{\eta} = \phi^L(\beta(\xi,y,t),y,t)$, is a (global) observer for the system (5.1).

Proof. Consider the (off-the-manifold) variables

$$z = \beta(\xi,y,t) - \phi(\eta,y,t), \qquad (5.6)$$

where $\beta(\cdot)$ is a continuously differentiable function such that (A1) holds. Note that $|z|$ represents the distance of the system trajectories from the manifold $\mathcal{M}$ defined in (5.3). The dynamics of z are given by

$$\dot{z} = \frac{\partial \beta}{\partial y}f_2(\eta,y,t) + \frac{\partial \beta}{\partial \xi}\alpha(\xi,y,t) + \frac{\partial \beta}{\partial t} - \frac{\partial \phi}{\partial y}f_2(\eta,y,t) - \frac{\partial \phi}{\partial \eta}f_1(\eta,y,t) - \frac{\partial \phi}{\partial t}.$$

Substituting the function $\alpha(\cdot)$ from (5.5) and noting that by assumption (A1) this function is well-defined, yields the dynamics (5.4). It follows from (A2) that the distance $|z|$ from the manifold $\mathcal{M}$ converges asymptotically to zero. Note, moreover, that $\mathcal{M}$ is invariant, i.e., if $z(t) = 0$ for some t, then $z(\tau) = 0$ for all $\tau > t$. Hence, by Definition 5.1, the system (5.2) with $\alpha(\cdot)$ given by (5.5) is a (global) observer for (5.1). □

Theorem 5.1 provides an implicit description of the observer dynamics (5.2) in terms of the mappings $\beta(\cdot)$, $\phi(\cdot)$ and $\phi^L(\cdot)$ which must then be selected to satisfy (A2). (Note, however, that the function $\alpha(\cdot)$ in (5.5) renders the manifold $\mathcal{M}$ invariant *for any* mappings $\beta(\cdot)$ and $\phi(\cdot)$.) As a result, the problem of constructing an observer for the system (5.1) is *reduced* to the problem of rendering the system (5.4) asymptotically stable by assigning the functions $\beta(\cdot)$, $\phi(\cdot)$ and $\phi^L(\cdot)$. This non-standard stabilisation problem can be extremely difficult to solve, since, in general, it relies on the solution of a set

5.2 Reduced-order Observers

of partial differential equations (or inequalities)[7]. However, in many cases of practical interest, these equations turn out to be solvable, as demonstrated in the following examples.

Example 5.1 (Range estimation). Consider the problem of estimating the range of an object moving in the three-dimensional space by observing the motion of its projected feature on the two-dimensional image space of a camera[8].

The motion of an object undergoing rotation, translation and linear deformation can be described by the affine system

$$\begin{bmatrix} \dot{x}_1 \\ \dot{x}_2 \\ \dot{x}_3 \end{bmatrix} = \begin{bmatrix} a_{11}(t) & a_{12}(t) & a_{13}(t) \\ a_{21}(t) & a_{22}(t) & a_{23}(t) \\ a_{31}(t) & a_{32}(t) & a_{33}(t) \end{bmatrix} \begin{bmatrix} x_1 \\ x_2 \\ x_3 \end{bmatrix} + \begin{bmatrix} b_1(t) \\ b_2(t) \\ b_3(t) \end{bmatrix}, \quad (5.7)$$

where $[x_1, x_2, x_3]^\top \in \mathbb{R}^3$ are the unmeasurable co-ordinates of the object in an inertial reference frame with x_3 being perpendicular to the camera image space, as shown in Figure 5.1. Using the perspective (or "pinhole") model for the camera, the measurable co-ordinates on the image space are given by

$$y = [y_1, y_2]^\top = l \left[\frac{x_1}{x_3}, \frac{x_2}{x_3} \right]^\top, \quad (5.8)$$

where l is the focal length of the camera, *i.e.*, the distance between the camera and the origin of the image-space axes. Without loss of generality, we assume that $l = 1$. The design objective is to reconstruct the co-ordinates x_1, x_2 and x_3 from measurements of the image-space co-ordinates y_1 and y_2.

Let $\eta = 1/x_3$ and rewrite the system (5.7) in the (η, y) co-ordinates, *i.e.*,

$$\begin{aligned} \dot{\eta} &= h_1(y,t)\eta - b_3(t)\eta^2, \\ \dot{y} &= h_2(y,t) + g(y,t)\eta, \end{aligned} \quad (5.9)$$

where $y = [y_1, y_2]^\top \in \mathbb{R}^2$ are the measurable co-ordinates on the image space, $\eta \in \mathbb{R}$ is the inverse range of the object, and

$$h_1(y,t) = -\left(a_{31}(t)y_1 + a_{32}(t)y_2 + a_{33}(t)\right),$$

$$h_2(y,t) = \begin{bmatrix} a_{11}(t) - a_{33}(t) & a_{12}(t) \\ a_{21}(t) & a_{22}(t) - a_{33}(t) \end{bmatrix} y + \begin{bmatrix} a_{13}(t) \\ a_{23}(t) \end{bmatrix} - yy^\top \begin{bmatrix} a_{31}(t) \\ a_{32}(t) \end{bmatrix},$$

$$g(y,t) = \begin{bmatrix} b_1(t) - b_3(t)y_1 \\ b_2(t) - b_3(t)y_2 \end{bmatrix}.$$

---

[7] This is in contrast with the approach in [108, 119], where, due to the fewer degrees of freedom, these PDEs can be (approximately) solved under certain conditions.

[8] Previous solutions to this problem, which typically arises in machine vision as well as target tracking, can be found in [82, 43, 50, 93].

5 Nonlinear Observer Design

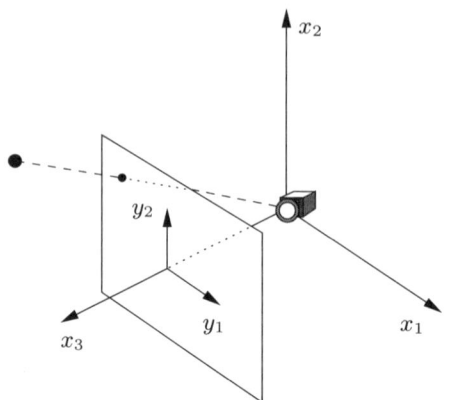

Fig. 5.1. Diagram of the perspective vision system.

It is assumed that the signals $a_{ij}(t)$, $b_i(t)$ and $y(t)$ are bounded, $a_{ij}(t)$ and $b_i(t)$ are differentiable, and the "instantaneous observability" condition[9]

$$|g(y,t)| \geq \epsilon > 0 \tag{5.10}$$

holds, for some ϵ and for all t and y.

For simplicity let $\phi(\eta, y, t) = \varepsilon(y,t)\eta$, where $\varepsilon(\cdot) \neq 0$ is a function to be determined, and consider an observer of the form given in Theorem 5.1, namely

$$\dot{\xi} = -\left(\frac{\partial \beta}{\partial \xi}\right)^{-1} \left(\frac{\partial \beta}{\partial y}\left(h_2(y,t) + g(y,t)\hat{\eta}\right) + \frac{\partial \beta}{\partial t} - \frac{\partial \varepsilon}{\partial y}\left(h_2(y,t) + g(y,t)\hat{\eta}\right)\hat{\eta}\right.$$
$$\left. - \varepsilon(y,t)\left(h_1(y,t)\hat{\eta} - b_3(t)\hat{\eta}^2\right) - \frac{\partial \varepsilon}{\partial t}\hat{\eta}\right), \tag{5.11}$$
$$\hat{\eta} = \varepsilon(y,t)^{-1}\beta(\xi, y, t).$$

From (5.4) the dynamics of the error $z = \beta(\xi, y, t) - \varepsilon(y,t)\eta = \varepsilon(y,t)(\hat{\eta} - \eta)$ are given by

$$\dot{z} = -\frac{\partial \beta}{\partial y}g(y,t)\left(\hat{\eta} - \eta\right) + \left(\frac{\partial \varepsilon}{\partial y}h_2(y,t) + \varepsilon(y,t)h_1(y,t)\right)(\hat{\eta} - \eta)$$
$$+ \left(\frac{\partial \varepsilon}{\partial y}g(y,t) - \varepsilon(y,t)b_3(t)\right)(\hat{\eta}^2 - \eta^2)$$

---

[9] The inequality (5.10) is a sufficient condition for observability of the system (5.9). In particular, it is obvious that when $g(y,t) = 0$, for all t, the state η becomes unobservable from the output y. For $b_3(t) \neq 0$ this happens when the projected feature is at the point $(y_1, y_2) = (b_1(t)/b_3(t), b_2(t)/b_3(t))$, which is known as the *focus of expansion* [82].

5.2 Reduced-order Observers

$$= -\left(\frac{\partial \beta}{\partial y}g(y,t) - \frac{\partial \varepsilon}{\partial y}h_2(y,t) - \varepsilon(y,t)h_1(y,t)\right)\varepsilon(y,t)^{-1}z$$
$$+ \left(\frac{\partial \varepsilon}{\partial y}g(y,t) - \varepsilon(y,t)b_3(t)\right)(\hat{\eta}^2 - \eta^2). \tag{5.12}$$

The observer design problem is now reduced to finding functions $\beta(\cdot)$ and $\phi(\cdot) = \varepsilon(\cdot)\eta$, with $\varepsilon(\cdot) \neq 0$, that satisfy assumptions (A1) and (A2) of Theorem 5.1. In view of (5.12) this can be achieved by solving the PDEs

$$\frac{\partial \beta}{\partial y}g(y,t) - \frac{\partial \varepsilon}{\partial y}h_2(y,t) - \varepsilon(y,t)h_1(y,t) = \kappa(y,t)\varepsilon(y,t), \tag{5.13}$$

$$\frac{\partial \varepsilon}{\partial y}g(y,t) - \varepsilon(y,t)b_3(t) = 0, \tag{5.14}$$

for some $\kappa(\cdot)$. From (5.14) we obtain the solution

$$\varepsilon(y,t) = -\frac{1}{|g(y,t)|},$$

which by (5.10) is well-defined and nonzero for all y and t. Let

$$\kappa(y,t) = \lambda|g(y,t)|^3 - \frac{\partial \varepsilon}{\partial y}h_2(y,t)\varepsilon(y,t)^{-1} - h_1(y,t)$$

and note that, by (5.10) and the fact that the signals $a_{ij}(t)$, $b_i(t)$ and $y(t)$ are bounded, there exists $\lambda > 0$ (sufficiently large) such that $\kappa(\cdot) \geq \underline{\kappa}$, for some constant $\underline{\kappa} > 0$. The PDE (5.13) is now reduced to

$$\frac{\partial \beta}{\partial y}g(y,t) = -\lambda|g(y,t)|^2,$$

which can be solved for $\beta(\cdot)$ yielding

$$\beta(\xi, y, t) = c(\xi, t) + \frac{\lambda}{2}\left((y_1^2 + y_2^2)b_3(t) - 2b_1(t)y_1 - 2b_2(t)y_2\right),$$

where $c(\cdot)$ is a free function. Selecting $c(\xi,t) = \xi$ ensures that assumption (A1) is satisfied. Substituting the above expressions into (5.12) yields the system

$$\dot{z} = -\kappa(y,t)z,$$

which has a uniformly (globally) asymptotically stable equilibrium at $z = 0$, hence assumption (A2) holds.

The observer (5.11) has been simulated using the perspective system[10]

$$\begin{bmatrix} \dot{x}_1 \\ \dot{x}_2 \\ \dot{x}_3 \end{bmatrix} = \begin{bmatrix} -0.2 & 0.4 & -0.6 \\ 0.1 & -0.2 & 0.3 \\ 0.3 & -0.4 & 0.4 \end{bmatrix} \begin{bmatrix} x_1 \\ x_2 \\ x_3 \end{bmatrix} + \begin{bmatrix} 0.5 \\ 0.25 \\ 0.3 \end{bmatrix},$$

---

[10] The numerical values are taken from [43, 50], see also [93] for a comparison.

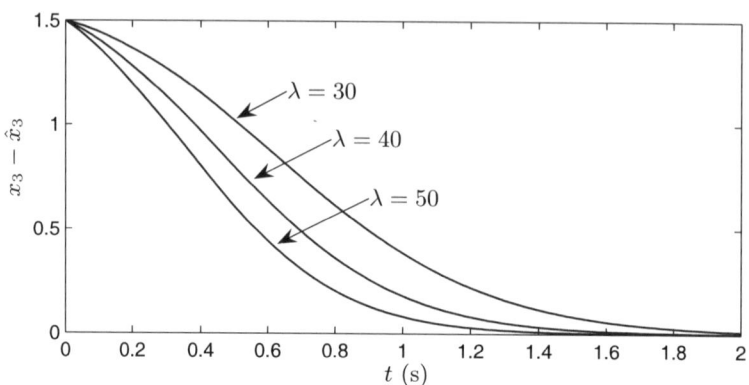

Fig. 5.2. Time history of the range estimation error for different values of λ.

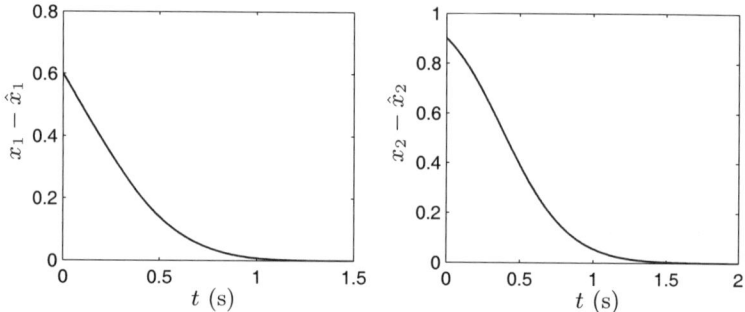

Fig. 5.3. Time history of the observation errors for $\lambda = 50$.

with the initial conditions $x_1(0) = 1$, $x_2(0) = 1.5$, $x_3(0) = 2.5$ and $\hat{\eta}(0) = 1$. Note that, since the b_is are constant, the functions $\beta(\cdot)$ and $\varepsilon(\cdot)$ do not depend explicitly on time, *i.e.*, $\frac{\partial \beta}{\partial t} = 0$ and $\frac{\partial \varepsilon}{\partial t} = 0$.

Figure 5.2 shows the time history of the range estimation error $x_3 - \hat{x}_3 = x_3 - 1/\hat{\eta}$ for different values of λ, namely $\lambda = 30$, $\lambda = 40$ and $\lambda = 50$. Note that the convergence rate can be arbitrarily increased simply by increasing the parameter λ. (However, the sensitivity to noise also increases in this way.) Finally, note that from (5.8) and the definition of η the estimates of the coordinates x_1, x_2 are given by $\hat{x}_1 = y_1/\hat{\eta}$, $\hat{x}_2 = y_2/\hat{\eta}$, respectively. Figure 5.3 shows the time histories of the observation errors $x_1 - \hat{x}_1$ and $x_2 - \hat{x}_2$ for $\lambda = 50$. ∎

Note 5.2. In [93] a *semi-global* observer for the perspective system (5.9) has been obtained using the same procedure but with an identity mapping $\phi(\eta, y, t) = \varepsilon(y, t)\eta$ (*i.e.*, with $\varepsilon(y, t) = 1$). In that case the error dynamics (5.12) reduce to

$$\dot{z} = -\left(\frac{\partial \beta}{\partial y} g(y, t) - h_1(y, t) + 2b_3(t)\eta\right) z - b_3(t)z^2.$$

5.2 Reduced-order Observers

Hence, selecting $\beta(\xi, y, t) = \xi - \frac{\lambda}{2}\left((y_1^2 + y_2^2)b_3(t) - 2b_1(t)y_1 - 2b_2(t)y_2\right)$ yields

$$\dot{z} = -\kappa(y, t)z - b_3(t)z^2,$$

where $\kappa(y, t) = \lambda |g(y, t)|^2 - h_1(y, t) + 2b_3(t)\eta$, and λ can be selected as before to ensure $\kappa(\cdot) \geq \delta > 0$. As a result, the origin $z = 0$ is a uniformly locally asymptotically stable equilibrium with a region of attraction containing the invariant set

$$B = \{z \in \mathbb{R} : |z| < \delta / \max |b_3(t)|\}.$$

The semi-global property follows by noting that, for any set of initial conditions $z(0)$, there exists δ, hence λ, sufficiently large such that $z(0) \in B$. ◁

Example 5.2 (Magnetic levitation). Consider again the magnetic levitation system of Examples 1.1 and 2.4, which is described by equations (1.7), and suppose that only ξ_1, *i.e.*, the position of the ball, is measurable. The objective is to design an asymptotic observer for the unmeasured states ξ_2 and ξ_3. Note that the system (1.7) can be rewritten in the form (5.1), namely

$$\dot{\eta}_1 = \frac{1}{2k}\eta_2^2 - mg,$$

$$\dot{\eta}_2 = -\frac{1}{k}R_2(1-y)\eta_2 + w,$$

$$\dot{y} = \frac{1}{m}\eta_1,$$

where $\eta = [\eta_1, \eta_2]^\top = [\xi_2, \xi_3]^\top$ and $y = \xi_1$.

The first step is to define a mapping $\phi(\eta, y)$ that enables to render the manifold $\mathcal{M}$ attractive. In contrast with the previous example, this cannot be achieved with an affine mapping $\phi(\eta, y) = \varepsilon(y)\eta$ due to the fact that the quadratic nonlinearity η_2^2 does not appear in the dynamics of the output. However, a solution can be obtained by applying *overparameterisation* and taking $\phi(\eta, y) = [\eta_1, \eta_2, \eta_2^2]$. Note that the observer (5.5) is third-order and the resulting error dynamics in (5.4) are given by

$$\dot{z} = -\frac{\partial \beta}{\partial y}\frac{1}{m}\begin{bmatrix}1 & 0 & 0\end{bmatrix}z + \begin{bmatrix} 0 & 0 & \frac{1}{2k} \\ 0 & -\frac{1}{k}R_2(1-y) & 0 \\ 0 & 2w & -\frac{2}{k}R_2(1-y) \end{bmatrix}z.$$

Hence, selecting

$$\beta(\xi, y) = \xi + \begin{bmatrix} \lambda_1 \\ 0 \\ \lambda_2/k \end{bmatrix} my,$$

where $\lambda_1 > 0$, $\lambda_2 > 0$ are constants, ensures that assumption (A1) is satisfied and the error dynamics become

100   5 Nonlinear Observer Design

$$\dot{z} = -\begin{bmatrix} \lambda_1 & 0 & -\dfrac{1}{2k} \\ 0 & \dfrac{1}{k}R_2(1-y) & 0 \\ \dfrac{\lambda_2}{k} & -2w & \dfrac{2}{k}R_2(1-y) \end{bmatrix} z. \tag{5.15}$$

Assume now that w (which corresponds to the capacitor voltage) is uniformly bounded, i.e., there exists a constant $c > 0$ such that $|w(t)| < c$, for all t, and, moreover, $1 - y(t) \geq \delta > 0$. (Physically, the last assumption implies that the ball remains strictly below the magnet.) Then the zero equilibrium of the system (5.15) is uniformly globally asymptotically stable. This is established with the Lyapunov function

$$V(z) = \lambda_2 z_1^2 + a z_2^2 + \frac{1}{2} z_3^2,$$

with $a \triangleq c^2 k^2/(R_2^2 \delta^2)$, whose time-derivative along the trajectories of (5.15) satisfies

$$\dot{V} = -2\lambda_1 \lambda_2 z_1^2 - \frac{2a}{k} R_2(1-y) z_2^2 + 2w z_2 z_3 - \frac{2}{k} R_2(1-y) z_3^2$$

$$\leq -2\lambda_1 \lambda_2 z_1^2 - \frac{a}{k} R_2(1-y) z_2^2 - \frac{1}{k} R_2(1-y) z_3^2,$$

hence, assumption (A2) holds. ∎

5.3 Systems with Monotonic Nonlinearities Appearing in the Output Equation

The generality of Theorem 5.1 comes at a price: it does not provide a constructive way of finding a function $\beta(\cdot)$ that satisfies assumption (A2). In this section we attempt to give a better insight into the design of this function by studying a more specific class of systems. In particular, we show that the presence of monotonic nonlinearities in the output dynamics can be exploited to stabilise the zero equilibrium of the observer error system even in the presence of general, i.e., possibly nonmonotonic, nonlinearities.

From a passivity point of view, this approach ensures that the "shortage" of passivity due to the nonmonotonic terms is compensated for by an "excess" of passivity due to output injection in the observer error dynamics. Additionally, we show that for systems with *finite escape time* the output injection term can be, in some cases, shaped so that the observer error converges at the time of escape.

Consider a class of nonlinear systems described by equations of the form

$$\begin{aligned} \dot{\eta} &= F_1(y)\gamma(Cy + H\eta) + \delta(y, H\eta) + g_1(y, u), \\ \dot{y} &= F_2(y)\gamma(Cy + H\eta) + g_2(y, u), \end{aligned} \tag{5.16}$$

5.3 Systems with Monotonic Nonlinearities Appearing in the Output Equation

where $\eta \in \mathbb{R}^n$ is the unmeasured part of the state, $y \in \mathbb{R}^m$ is the measurable output, $u \in \mathbb{R}^r$ is the input, and $\gamma(\cdot) \in \mathbb{R}^p$ is a vector with each element a nonlinear function of a linear combination of the states, i.e.,

$$\gamma_i = \gamma_i(C_i y + H_i \eta),$$

where C_i and H_i denote the rows of C and H, respectively. It is assumed that the functions $\gamma(\cdot)$ and $\delta(\cdot)$ are continuously differentiable and that each $\gamma_i(\cdot)$ is nondecreasing, i.e.,

$$(a - b)(\gamma_i(a) - \gamma_i(b)) \geq 0, \tag{5.17}$$

for all $a, b \in \mathbb{R}$.

For simplicity we fix the mapping $\phi(\cdot)$ in Definition 5.1 to be the identity and choose $p = n$, hence the problem considered is to find a reduced-order observer of the form

$$\dot{\xi} = \alpha(\xi, y), \qquad \hat{\eta} = \beta(\xi, y), \tag{5.18}$$

with $\xi \in \mathbb{R}^n$.

Note that, when $\delta(y, H\eta) \equiv 0$, the system (5.16) can be rewritten in the form

$$\begin{aligned}\dot{\eta} &= A_1(y)\eta + G_1(y)\bar{\gamma}(\bar{C}y + \bar{H}\eta) + g_1(y, u), \\ \dot{y} &= A_2(y)\eta + G_2(y)\bar{\gamma}(\bar{C}y + \bar{H}\eta) + g_2(y, u),\end{aligned} \tag{5.19}$$

where $G_1(y) = [F_1(y), -A_1(y)]$, $G_2(y) = [F_2(y), -A_2(y)]$ and

$$\bar{\gamma}(\bar{C}y + \bar{H}\eta) = \begin{bmatrix} \gamma(Cy + H\eta) \\ \eta \end{bmatrix}.$$

Note 5.3. In comparison with the class of systems considered in [12], the system (5.16) contains an additional nonlinearity $\delta(\cdot)$, which is not necessarily monotonic, and the matrices F_1 and F_2 generally depend on the output y. It should be mentioned that the result of [12] relies strongly on the matrices F_1 and F_2 being constant. Moreover, the system dynamics must include a linear (detectable) part which plays an essential role in the stabilisation of the observer error dynamics. In contrast, the construction in this chapter does not require a linear part, since stabilisation of the observer error dynamics is achieved by exploiting the monotonic nonlinearities contained in the vector $\gamma(\cdot)$—which may also include linear terms. Finally, note that, if F_1, F_2, A_1 and A_2 are constant, then (5.19) coincide with the systems considered in [12]. ◁

Example 5.3. To motivate the proposed approach (and highlight the differences from the one in [12]), consider a simple example described by the two-dimensional system

$$\begin{aligned}\dot{\eta} &= \eta^2 + u, \\ \dot{y} &= \eta + \eta^3 - y,\end{aligned} \tag{5.20}$$

and the problem of constructing a (reduced-order) observer for the unmeasured state η. In an attempt to apply the methodology in [12], rewrite the system in the form

$$\dot{\eta} = A_1\eta + G_1\gamma(\eta) + u,$$
$$\dot{y} = A_2\eta + G_2\gamma(\eta) - y,$$

where $\gamma(\eta) = [\eta^3, \eta + \eta^2 + \eta^3]^\top$ satisfies the monotonicity condition (5.17), and $A_1 = -1$, $G_1 = [-1, 1]$, $A_2 = 1$, and $G_2 = [1, 0]$. Defining the error $z = \xi + By - \eta$ and the observer dynamics

$$\dot{\xi} = (A_1 - BA_2)(\xi + By) + (G_1 - BG_2)\gamma(\xi + By) + u + By$$

yields the error system

$$\dot{z} = (A_1 - BA_2)z + (G_1 - BG_2)(\gamma(\eta + z) - \gamma(\eta)),$$

which has an asymptotically stable equilibrium at zero if there exist constants $\nu_1 > 0$, $\nu_2 > 0$ and B such that[11]

$$A_1 - BA_2 \leq 0, \qquad G_1 - BG_2 = [-\nu_1, -\nu_2].$$

It can be readily seen that the last condition cannot be satisfied, since it requires $\nu_2 = -1$.

The method in [12] fails in this example because it is not possible to assign the "output injection gain" B to render the error system *passive* with respect to *each one* of the elements of the vector $\gamma(\eta + z) - \gamma(\eta)$, which are passive by condition (5.17). Obviously, this passivity requirement becomes more restrictive as the dimension of the vector $\gamma(\cdot)$ increases. However, since passivity is not a necessary condition for stability, there may still exist a B such that the zero equilibrium of the error system is asymptotically stable.

It is now shown that such a solution can be constructed by means of Theorem 5.1. Towards this end, let $\beta(\xi, y) = \xi + f(y)$, which satisfies condition (A1), and consider the observer (5.5), which in this case becomes

$$\dot{\xi} = \alpha(\xi, y) = (\xi + f(y))^2 - \frac{\partial f}{\partial y}\left(\xi + f(y) + (\xi + f(y))^3\right) + u + \frac{\partial f}{\partial y}y.$$

Substituting into the dynamics of the error $z = \hat{\eta} - \eta = \xi + f(y) - \eta$ yields the system (5.4), which can be written as

$$\dot{z} = \left((\eta + z)^2 - \eta^2\right) - \frac{\partial f}{\partial y}\left(z + (\eta + z)^3 - \eta^3\right)$$
$$= \left(2\eta + z - \frac{\partial f}{\partial y}(1 + 3\eta^2 + 3\eta z + z^2)\right)z$$
$$\triangleq \lambda(\eta, y, z)z. \tag{5.21}$$

---
[11] See equation (25) in [12].

5.3 Systems with Monotonic Nonlinearities Appearing in the Output Equation

The aim now is to find $f(y)$ such that $\lambda(\eta, y, z) < 0$ for all y, η and z. From the inequalities

$$1 + 3\eta^2 + 3\eta z + z^2 \geq 1, \qquad \left|\frac{2\eta + z}{1 + 3\eta^2 + 3\eta z + z^2}\right| \leq \frac{1}{\sqrt{3}},$$

the required condition holds if

$$\frac{\partial f}{\partial y} > \frac{1}{\sqrt{3}}.$$

Hence, selecting $f(y) = By$, with $B > 1/\sqrt{3}$, ensures that the zero equilibrium of the error system (5.21) is uniformly globally asymptotically stable, hence assumption (A2) of Theorem 5.1 holds.

Notice that for $u(t) = 0$ and $\eta(0) > 0$ the states of the system (5.20) escape to infinity in finite time $t_e = 1/\eta(0)$. Nevertheless, it is possible to prove finite-time convergence of the observer error. To this end, define a scaled time variable τ such that $\frac{d\tau}{dt} = 1 + \eta^2$, which implies

$$\tau = \int_0^t \left(1 + \eta(\zeta)^2\right) d\zeta,$$

and note that τ tends to ∞ as t tends to t_e. From (5.21), the dynamics of the observer error with respect to the new time variable τ can be written as

$$\frac{dz}{d\tau} = \dot{z}\frac{dt}{d\tau} = \frac{\lambda(\eta, y, z)}{1 + \eta^2} z.$$

Note that the term $\lambda(\eta, y, z)/(1+\eta^2)$ is strictly negative even when $\eta(\tau) \to \infty$, hence

$$\lim_{\tau \to \infty} z(\tau) = \lim_{\tau \to \infty} \left(\hat{\eta}(\tau) - \eta(\tau)\right) = 0,$$

which, in turn, implies $\lim_{t \to t_e} \left(\hat{\eta}(t) - \eta(t)\right) = 0$, i.e., the manifold $\mathcal{M}$ in (5.3) is reached in finite time (namely at the time of escape t_e). ∎

The following theorem provides a generalisation of the above design for the class of systems (5.16).

Theorem 5.2. *Consider the system (5.16) with states $\eta(t)$ and $y(t)$ maximally defined in $[0, T)$, for some $T > 0$, and define two matrices $\Gamma(\eta, y, z)$ and $\Delta(\eta, y, z)$ such that*

$$\Gamma(\eta, y, z)Hz = \gamma(Cy + H(\eta + z)) - \gamma(Cy + H\eta),$$
$$\Delta(\eta, y, z)Hz = \delta(y, H(\eta + z)) - \delta(y, H\eta).$$

Suppose that there exist a positive-definite matrix P, a nonnegative function $\rho(\eta, y, z)$, with $\rho(\eta, y, z) \neq 0$ for $z \neq 0$, and a left-invertible (with respect to its first argument) function $\beta(\xi, y) : \mathbb{R}^m \times \mathbb{R}^n \to \mathbb{R}^n$, with $\det(\frac{\partial \beta}{\partial \xi}) \neq 0$, that satisfy the matrix inequality

$$\Lambda(\eta, y, z)^\top P + P\Lambda(\eta, y, z) < -\rho(\eta, y, z)I, \quad (5.22)$$

for all y, η and z, where

$$\Lambda(\eta, y, z) = \tilde{F}(\xi, y)\Gamma(\eta, y, z)H + \Delta(\eta, y, z)H,$$

$$\tilde{F}(\xi, y) = F_1(y) - \frac{\partial \beta}{\partial y} F_2(y).$$

Then the reduced-order observer

$$\dot{\xi} = \left(\frac{\partial \beta}{\partial \xi}\right)^{-1} \Bigg(F_1(y)\gamma(Cy + H\hat{\eta}) + \delta(y, H\hat{\eta}) + g_1(y, u)$$

$$- \frac{\partial \beta}{\partial y} \left(F_2(y)\gamma(Cy + H\hat{\eta}) + g_2(y, u) \right) \Bigg), \quad (5.23)$$

$$\hat{\eta} = \beta(\xi, y)$$

is such that $|\hat{\eta} - \eta|$ is bounded. Moreover, if $\rho(\eta, y, z) \geq \rho_0$ for some $\rho_0 > 0$, then $|\hat{\eta} - \eta|$ is integrable in $[0, T)$. If, in addition, $T = \infty$, then

$$\lim_{t \to \infty} (\hat{\eta}(t) - \eta(t)) = 0.$$

Proof. Define the error variable $z = \hat{\eta} - \eta = \beta(\xi, y) - \eta$, whose dynamics are described by the equation

$$\dot{z} = \frac{\partial \beta}{\partial \xi}\alpha(\xi, y) + \frac{\partial \beta}{\partial y}(F_2(y)\gamma(Cy + H\eta) + g_2(y, u))$$

$$- F_1(y)\gamma(Cy + H\eta) - \delta(y, H\eta) - g_1(y, u).$$

Assigning the observer dynamics as in (5.23), and noting that $\hat{\eta} = \eta + z$, yields the error system

$$\dot{z} = \tilde{F}(\xi, y)\Gamma(\eta, y, z)Hz + \Delta(\eta, y, z)Hz = \Lambda(\eta, y, z)z, \quad (5.24)$$

which, from (5.22), has a uniformly globally stable equilibrium at $z = 0$. Moreover, if $\rho(\eta, y, z) \geq \rho_0 > 0$, then we have that $\dot{V} \leq -\rho_0|z|^2$, which implies that $|z(t)| \leq k|z(0)|\exp(-\lambda t)$, for all $t \in [0, T)$ and for some positive constants k and λ. Integrating both sides yields $\int_0^T |z(t)|dt \leq \frac{k}{\lambda}|z(0)|$, hence the claim. □

It is apparent, from the construction given in the proof of Theorem 5.2, that the observer problem is reduced to the problem of finding a function $\beta(\cdot)$ (with the invertibility properties assumed in the theorem) that solves the (partial) differential inequality (5.22). In the special case of the class of systems (5.19) this inequality reduces to a parameterised matrix inequality that depends only on the output, as the following statement shows.

5.3 Systems with Monotonic Nonlinearities Appearing in the Output Equation

Corollary 5.1. *Consider the system (5.19) with states $y(t)$ and $\eta(t)$ defined for all $t \geq 0$, where $\bar{\gamma}(\cdot)$ satisfies the condition (5.17). Suppose that there exist a positive-definite matrix P, a function $\beta(y)$, a constant $\nu > 0$ and a diagonal matrix $N > 0$ that satisfy the matrix inequality*

$$\begin{bmatrix} \tilde{A}(y)^\top P + P\tilde{A}(y) + \nu I & P\tilde{G}(y) + \bar{H}^\top N \\ \tilde{G}(y)^\top P + N\bar{H} & 0 \end{bmatrix} \leq 0, \quad (5.25)$$

where

$$\tilde{A}(y) = A_1(y) - \frac{\partial \beta}{\partial y} A_2(y), \qquad \tilde{G}(y) = G_1(y) - \frac{\partial \beta}{\partial y} G_2(y).$$

Then there exists a reduced-order observer of the form (5.18) such that

$$\lim_{t \to \infty} (\hat{\eta}(t) - \eta(t)) = 0.$$

Proof. To begin with, note that the system (5.19) is a special instance of the system (5.16) with $F_2(y) = [G_2(y), A_2(y)]$, $F_1(y) = [G_1(y), A_1(y)]$ and

$$\gamma(Cy + H\eta) = \begin{bmatrix} \bar{\gamma}(\bar{C}y + \bar{H}\eta) \\ \eta \end{bmatrix}.$$

Define a (diagonal) matrix $\bar{\Gamma}(\eta, y, z)$ such that

$$\bar{\Gamma}(\eta, y, z)\bar{H}z = \bar{\gamma}(\bar{C}y + \bar{H}(\eta + z)) - \bar{\gamma}(\bar{C}y + \bar{H}\eta),$$

and note that (using the notation of Theorem 5.2)

$$\Gamma(\eta, y, z) = \begin{bmatrix} \bar{\Gamma}(\eta, y, z) & 0_{p \times n} \\ 0_{n \times p} & I \end{bmatrix}, \quad H = \begin{bmatrix} \bar{H} \\ I \end{bmatrix}.$$

The matrix inequality (5.22) can now be rewritten as

$$\Xi^\top P + P\Xi < 0, \quad (5.26)$$

where

$$\Xi = \tilde{G}(y)\bar{\Gamma}(\eta, y, z)\bar{H} + \tilde{A}(y).$$

It remains to show that (5.25) implies (5.26), hence the result in Theorem 5.2 is applicable. This follows directly from the fact that (5.25) implies

$$\tilde{A}(y)^\top P + P\tilde{A}(y) < 0,$$
$$P\tilde{G}(y) = -\bar{H}^\top N,$$

and that, by condition (5.17), $z^\top \bar{\Gamma}(\eta, y, z)\bar{H}z \geq 0$, for all y, η and z. Notice that in this case we have replaced $\beta(\xi, y)$ with $\xi + \beta(y)$, so the invertibility conditions are automatically satisfied. □

Note 5.4. In the special case in which the matrices A_1, A_2, G_1 and G_2 are constant, condition (5.25) reduces to the LMI proposed in [12] by selecting $\beta(y) = By$ with B constant. ◁

The result in Theorem 5.2 is based on the assumption that the system trajectories exist for all times. The following corollary shows that for a system with finite escape time (such as the one considered in Example 5.3) the observer can be made to converge in finite time.

Corollary 5.2. *Consider the system (5.16) with states $y(t)$ and $\eta(t)$ maximally defined in $[0, t_e)$, for some $t_e = t_e(y(0), \eta(0)) > 0$. Suppose that the assumptions of Theorem 5.2 hold and, moreover, there exists a function $\mu(\eta, y) \geq 1$ and a constant $\nu > 0$ such that*

$$\frac{1}{\mu(\eta, y)} \left(\Lambda(\eta, y, z)^\top P + P\Lambda(\eta, y, z) \right) \leq -\nu I, \tag{5.27}$$

for some positive-definite matrix P, and along any trajectory

$$\lim_{t \to t_e} \int_0^t \mu(\eta(\zeta), y(\zeta)) d\zeta = \infty. \tag{5.28}$$

Then there exists a reduced-order observer of the form (5.18) such that

$$\lim_{t \to t_e} (\hat{\eta}(t) - \eta(t)) = 0. \tag{5.29}$$

Proof. Define the scaled time variable

$$\tau = \int_0^t \mu(\eta(\zeta), y(\zeta)) d\zeta$$

and note that, from (5.28), τ tends to ∞ as t tends to t_e. Following the construction in the proof of Theorem 5.2, the observer error dynamics can be rewritten as

$$\dot{z} = \frac{dz}{d\tau} \frac{d\tau}{dt} = \Lambda(\eta, y, z)z,$$

which implies

$$\frac{dz}{d\tau} = \frac{1}{\mu(\eta, y)} \Lambda(\eta, y, z) z.$$

Note that, from (5.27), the above system has a uniformly globally asymptotically stable equilibrium at the origin, with the Lyapunov function $V(z) = z^\top P z$, which implies that

$$\lim_{\tau \to \infty} (\hat{\eta}(\tau) - \eta(\tau)) = 0,$$

hence (5.29) holds. □

5.4 Mechanical Systems with Two Degrees of Freedom

In this section we use the foregoing approach to construct globally asymptotically convergent observers for general Euler–Lagrange systems with two degrees of freedom (2DOF), where the objective is to estimate the velocities by measuring the positions. This is a challenging problem since the dynamics involved contain nonmonotonic (i.e., quadratic) nonlinearities.

Following the proposed methodology, we first construct an output injection function $\beta(\cdot)$ by solving a set of partial differential equations and then ensure the convergence of the observer error by means of a quadratic Lyapunov function (see Theorem 5.2). As an example we consider the well-known "ball and beam" system.

Note 5.5. An intrinsic *local* observer for this class of systems has been developed in [4] based on the Riemannian structure associated to the inertia matrix, while a global observer has been proposed in [28] albeit for a narrower class of systems that can be rendered linear in the unmeasured states via a transformation. ◁

5.4.1 Model Description

Consider a 2DOF mechanical system whose dynamics are described by the Euler–Lagrange equations

$$\frac{d}{dt}\frac{\partial \mathcal{L}}{\partial \dot{q}}(q,\dot{q}) - \frac{\partial \mathcal{L}}{\partial q}(q,\dot{q}) = \tau - F(q,\dot{q}), \tag{5.30}$$

where $q = [q_1, q_2]^\top \in \mathbb{R}^2$ is the vector of the (measurable) joint position variables, $\tau \in \mathbb{R}^2$ the vector of the external torques and $F(q,\dot{q})$ represents the dissipative forces. The Lagrangian function $\mathcal{L}(q,\dot{q}) = \mathcal{T}(q,\dot{q}) - \mathcal{V}(q)$ is defined as the difference between the kinetic energy $\mathcal{T}(q,\dot{q}) = \frac{1}{2}\dot{q}^\top M(q)\dot{q}$ and the potential energy $\mathcal{V}(q)$. We consider a two-body system whose inertia matrix has the form

$$M(q) = \begin{bmatrix} m_{11}(q_2) & m_{12}(q_2) \\ m_{12}(q_2) & m_{22} \end{bmatrix}, \tag{5.31}$$

with the standard assumption $M(q) = M(q)^\top > 0$, for all q. The Euler–Lagrange equations (5.30) can also be written as

$$M(q)\ddot{q} + C(q,\dot{q})\dot{q} + G(q) = \tau - F(q,\dot{q}), \tag{5.32}$$

where the vector $G(q)$ accounts for the potential forces and the matrix $C(q,\dot{q})$ represents the Coriolis and centrifugal forces and is given by[12]

$$C(q,\dot{q}) = \begin{bmatrix} \frac{1}{2}m_{11}'(q_2)\dot{q}_2 & \frac{1}{2}m_{11}'(q_2)\dot{q}_1 + m_{12}'(q_2)\dot{q}_2 \\ -\frac{1}{2}m_{11}'(q_2)\dot{q}_1 & 0 \end{bmatrix},$$

---
[12] The Coriolis–centrifugal force matrix is derived using the Christoffel symbols [196].

where $m_{i,j}{}'$ denotes the first derivative of $m_{i,j}$ with respect to q_2. Note that the matrix $C(q,\dot{q})$ is linear in the second argument, i.e., the following property holds
$$C(q, \eta + z) = C(q, \eta) + C(q, z), \tag{5.33}$$
for all q, η and z. Defining the states $y = q$ and $\eta = \dot{q}$, the system (5.32) can be rewritten in the form (5.1), namely
$$\begin{aligned} \dot{y} &= \eta, \\ \dot{\eta} &= -M(y)^{-1}\left(C(y,\eta)\eta + F(y,\eta) + G(y) - \tau\right). \end{aligned} \tag{5.34}$$

5.4.2 Observer Design

We now proceed to the design of a reduced-order observer of the form (5.2) for the velocity vector η using the result in Section 5.1. Let $\phi(\eta, y, t) = T(y)\eta$, where $T(y)$ is a 2×2 transformation matrix to be determined, and define the error variable
$$z = \beta(\xi, y) - T(y)\eta = T(y)\left(\hat{\eta} - \eta\right),$$
whose dynamics are given by
$$\dot{z} = \frac{\partial \beta}{\partial \xi}\dot{\xi} + \frac{\partial \beta}{\partial y}\eta - \bar{C}(y,\eta)\eta + T(y)M(y)^{-1}\left(F(y,\eta) + G(y) - \tau\right),$$
where $\bar{C}(y,\eta) = \dot{T}(y) - T(y)M(y)^{-1}C(y,\eta)$.

Assume for now that the Jacobian $\frac{\partial \beta}{\partial \xi}$ is invertible—this will be verified by the choice of the function $\beta(\cdot)$. Selecting the observer dynamics (5.2) as in Theorem 5.1, namely
$$\dot{\xi} = -\left(\frac{\partial \beta}{\partial \xi}\right)^{-1}\left(\frac{\partial \beta}{\partial y}\hat{\eta} - \bar{C}(y,\hat{\eta})\hat{\eta} + T(y)M(y)^{-1}\left(F(y,\hat{\eta}) + G(y) - \tau\right)\right),$$
and using the fact that $\bar{C}(y,\eta)$ satisfies the linearity property (5.33) yields the error dynamics
$$\begin{aligned} \dot{z} = &-\frac{\partial \beta}{\partial y}L(y)z + \bar{C}(y, L(y)z)\hat{\eta} + \bar{C}(y,\eta)L(y)z \\ &- T(y)M(y)^{-1}\left(F(y, \eta + L(y)z) - F(y,\eta)\right), \end{aligned} \tag{5.35}$$
where $L(y) = T(y)^{-1}$.

The transformation matrix $T(y)$ is selected to satisfy the equation[13] $M(y) = T(y)^\top T(y)$, which yields

---

[13] This factorisation of the inertia matrix $M(y)$ is known as Cholesky factorisation and has been introduced in [81].

5.4 Mechanical Systems with Two Degrees of Freedom

$$T(y) = \begin{bmatrix} \sqrt{m_{11}(y_2)} & \dfrac{m_{12}(y_2)}{\sqrt{m_{11}(y_2)}} \\ 0 & \dfrac{r(y_2)}{m_{11}(y_2)} \end{bmatrix},$$

where

$$r(y_2) = \sqrt{m_{11}(y_2)(m_{22}m_{11}(y_2) - m_{12}(y_2)^2)}.$$

Note that, since the inertia matrix is positive-definite, $m_{11}(y_2) > 0$ and $m_{22}m_{11}(y_2) - m_{12}(y_2)^2 > 0$, hence the matrix $T(y)$ is invertible. The above selection of $T(y)$ is such that the matrix

$$\bar{C}(y,\eta)L(y) = \left[\dot{T}(y) - L(y)^\top C(y,\eta)\right] L(y)$$

is *skew-symmetric*, i.e., $\bar{C}(y,\eta)L(y) + L(y)^\top \bar{C}(y,\eta)^\top = 0$. In particular,

$$\bar{C}(y,\eta)L(y) = \frac{m_{11}{}'(y_2)}{2r(y_2)} (m_{11}(y_2)\eta_1 + m_{12}(y_2)\eta_2) \begin{bmatrix} 0 & -1 \\ 1 & 0 \end{bmatrix}.$$

Substituting $\hat{\eta} = L(y)\beta(\xi,y)$ and rearranging the second term on the right-hand side of (5.35) yields the error dynamics

$$\dot{z} = -\Gamma(y,\xi)z + \bar{C}(y,\eta)L(y)z - L(y)^\top \left(F(y,\eta + L(y)z) - F(y,\eta)\right), \quad (5.36)$$

where

$$\Gamma(y,\xi) = \begin{bmatrix} \gamma_{11}(y,\xi) & \gamma_{12}(y,\xi) \\ \gamma_{21}(y,\xi) & \gamma_{22}(y,\xi) \end{bmatrix}$$

$$= \begin{bmatrix} \dfrac{1}{\sqrt{m_{11}(y_2)}} \dfrac{\partial \beta_1}{\partial y_1} + \dfrac{m_{11}{}'(y_2)\beta_2(\xi,y)}{2r(y_2)} & \dfrac{m_{11}(y_2)}{r(y_2)} \dfrac{\partial \beta_1}{\partial y_2} - \dfrac{m_{12}(y_2)}{r(y_2)} \dfrac{\partial \beta_1}{\partial y_1} \\ \dfrac{1}{\sqrt{m_{11}(y_2)}} \dfrac{\partial \beta_2}{\partial y_1} - \dfrac{m_{11}{}'(y_2)\beta_1(\xi,y)}{2r(y_2)} & \dfrac{m_{11}(y_2)}{r(y_2)} \dfrac{\partial \beta_2}{\partial y_2} - \dfrac{m_{12}(y_2)}{r(y_2)} \dfrac{\partial \beta_2}{\partial y_1} \end{bmatrix}.$$

We now shape the mapping $\beta(\xi,y) = [\beta_1(\xi,y), \beta_2(\xi,y)]^\top$ so that the system (5.36) has a uniformly globally asymptotically stable equilibrium at the origin. To this aim, we need the following assumption[14].

Assumption 5.1. *There exists a constant $\nu \in \mathbb{R}$ such that the dissipative forces vector $F(y,\eta)$ satisfies the property*

$$z^\top L(y)^\top \frac{\partial F(y,\eta)}{\partial \eta} L(y) z \geq \nu z^\top z, \tag{5.37}$$

for all y, η and z.

---

[14] Note that Assumption 5.1 is satisfied for $F(y,\eta) = \tilde{F}(y)\eta$, if $\tilde{F}(y)$ is positive-definite. This is the case, for instance, if the dissipative forces are due to viscous friction.

Consider the quadratic Lyapunov function $V(z) = \frac{1}{2}z^\top z$ whose time-derivative along the trajectories of (5.36) satisfies

$$\begin{aligned}\dot V(z) &= -\gamma_{11}(y,\xi)z_1^2 - (\gamma_{12}(y,\xi)+\gamma_{21}(y,\xi))\,z_1 z_2 - \gamma_{22}(y,\xi)z_2^2 \\ &\quad - z^\top L(y)^\top \left(F(y,\eta+L(y)z) - F(y,\eta)\right) \\ &\leq -\gamma_{11}(y,\xi)z_1^2 - (\gamma_{12}(y,\xi)+\gamma_{21}(y,\xi))\,z_1 z_2 - \gamma_{22}(y,\xi)z_2^2 \\ &\quad - \nu z^\top z, \end{aligned} \qquad (5.38)$$

where we have used (5.37) combined with the Mean Value Theorem, namely

$$F(y,\eta+z) - F(y,\eta) = \left.\frac{\partial F(y,\eta)}{\partial \eta}\right|_{\eta=\eta^*} z,$$

where η^* is a convex combination of η and z. It remains to find functions $\beta_1(\cdot)$ and $\beta_2(\cdot)$ that solve the partial differential (in)equalities

$$\gamma_{11}(y,\xi) + \nu > 0,$$
$$\gamma_{12}(y,\xi) + \gamma_{21}(y,\xi) = 0,$$
$$\gamma_{22}(y,\xi) + \nu > 0,$$

and guarantee invertibility of the matrix $\frac{\partial \beta}{\partial \xi}$ and left-invertibility of $\beta(\cdot)$ with respect to its first argument.

We consider two different cases: (1) the inertia matrix $M(y)$ has the general form (5.31) and its entries are *bounded*; and (2) the inertia matrix $M(y)$ is *diagonal*, i.e., $m_{12}(y_2) = 0$.

5.4.3 Non-diagonal Inertia Matrix with Bounded Entries

Assume that the inertia matrix satisfies the property

$$M_{\min} I_2 \leq M(y) \leq M_{\max} I_2, \qquad (5.39)$$

for all y and for some constants $M_{\max} \geq M_{\min} > 0$, and that $m_{11}'(y_2)$ is bounded for all y_2. Selecting $\beta_1(\cdot)$ and $\beta_2(\cdot)$ as

$$\beta_1(\xi,y) = \sqrt{m_{11}(y_2)}\left[k_1\left(y_1 + \int_{y_{20}}^{y_2} \frac{m_{12}(\zeta)}{m_{11}(\zeta)}\,d\zeta\right)(k_2 \pi + \xi_2^2) + \xi_1\right], \quad (5.40)$$

$$\beta_2(\xi,y) = k_2 \arctan(y_2) + \xi_2, \qquad (5.41)$$

leads to $\gamma_{12}(y,\xi) + \gamma_{21}(y,\xi) = 0$ and

$$\gamma_{11}(y,\xi) \geq -c_2\left(k_2 \arctan(y_2) + \xi_2\right) + k_1\left(\xi_2^2 + k_2 \pi\right),$$

$$\gamma_{22}(y,\xi) \geq \frac{k_2 c_1}{1 + y_2^2},$$

5.4 Mechanical Systems with Two Degrees of Freedom

where the constants c_1 and c_2 are such that $c_1 \leq m_{11}(y_2)/r(y_2)$ and $c_2 \geq |m_{11}'(y_2)/(2r(y_2))|$. Note that the existence of such constants is guaranteed by the property (5.39) and the boundedness of $m_{11}'(y_2)$. The positivity of $\gamma_{11}(\cdot)$ and $\gamma_{22}(\cdot)$ is ensured by picking the "observer gains" k_1 and k_2 such that $k_1 \geq c_2$ and $k_2 > \frac{1}{2\pi}$.

Finally, note that $\beta(\xi, y)$ is invertible with respect to ξ and, moreover,

$$\frac{\partial \beta}{\partial \xi} = \begin{bmatrix} \sqrt{m_{11}(y_2)} & 2k_1\xi_2\sqrt{m_{11}(y_2)}\left(\int_{y_{20}}^{y_2} \frac{m_{12}(\zeta)}{m_{11}(\zeta)}\,d\zeta + y_1\right) \\ 0 & 1 \end{bmatrix},$$

which is invertible for any y_2.

The result is summarised in the following proposition.

Proposition 5.1. *Consider the system (5.36), where $\beta_1(\cdot)$ and $\beta_2(\cdot)$ are given by (5.40) and (5.41), respectively, and assume that (5.39) holds. Then the zero equilibrium of the system (5.36) is:*

(i) uniformly globally exponentially stable, if Assumption 5.1 holds with $\nu > 0$;
(ii) uniformly globally stable, if Assumption 5.1 holds with $\nu = 0$;
(iii) uniformly globally asymptotically stable, if Assumption 5.1 holds with $\nu = 0$ and $\frac{y_2(t)}{\epsilon + \sqrt{t}}$ is bounded for all $t > 0$ and for some $\epsilon > 0$.

Proof. We only prove (iii), since (i) and (ii) follow directly from (5.38). Set $\nu = 0$ in (5.38) and note that as y_2 tends to infinity, the function $\gamma_{22}(y, \xi)$ tends to zero. Since from (5.38)

$$|z(t)| \leq \sqrt{2V(t_0)} \exp\left(-\frac{1}{2}\int_{t_0}^{t} [\min\{\gamma_{11}(y(\tau), \xi(\tau)), \gamma_{22}(y(\tau), \xi(\tau))\} + \nu]\,d\tau\right),$$

the convergence of $|z(t)|$ to zero is guaranteed if

$$\lim_{t\to\infty} \int_{t_0}^{t} \gamma_{22}(y(\tau), \xi(\tau))\,d\tau = \lim_{t\to\infty} \int_{t_0}^{t} \frac{k_2 c_1}{1 + y_2(\tau)^2}\,d\tau = \infty.$$

The above condition is satisfied when $\lim_{t\to\infty} |y_2(t)|/\sqrt{t} \leq d$, for some finite d, namely when $y_2(t)$ tends to infinity not faster than $\sqrt{t}$, yielding $|z(t)| \leq \sqrt{2V(t_0)/(1+t)^{k_2 c_1}}$, hence the claim. □

5.4.4 Diagonal Inertia Matrix

This is a special case that is encountered, for instance, in the "ball and beam" system as well as in Cartesian manipulators. In this case, we relax condition (5.39) and replace it with the following assumption.

Assumption 5.2. There exists a $\mathcal{C}^r$ function $f_2 : \mathbb{R} \to \mathbb{R}$, $r \geq 1$, such that

$$f_2'(y_2) > \epsilon\sqrt{m_{22}} \quad \text{and} \quad m_{11}'(y_2)f_2(y_2) \geq 0,$$

where ϵ is a positive constant such that $\nu \geq -\epsilon$.[15]

Selecting the functions $\beta_1(\xi, y)$ and $\beta_2(\xi, y)$ as

$$\beta_1(\xi, y) = \sqrt{m_{11}(y_2)}\left(k_1 y_1\left(1 + \xi_2^2\right) + \xi_1\right), \tag{5.42}$$
$$\beta_2(\xi, y) = f_2(y_2) + \xi_2 \tag{5.43}$$

guarantees that $\gamma_{12}(y, \xi) + \gamma_{21}(y, \xi) = 0$ and

$$\gamma_{11}(y, \xi) \geq -c_2 \xi_2 + k_1\left(1 + \xi_2^2\right),$$
$$\gamma_{22}(y, \xi) = \frac{1}{\sqrt{m_{22}}} f_2'(y_2) > \epsilon > 0,$$

where $c_2 \geq |m_{11}'(y_2)/(2r(y_2))| > 0$ and we have used Assumption 5.2. Choosing the "observer gain" k_1 such that

$$k_1 > \frac{\epsilon + \sqrt{\epsilon^2 + c_2^2}}{2}$$

ensures that $\gamma_{11}(y, \xi) > \epsilon$ for all y and ξ, hence the derivative of the Lyapunov function given in (5.38) is negative-definite. Finally, note that the Jacobian matrix $\frac{\partial \beta}{\partial \xi}$ is invertible, in particular

$$\frac{\partial \beta}{\partial \xi} = \begin{bmatrix} \sqrt{m_{11}(y_2)} & 2k_2 y_1 \xi_2 \sqrt{m_{11}(y_2)} \\ 0 & 1 \end{bmatrix},$$

and $\beta(\xi, y)$ is invertible with respect to ξ. The result is summarised in the following statement.

Proposition 5.2. *Consider the system (5.36), where $\beta_1(\cdot)$ and $\beta_2(\cdot)$ are given by (5.42) and (5.43) respectively, and suppose that Assumptions 5.1 and 5.2 hold. Then the system (5.36) has a uniformly globally exponentially stable equilibrium at the origin.*

5.5 Example: Ball and Beam

Consider the "ball and beam" system[16] depicted in Figure 5.4, which is described by equations of the form (5.34), where the states y_1 and y_2 correspond to the position of the ball and the angle of the beam, respectively, $F(y, \eta) = 0$,

---
[15] Note that in this case ν may be negative.
[16] See [67, 72] for details on this problem.

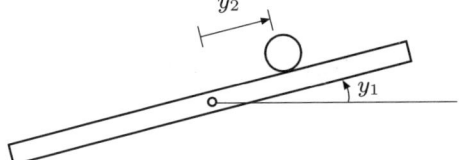

Fig. 5.4. Ball and beam.

$$M(y) = \begin{bmatrix} B + my_2^2 & 0 \\ 0 & A \end{bmatrix}, \quad C(y, \eta) = \begin{bmatrix} my_2\eta_2 & my_2\eta_1 \\ -my_2\eta_1 & 0 \end{bmatrix}$$

and

$$G(y) = \begin{bmatrix} mgy_2 \cos(y_1) \\ mg \sin(y_1) \end{bmatrix},$$

where $A = \frac{J_b}{R^2} + m$, $B = J + J_b$, m, R and J_b are the mass, radius and moment of inertia of the ball, respectively, J is the moment of inertia of the beam, g is the gravitational acceleration, and τ is the applied torque.

Following the observer design presented in Section 5.4.4, the functions $\beta_1(\xi, y)$ and $\beta_2(\xi, y)$ are obtained from (5.42) and (5.43) as

$$\beta_1(\xi, y) = \sqrt{B + my_2^2} \left(k_1 y_1 \left(1 + \xi_2^2 \right) + \xi_1 \right),$$
$$\beta_2(\xi, y) = k_2 \left(\tfrac{1}{3} y_2^3 + y_2 \right) + \xi_2,$$

where we have selected the function $f_2(y_2)$ to satisfy Assumption 5.2. Note that the derivative of the Lyapunov function $V(z) = \tfrac{1}{2} z^\top z$ satisfies

$$\dot{V}(z) \leq -\left(k_1 \left(1 + \xi_2^2 \right) - c_2 |\xi_2| \right) z_1^2 - k_2 \left(1 + y_2^2 \right) z_2^2 \quad (5.44)$$

and is negative-definite provided $k_1 \geq c_2$.

Simulations of the *unforced* (i.e., $\tau = 0$) ball and beam system and the observer described above have been carried out with the parameter values[17] $m = 0.05$ kg, $R = 0.01$ m, $J_b = 2 \times 10^{-6}$ kg m^2, $J = 0.02$ kg m^2, and $g = 9.81$ m/s^2. The response of the system states to the initial conditions $(y_{10}, y_{20}, x_{10}, x_{20}) = (0.1, 0, 0.5, 0.5)$ is shown in Figure 5.5. The time histories of the observer error variables $\hat{\eta}_1 - \eta_1$ and $\hat{\eta}_2 - \eta_2$, for different values of the parameters k_1 and k_2, are shown in Figure 5.6. Observe that the errors converge asymptotically to zero at a rate that increases with k_1 and k_2. From (5.44) it is clear that incrementing k_1 leads to a higher decay rate of the observer error z_1, whereas incrementing k_2 leads mainly to an increment of the decay rate for the observer error z_2, which corresponds to the angular velocity of the beam.

---

[17] We use the same parameter values as in [72].

114   5 Nonlinear Observer Design

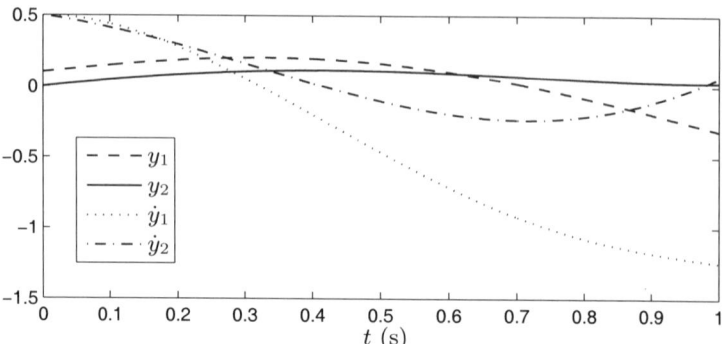

Fig. 5.5. Unforced trajectories of the ball and beam.

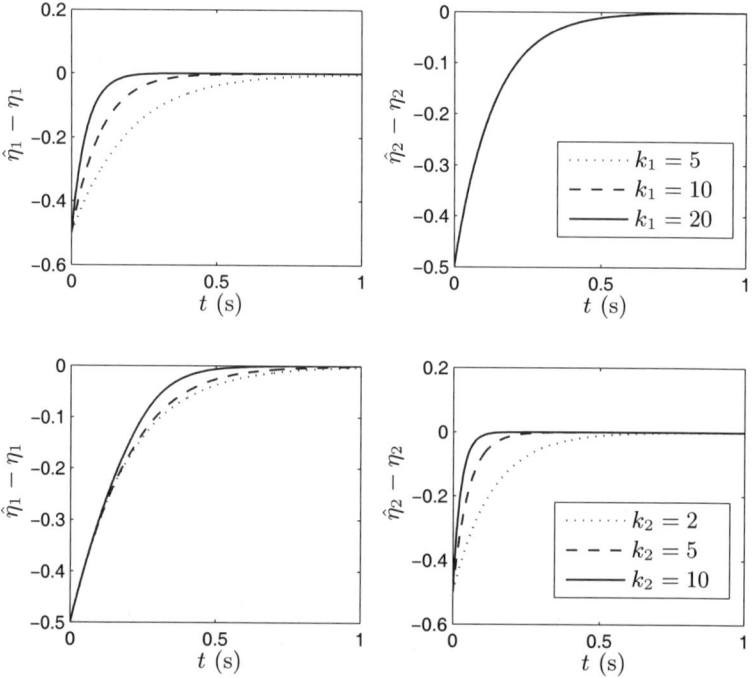

Fig. 5.6. Time histories of the observer errors $\hat{\eta}_1 - \eta_1$ and $\hat{\eta}_2 - \eta_2$ for fixed $k_2 = 2$ (top) and for fixed $k_1 = 5$ (bottom).

6
Robust Stabilisation via Output Feedback

In this chapter the I&I methodology is applied to the problem of asymptotic stabilisation by means of output feedback. It is known that for general nonlinear systems the combination of an exponentially convergent observer with a globally stabilising, state feedback control law may lead to instability and even finite escape time[1]. One way to overcome this problem (without modifying the observer design) is to construct state feedback controllers that ensure the closed-loop system is either input-to-state stable or $\mathcal{L}_p$ stable with respect to the state estimation error. A less conservative condition can be obtained by rewriting the closed-loop dynamics as a system with a globally asymptotically stable equilibrium perturbed by an *integrable* term[2].

The approach used in the present chapter follows the results of Chapters 3 and 4 and consists in writing the closed-loop system as a *cascaded* or *feedback* interconnection of two systems possessing specific input–output and asymptotic properties. Hence, unlike some existing results, the method does not rely on the construction of a Lyapunov function for the closed-loop system. Moreover, from an adaptive control viewpoint, it treats unknown parameters and unmeasured states in a uniform manner.

The methodology is further extended to study systems with *unstructured* uncertainties, *i.e.*, with unknown functions appearing in the system equations. This is mainly achieved by using the reduced-order observer design described in Chapter 5 and expressing the augmented system as a feedback interconnection of input-to-state stable systems and applying a small-gain argument.

The method is illustrated by means of several academic examples and it is applied to the design of a globally asymptotically stabilising controller for the benchmark *translational oscillator/rotational actuator* (TORA) system, where only the rotation angle is measurable.

---

[1] See the examples in [115, 204, 217, 13, 8].
[2] Such conditions can be found, for instance, in [58] and [177]. See also [8] for a survey.

6 Robust Stabilisation via Output Feedback

Note 6.1. The problem of asymptotic stabilisation by means of output feedback for general nonlinear systems has been widely studied in recent years. Following the fundamental existence result presented in [218, 215, 195] (see also [178] for an extension to the global case) several attempts have been made to construct explicit output feedback algorithms for classes of nonlinear systems. In particular, the class of systems that are linear in the unmeasured states has received special attention, see [135, 137, 123, 25, 26, 59, 58, 212, 84, 180, 122, 121, 179, 90, 9, 10], and references therein. ◁

6.1 Introduction

The results of Chapters 3 and 4 can be straightforwardly extended to the output feedback stabilisation problem, *i.e.*, to the problem of finding a dynamic stabilising control law that does not depend on the unmeasured states. To this end, consider the system

$$\begin{aligned} \dot{\eta} &= f(\eta, y, u), \\ \dot{y} &= h(\eta, y, u), \end{aligned} \tag{6.1}$$

with state $(\eta, y) \in \mathbb{R}^n \times \mathbb{R}^r$ and input $u \in \mathbb{R}^m$, and suppose that only the state y is available for measurement. Note that the system (6.1) may also include unknown parameters, *i.e.*, equations of the form $\dot{\eta}_i = 0$. The output feedback regulation problem is to find a dynamic output feedback control law described by equations of the form

$$\begin{aligned} \dot{\hat{\eta}} &= w(y, \hat{\eta}), \\ u &= v(y, \hat{\eta}), \end{aligned} \tag{6.2}$$

with $\hat{\eta} \in \mathbb{R}^n$, such that all trajectories of the closed-loop system (6.1), (6.2) are bounded and

$$\lim_{t \to \infty} y(t) = y^*, \tag{6.3}$$

for some $y^* \in \mathbb{R}^r$.

Note that, in contrast with the state feedback stabilisation results of Chapters 2 and 3, in this case it is only required that the output y converges to a set-point and that η remains bounded. The reason for this is that it may not be possible to drive the whole state (η, y) to a desired equilibrium. This is the case, for instance, when the vector η contains unknown parameters. However, it is often possible to establish global stability of a (partially specified) equilibrium (with $y = y^*$).

Following the ideas of Section 3.1, suppose that there exists a full-information control law $u = v(\eta, y)$ such that all trajectories of the closed-loop system

$$\begin{aligned} \dot{\eta} &= f(\eta, y, v(\eta, y)), \\ \dot{y} &= h(\eta, y, v(\eta, y)), \end{aligned} \tag{6.4}$$

are bounded and (6.3) holds. Consider now the system

$$\dot{\eta} = f(\eta, y, v(\hat{\eta} + \beta(y), y)),$$
$$\dot{y} = h(\eta, y, v(\hat{\eta} + \beta(y), y)), \tag{6.5}$$
$$\dot{\hat{\eta}} = w(y, \hat{\eta}),$$

with extended state $(\eta, y, \hat{\eta})$, where $\hat{\eta} \in \mathbb{R}^n$. The output feedback stabilisation problem boils down to finding functions $\beta(\cdot)$ and $w(\cdot)$ such that the system (6.5) has all trajectories bounded and is asymptotically immersed into the system (6.4).

6.2 Linearly Parameterised Systems

As in Section 3.1, a first approach is to consider linearly parameterised, control affine systems of the form

$$\begin{aligned} \dot{\eta} &= f_0(y) + f_1(y)\eta + g_1(y)u, \\ \dot{y} &= h_0(y) + h_1(y)\eta + g_2(y)u, \end{aligned} \tag{6.6}$$

with state $(\eta, y) \in \mathbb{R}^n \times \mathbb{R}^r$, output y and input $u \in \mathbb{R}^m$.

Theorem 6.1. *Consider the system (6.6) and assume the following hold.*

(A1) There exists a full-information control law $u = v(\eta, y)$ such that all trajectories of the closed-loop system

$$\begin{aligned} \dot{\eta} &= f^*(\eta, y) \triangleq f_0(y) + f_1(y)\eta + g_1(y)v(\eta, y), \\ \dot{y} &= h^*(\eta, y) \triangleq h_0(y) + h_1(y)\eta + g_2(y)v(\eta, y) \end{aligned} \tag{6.7}$$

are bounded and satisfy (6.3).
(A2) There exists a mapping $\beta(\cdot)$ such that all trajectories of the system

$$\begin{aligned} \dot{z} &= \left[f_1(y) - \frac{\partial \beta}{\partial y} h_1(y) \right] z, \\ \dot{\eta} &= f^*(\eta, y) + g_1(y) \left(v(\eta + z, y) - v(\eta, y) \right), \\ \dot{y} &= h^*(\eta, y) + g_2(y) \left(v(\eta + z, y) - v(\eta, y) \right) \end{aligned} \tag{6.8}$$

are bounded and satisfy

$$\lim_{t \to \infty} \left[v(\eta(t) + z(t), y(t)) - v(\eta(t), y(t)) \right] = 0.$$

Then there exists a dynamic output feedback control law described by equations of the form (6.2) such that all trajectories of the closed-loop system (6.1), (6.2) are bounded and satisfy (6.3).

6 Robust Stabilisation via Output Feedback

Proof. Similarly to Theorem 2.1, the proof consists in constructing a function $w(y, \hat{\eta})$ so that the closed-loop system (6.6), (6.2) is transformed into (6.8). To this end, let

$$z = \hat{\eta} - \eta + \beta(y)$$

and note that the dynamics of the off-the-manifold variable z are given by

$$\dot{z} = w(y, \hat{\eta}) - f_0(y) - f_1(y)(\hat{\eta} + \beta(y) - z) - g_1(y)v(\hat{\eta} + \beta(y), y)$$
$$+ \frac{\partial \beta}{\partial y}[h_0(y) + h_1(y)(\hat{\eta} + \beta(y) - z) + g_2(y)v(\hat{\eta} + \beta(y), y)].$$

Selecting

$$w(y, \hat{\eta}) = f_0(y) + f_1(y)(\hat{\eta} + \beta(y)) + g_1(y)v(\hat{\eta} + \beta(y), y)$$
$$- \frac{\partial \beta}{\partial y}[h_0(y) + h_1(y)(\hat{\eta} + \beta(y)) + g_2(y)v(\hat{\eta} + \beta(y), y)]$$

yields the first of equations (6.8), while the other two equations are obtained from (6.6) setting $u = v(\hat{\eta} + \beta(y), y) = v(\eta + z, y)$ and adding and subtracting the term $v(\eta, y)$. Hence, by assumptions (A1) and (A2), all trajectories of the closed-loop system (6.1), (6.2) are bounded and satisfy (6.3). □

Example 6.1. Consider the two-dimensional nonlinear system

$$\begin{aligned} \dot{\eta} &= \eta + y, \\ \dot{y} &= \eta(y^2 + 1) + u, \end{aligned} \qquad (6.9)$$

with input u and output y, where η is an unmeasured state, and the problem of stabilising the origin $(\eta, y) = (0, 0)$ by output feedback. It is interesting to note that the zero dynamics are described by the equation $\dot{\eta} = \eta$, hence the system (6.9) is not minimum-phase.

To begin with, note that (A1) is satisfied by setting

$$v(\eta, y) = -\eta(y^2 + 1) - k_1\eta - k_2 y,$$

with $k_1 > k_2 > 1$. Consider now the dynamic control law

$$\dot{\hat{\eta}} = w(y, \hat{\eta}),$$
$$u = v(\hat{\eta} + \beta(y), y) = -(\hat{\eta} + \beta(y))(y^2 + 1) - k_1(\hat{\eta} + \beta(y)) - k_2 y.$$

Selecting

$$w(y, \hat{\eta}) = -\frac{\partial \beta}{\partial y}[-k_1(\hat{\eta} + \beta(y)) - k_2 y] + \hat{\eta} + \beta(y) + y,$$

yields the system

$$\begin{aligned}
\dot{z} &= -\left[\frac{\partial \beta}{\partial y}(y^2+1) - 1\right]z, \\
\dot{\eta} &= \eta + y, \\
\dot{y} &= -k_1\eta - k_2 y - (y^2+1+k_1)z.
\end{aligned} \qquad (6.10)$$

A suitable selection for the function $\beta(\cdot)$ is given by

$$\beta(y) = \gamma\left(\frac{y^3}{3}+y\right) + (1+\gamma)\arctan(y),$$

with $\gamma > 0$, which is such that

$$\frac{\partial \beta}{\partial y}(y^2+1) - 1 = \gamma(y^2+1)^2 + \gamma.$$

This ensures that $(y^2(t)+1)z(t) \in \mathcal{L}_2$ and $k_1 z(t) \in \mathcal{L}_2$, hence all trajectories of the system (6.10) are bounded and η and y converge to zero. ∎

Example 6.2. Consider a linear nonminimum-phase system with input u and output y described by the transfer function

$$H(s) = -\frac{s-1}{s^2 + \theta s},$$

where θ is an unknown parameter. Defining $\eta_1 = \theta$, a state-space realisation of the form (6.1) is given by

$$\begin{aligned}
\dot{\eta}_1 &= 0, \\
\dot{\eta}_2 &= \eta_2 - (1+\eta_1)y, \\
\dot{y} &= \eta_2 - (1+\eta_1)y - u.
\end{aligned} \qquad (6.11)$$

Suppose that the system (6.11) is stabilisable, i.e., $1+\eta_1 = 1+\theta \neq 0$, and, moreover, that there exists a known constant ϵ such that $1+\eta_1 \geq \epsilon > 0$. The control objective is to regulate η_2 and y to zero. A full-information control law $v(\eta, y)$ that achieves this objective is given by

$$v(\eta, y) = \eta_2 - (1+\eta_1)y + k_1 \eta_2 + k_2 y,$$

where k_1 and k_2 are constants satisfying the inequalities

$$k_2 > 1, \qquad k_1 < -\frac{k_2}{\epsilon}.$$

Consider now a dynamic output feedback control law of the form

$$\begin{aligned}
\dot{\hat{\eta}} &= w(y, \hat{\eta}), \\
u &= v(\hat{\eta} + \beta(y), y).
\end{aligned}$$

It remains to find functions $\beta(\cdot)$ and $w(\cdot)$ such that all trajectories of the closed-loop system are bounded and the manifold $\hat{\eta} - \eta + \beta(y) = 0$ is attractive. To this end, consider the dynamics of the off-the-manifold co-ordinates $z = \hat{\eta} - \eta + \beta(y)$, namely

$$\dot{z} = w(\hat{\eta}, y) - \begin{bmatrix} 0 \\ \eta_2 - (1 + \eta_1) y \end{bmatrix} + \frac{\partial \beta}{\partial y} [\eta_2 - (1 + \eta_1) y - u],$$

with $\beta(\cdot) = [\beta_1(\cdot), \beta_2(\cdot)]^\top$. Selecting

$$w(y, \hat{\eta}) = \begin{bmatrix} 0 \\ \hat{\eta}_2 + \beta_2(y) - (1 + \hat{\eta}_1 + \beta_1(y)) y \end{bmatrix} - \frac{\partial \beta}{\partial y} [\hat{\eta}_2 + \beta_2(y) - (1 + \hat{\eta}_1 + \beta_1(y)) y - u]$$

yields

$$\dot{z} = \begin{bmatrix} \dfrac{\partial \beta_1}{\partial y} y & -\dfrac{\partial \beta_1}{\partial y} \\ \left(\dfrac{\partial \beta_2}{\partial y} - 1\right) y & -\left(\dfrac{\partial \beta_2}{\partial y} - 1\right) \end{bmatrix} z,$$

which suggests setting $\beta(y) = [\beta_1(y), \beta_2(y)] = \left[-\frac{1}{2} y^2, 2y\right]$, yielding

$$\dot{z} = \begin{bmatrix} -y^2 & y \\ y & -1 \end{bmatrix} z. \tag{6.12}$$

Note that the above selection does not imply convergence of z to zero. In fact, it can only be concluded that z is bounded and $y(t) z_1(t) - z_2(t) \in \mathcal{L}_2$. To show that this is sufficient to prove stability of the origin for the closed-loop system, define the variable $\tilde{\eta}_2 = \hat{\eta}_2 + \beta_2(y)$ and note that the closed-loop system can be written in the $(\tilde{\eta}_2, y)$ co-ordinates as

$$\dot{\tilde{\eta}}_2 = \tilde{\eta}_2 - (1 + \eta_1) y - z_2 + (y z_1 - z_2),$$
$$\dot{y} = -k_2 y - k_1 \tilde{\eta}_2 + (y z_1 - z_2),$$

i.e., as an exponentially stable system perturbed by the terms $y z_1 - z_2$ and z_2. Note now that $y z_1 - z_2$ converges to zero and z_2 converges to a constant. Hence all trajectories converge towards an equilibrium, but not necessarily such that $y = 0$. This situation can be rectified by adding an integrator to the control law, which is now defined as

$$\begin{aligned} \dot{\hat{\eta}} &= w(y, \hat{\eta}), \\ \dot{\chi} &= y, \\ u &= v(\hat{\eta} + \beta(y), y) + k_0 \chi. \end{aligned} \tag{6.13}$$

It is straightforward to show that the closed-loop system (6.11), (6.13) is such that η_2 and y converge to zero, provided the constants k_1, k_2 and k_0 are chosen appropriately. Figure 6.1 shows the response of the system (6.11), (6.13) with the initial conditions $y(0) = 2$, $\eta_2(0) = 1$, for $\eta_1 = 1$, $\epsilon = 1$, $k_1 = -3$, $k_2 = 2$ and $k_0 = -1$. ∎

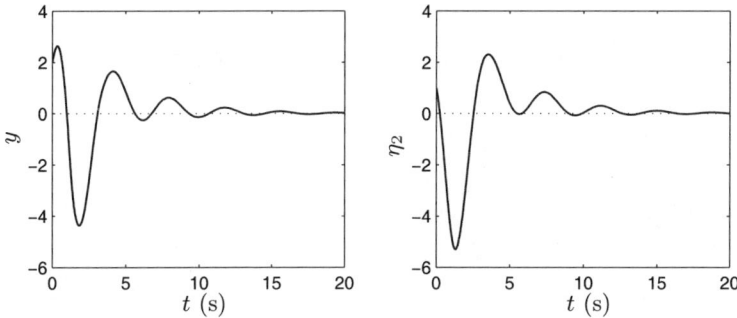

Fig. 6.1. Response of the system (6.11), (6.13).

6.3 Control Design Using a Separation Principle

The result in Theorem 6.1 relies on the existence of a function $\beta(\cdot)$ that stabilises the cascaded system (6.8). Note that, by construction, when $z = 0$, the (η, y)-subsystem in (6.8) has bounded trajectories, with y converging to the desired equilibrium. Therefore, it is natural to ask whether it is possible to neglect the (η, y)-subsystem and concentrate on the stabilisation of the z-subsystem alone. In the case of Theorem 6.1 this is not feasible because, even if z converges to zero exponentially, the term $v(\eta+z, y) - v(\eta, y)$ may *destabilise* the (η, y)-subsystem. This clearly poses a restriction on the function $v(\eta, y)$ (see, *e.g.*, Corollary 3.1).

The present section addresses this problem by establishing a condition under which the design of the full-information control law $u = v(\eta, y)$ can be *decoupled* from the design of the function $\beta(\cdot)$ and so the design for the z-subsystem can be tackled independently. In this sense it can be considered as a nonlinear counterpart of the separation principle used in linear systems. Interestingly, this result can be applied to systems that are not affine in the input.

To simplify the presentation we consider linearly parameterised systems described by equations of the form

$$\begin{aligned} \dot{\eta} &= f_0(y, u) + f_1(y, u)\eta, \\ \dot{y} &= h_0(y, u) + h_1(y, u)\eta, \end{aligned} \quad (6.14)$$

with state $(\eta, y) \in \mathbb{R}^n \times \mathbb{R}^r$, output y and input $u \in \mathbb{R}^m$. Note, however, that the result can be extended to systems that are nonlinear in the unmeasured states by using the general observer design of Chapter 5 and by replacing the observer error system (6.16) in Theorem 6.2 with (5.4).

Theorem 6.2. *Consider the system (6.14) and assume the following hold.*

(B1) There exists a full-information control law

$$u = v(\eta, y) \tag{6.15}$$

such that all trajectories of the closed-loop system (6.14), (6.15) are bounded and satisfy (6.3). Moreover, the system (6.14) with

$$u = v(\eta + z, y)$$

is globally bounded-input bounded-state stable with respect to the input z.
(B2) There exists a mapping $\beta(\cdot)$ such that the zero equilibrium of the system

$$\dot{z} = \left[f_1(y, u) - \frac{\partial \beta}{\partial y} h_1(y, u) \right] z \tag{6.16}$$

is globally stable, uniformly in y and u, and $z(t)$ is such that, for any η and y,

$$\lim_{t \to \infty} [v(\eta + z(t), y) - v(\eta, y)] = 0. \tag{6.17}$$

Then there exists a dynamic output feedback control law described by equations of the form (6.2) such that all trajectories of the closed-loop system (6.14), (6.2) are bounded and satisfy (6.3).

Proof. To begin with, consider a dynamic output feedback control law of the form (6.2), where the mapping $\beta(\cdot)$ is as in assumption (B2) and the function $w(\cdot)$ acts as a new control signal. Let

$$z = \hat{\eta} - \eta + \beta(y)$$

and note that the closed-loop system (6.14), (6.2) can be written in the η, y and z co-ordinates as

$$\begin{aligned}
\dot{\eta} &= f_0(y, v(\eta + z, y)) + f_1(y, v(\eta + z, y))\eta, \\
\dot{y} &= h_0(y, v(\eta + z, y)) + h_1(y, v(\eta + z, y))\eta, \\
\dot{z} &= w(y, \hat{\eta}) - f_0(y, v(\eta + z, y)) - f_1(y, v(\eta + z, y))(\hat{\eta} + \beta(y) - z) \\
&\quad + \frac{\partial \beta}{\partial y} \left[h_0(y, v(\eta + z, y)) + h_1(y, v(\eta + z, y))(\hat{\eta} + \beta(y) - z) \right].
\end{aligned}$$

It must be noted that, with the exception of the terms

$$f_1(y, v(\eta + z, y))z, \qquad \frac{\partial \beta}{\partial y} h_1(y, v(\eta + z, y))z,$$

all terms in the $\dot{z}$ equation are measurable. Therefore, setting

$$\begin{aligned}
w(y, \hat{\eta}) &= f_0(y, v(\eta + z, y)) + f_1(y, v(\eta + z, y))(\hat{\eta} + \beta(y)) \\
&\quad - \frac{\partial \beta}{\partial y} \left[h_0(y, v(\eta + z, y)) + h_1(y, v(\eta + z, y))(\hat{\eta} + \beta(y)) \right]
\end{aligned}$$

yields the system

6.3 Control Design Using a Separation Principle

$$\dot{\eta} = f_0(y, v(\eta + z, y)) + f_1(y, v(\eta + z, y))\eta,$$
$$\dot{y} = h_0(y, v(\eta + z, y)) + h_1(y, v(\eta + z, y))\eta,$$
$$\dot{z} = \left[f_1(y, v(\eta + z, y)) - \frac{\partial \beta}{\partial y} h_1(y, v(\eta + z, y)) \right] z.$$

As a result, by assumption (B2), the variable z remains bounded for all t and it is such that (6.17) holds. Hence, by assumption (B1), η and y are bounded for all t and (6.3) holds, which proves the claim. $\square$

Theorem 6.2 states that it is possible to solve the considered output feedback regulation problem provided that two subproblems are solvable. The former is a *robust* full-information stabilisation problem, the latter is a problem of *robust* stabilisation by output injection with a constraint on the ω-limit set of the state z. Note, moreover, that the output injection gain has to possess a special form, i.e., it is the Jacobian matrix of the mapping $\beta(y)$. This, in particular, poses a restriction when the number of outputs is larger than one.

Assumption (B2) can be replaced by the following (stronger) condition.

(B2′) *There exists a mapping $\beta(\cdot)$ such that the zero equilibrium of the system (6.16) is globally asymptotically stable, uniformly in y and u.*

If assumption (B2′) holds, then $\hat{\eta}$ can be used to construct an asymptotically converging estimate of the unmeasured states η, which is given by $\hat{\eta} + \beta(y)$. Note, however, that, to achieve the control goal, i.e., regulation of y, only the reconstruction of the full-information control law $v(\eta, y)$ is necessary.

Example 6.3. Consider the three-dimensional system described by equations of the form

$$\begin{aligned} \dot{x}_1 &= x_1 + x_2, \\ \dot{x}_2 &= x_3 + x_2^2, \\ \dot{x}_3 &= x_1 x_3^2 + x_1 + u, \\ y &= [x_2, x_3]^\mathsf{T}. \end{aligned} \quad (6.18)$$

Note that the zero dynamics of the system (6.18) are described by $\dot{x}_1 = x_1$, hence the system is not minimum-phase and cannot be stabilised using standard approaches[3]. On the contrary, it is now shown that using the result in Theorem 6.2 it is possible to achieve asymptotic stabilisation by dynamic output feedback.

To this end, note that a simple application of a backstepping procedure shows that there exists a full-information control law such that assumption (B1) holds. To verify assumption (B2), note that the system (6.16) reduces to

$$\dot{z} = \left[1 - \frac{\partial \beta}{\partial x_3} \left(1 + x_3^2 \right) \right] z.$$

---
[3] See, for instance, the method in [58].

Hence, the selection $\beta(y) = k\arctan(x_3)$, with $k > 1$, yielding $\frac{\partial \beta}{\partial x_3} = \frac{k}{1+x_3^2}$, is such that the system (6.16) in assumption (B2) reduces to

$$\dot{z} = -(k-1)z,$$

which has a (uniformly) globally asymptotically stable equilibrium at zero. ■

6.4 Systems with Monotonic Nonlinearities Appearing in the Output Equation

As mentioned in the introduction, invoking certainty equivalence and combining a globally convergent observer with a globally stabilising state feedback control law may result in an *unstable* closed-loop system. A condition for such an output feedback scheme to be stabilising is given in the following statement[4] for the class of systems (5.16), which we rewrite here for ease of reference

$$\begin{aligned}\dot{\eta} &= f_1(\eta, y, u) \triangleq F_1(y)\gamma(Cy + H\eta) + \delta(y, H\eta) + g_1(y, u),\\ \dot{y} &= f_2(\eta, y.u) \triangleq F_2(y)\gamma(Cy + H\eta) + g_2(y, u),\end{aligned} \quad (6.19)$$

with state $x = (\eta, y) \in \mathbb{R}^n \times \mathbb{R}^m$, output y and input $u \in \mathbb{R}^q$, where $f_1(\cdot)$ and $f_2(\cdot)$ are locally Lipschitz functions, which are zero at the origin, and H is a full-rank matrix.

Theorem 6.3. *Consider the system (6.19) and a reduced-order observer of the form (5.23) and such that Theorem 5.2 holds with $\rho(x,z)/|\Gamma(x,z)| \geq \rho_0$, for some $\rho_0 > 0$, and suppose that the following hold.*

(C1) There exists a state feedback control law $u = \phi(x)$ with $\phi(0) = 0$, and a C^1, positive-definite and radially unbounded function $V(x)$ such that

$$\frac{\partial V}{\partial x}(x) f(x, \phi(x)) \leq -\kappa(x),$$

for all $x \in \mathbb{R}^n \times \mathbb{R}^m$, and for some positive-definite function $\kappa(\cdot)$.
(C2) There exists a C^1, class $\mathcal{K}_\infty$ function $L(\cdot)$ such that[5]

$$|L'(V(\hat{x}))| \left|\frac{\partial V}{\partial x}(\hat{x}) k(\xi, y)\right| < M \quad (6.20)$$

for some $M > 0$, where $\hat{x} = [\hat{\eta}^\top, y^\top]^\top$ and $k(\xi, y) = -\left[F_1(y), \frac{\partial \beta}{\partial y} F_1(y)\right]^\top$.

Then the system (6.19) in closed loop with (5.23) and $u = \phi(\hat{x})$ has a globally asymptotically stable equilibrium at the origin.

---
[4] This idea has been introduced, in a general framework, in [177].
[5] We denote by $L'(\cdot)$ the derivative of $L(\cdot)$ with respect to its argument.

6.4 Systems with Monotonic Nonlinearities Appearing in the Output Equation

Proof. Define the observer error $z = \hat{\eta} - \eta$. Differentiating and substituting (5.23) yields the error system

$$\dot{z} = \big(\tilde{F}(y,\xi)\Gamma(x,z)H + \Delta(x,z)H\big)z,$$

which, by (5.22), has a uniformly globally stable equilibrium at the origin. The system (6.19) can be written in the $(\hat{\eta}, y)$ co-ordinates as

$$\begin{aligned}\dot{\hat{\eta}} &= F_1(y)\gamma(Cy + H\hat{x}) + \delta(y, H\hat{\eta}) + g_1(y,u) - \tfrac{\partial \beta}{\partial y}F_1(y)\varepsilon(t),\\ \dot{y} &= F_2(y)\gamma(Cy + H\hat{\eta}) + g_2(y,u) - F_1(y)\varepsilon(t),\end{aligned} \qquad (6.21)$$

where

$$\varepsilon(t) = \Gamma(x,z)Hz = \gamma(Cy + H(\eta + z)) - \gamma(Cy + H\eta).$$

To obtain an integral bound on $\varepsilon(t)$, define the variable $\tilde{z} = Hz \in \mathbb{R}^p$, whose dynamics are given by

$$\dot{\tilde{z}} = \frac{d(Hz)}{dt} = \big(H\tilde{F}(y,\xi)\Gamma(x,z) + H\Delta(x,z)\big)\tilde{z}, \qquad (6.22)$$

and consider again the matrix inequality (5.22) which can be written as[6]

$$\big(\tilde{F}\Gamma H + \Delta H\big)^\top P + P\big(\tilde{F}\Gamma H + \Delta H\big) \leq -\rho I.$$

Multiplying from the left with HP^{-1} and from the right with $P^{-1}H^\top$ yields

$$HP^{-1}H^\top\big(H\tilde{F}\Gamma + H\Delta\big)^\top + \big(H\tilde{F}\Gamma + H\Delta\big)HP^{-1}H^\top \\ \leq -\rho HP^{-2}H^\top.$$

Note that the matrix $HP^{-1}H^\top$ is positive-definite. Multiplying both sides with $X \triangleq (HP^{-1}H^\top)^{-1}$ yields

$$\big(H\tilde{F}\Gamma + H\Delta\big)^\top X + X\big(H\tilde{F}\Gamma + H\Delta\big) \leq -\rho Q, \qquad (6.23)$$

where $Q \triangleq XHP^{-2}H^\top X$. In view of (6.23) and the dynamics (6.22), it is convenient to define the norm

$$|\tilde{z}|_X \triangleq \sqrt{\tilde{z}^\top X \tilde{z}}, \qquad (6.24)$$

which is related to the 2-norm by the inequalities

$$\sqrt{\lambda_{\min}(X)}|\tilde{z}| \leq |\tilde{z}|_X \leq \sqrt{\lambda_{\max}(X)}|\tilde{z}|.$$

Differentiating (6.24) along the trajectories of (6.22) and using (6.23) yields

$$\frac{d}{dt}|\tilde{z}|_X \leq -\frac{\rho \tilde{z}^\top Q \tilde{z}}{2|\tilde{z}|_X} \leq -c_1|\tilde{z}|, \qquad (6.25)$$

---

[6] To improve readability the arguments of the functions are dropped.

where $c_1 = (\rho/2)\lambda_{\min}(Q)/\sqrt{\lambda_{\max}(X)}$. Note now that, by definition of ε and $\tilde{z}$, we have
$$|\varepsilon| = |\Gamma\tilde{z}| \leq |\Gamma||\tilde{z}|.$$

From (6.25) and the above inequality it follows that
$$\frac{d}{dt}|\tilde{z}|_x \leq -\frac{c_1}{|\Gamma|}|\varepsilon| \leq -c_2|\varepsilon|_1,$$

where $c_2 = c_1/(\sqrt{p}|\Gamma|) > \rho_0$. Integrating both sides from 0 to T we obtain
$$\int_0^T |\varepsilon(s)|ds \leq \frac{1}{c_2}\left(|\tilde{z}(0)|_x - |\tilde{z}(T)|_x\right),$$

hence the observer (5.23) is such that $|\varepsilon(t)|$ is integrable in $[0,T)$. The proof is completed by considering the derivative of $L(V(\hat{x}))$ along the trajectories of (6.21). □

Example 6.4. Consider the two-dimensional system
$$\begin{aligned}\dot{\eta} &= \psi(y)\eta^2 + u, \\ \dot{y} &= \eta + \eta^3 - y,\end{aligned} \tag{6.26}$$

with input u and output y, which is of the form (6.19) with $\delta(y,\eta) = \psi(y)\eta^2$. From (5.23), a reduced-order observer for the system (6.26) is given by
$$\dot{\xi} = \psi(y)\hat{\eta}^2 + u - \frac{\partial\beta}{\partial y}\left(\hat{\eta} + \hat{\eta}^3 - y\right),$$
$$\hat{\eta} = \xi + \beta(y),$$

where, for simplicity, we have taken $\frac{\partial\hat{\eta}}{\partial\xi} = 1$. The dynamics of the observer error $z = \hat{\eta} - \eta$ are given by $\dot{z} = \Lambda(\eta, y, z)z$, where
$$\Lambda(\eta, y, z) = \psi(y)(2\eta + z) - \frac{\partial\beta}{\partial y}\left(1 + 3\eta^2 + 3\eta z + z^2\right).$$

The design of the observer is completed by selecting the function $\beta(y)$ so that
$$\psi(y)(2\eta + z) - \frac{\partial\beta}{\partial y}\left(1 + 3\eta^2 + 3\eta z + z^2\right) \leq -\rho(\eta, y, z), \tag{6.27}$$

where $\rho(\eta, y, z) = \rho_0\left(1 + 3\eta^2 + 3\eta z + z^2\right)$, for some $\rho_0 > 0$. From the inequalities
$$1 + 3\eta^2 + 3\eta z + z^2 \geq 1, \quad \left|\frac{2\eta + z}{1 + 3\eta^2 + 3\eta z + z^2}\right| \leq \frac{1}{\sqrt{3}},$$

a possible solution is
$$\frac{\partial\beta}{\partial y} = \rho_0 + a\psi(y)^2 + b,$$

6.4 Systems with Monotonic Nonlinearities Appearing in the Output Equation

where a and b are positive constants such that $4ab \geq 1/3$, yielding the function

$$\beta(y) = \rho_0 y + a \int_0^y \psi(w)^2 dw + by.$$

Note now that a globally stabilising state feedback control law for the system (6.26) is given by

$$u = -\psi(y)\eta^2 - y\left(1 + \eta^2\right) - \eta,$$

which satisfies assumption (C1) of Theorem 6.3 with

$$V(x) = \frac{1}{2}y^2 + \frac{1}{2}\eta^2.$$

Moreover, the function $L(V) = \log(1 + V)$ is such that condition (6.20) becomes

$$\left| \frac{y + \frac{\partial \beta}{\partial y}\hat{\eta}}{1 + \frac{1}{2}(y^2 + \hat{\eta}^2)} \right| < M.$$

The existence of an upper bound M depends on the properties of $\frac{\partial \beta}{\partial y}$ and hence of $\psi(y)$. For instance, the foregoing inequality does not hold if $\psi(y)$ is quadratic (but it does hold if $\psi(y)$ is constant, which is the case considered in Example 5.3). ∎

As demonstrated by Example 6.4, the main difficulty in applying Theorem 6.3 to the class of systems (6.19) is that the term (6.20) may be of high order with respect to y due to the output injection term $\frac{\partial \beta}{\partial y}$ and the dependence of $F_1(\cdot)$ on y, and therefore it may be impossible to find a scaling function $L(\cdot)$ such that this term is bounded[7]. To overcome this difficulty we propose an alternative factorisation using the actual states of the system, rather than the estimated states as in (6.21), and we apply it on a problem, similar to the one in Example 6.4, which does not satisfy the boundedness condition in Theorem 6.3.

Theorem 6.4. *Consider the system (6.19) and a reduced-order observer of the form (5.23) and such that Theorem 5.2 holds with $\rho(\cdot) \geq \rho_0$ for some $\rho_0 > 0$, and suppose that the following hold.*

(D1) There exists a state feedback control law $u = \phi(x)$ with $\phi(0) = 0$, and a C^1, positive-definite and radially unbounded function $V(x)$ such that

$$\frac{\partial V}{\partial x}(x) f(x, \phi(x)) \leq -\kappa(x),$$

for all $x \in \mathbb{R}^n \times \mathbb{R}^m$, and for some positive-definite function $\kappa(\cdot)$.

---

[7] An additional complication is that $\frac{\partial \beta}{\partial y}$ may also depend on the observer state ξ. Note that for the special class of systems considered in [12] this is not an issue since both F_1 and $\frac{\partial \beta}{\partial y}$ are constant.

(D2) There exists a C^1, class $\mathcal{K}_\infty$ function $L(\cdot)$ such that

$$|L'(V(x))|\left|\frac{\partial V}{\partial x}(x)G(x,z)\right| \qquad (6.28)$$

is bounded, where the vector $G(x,z)$ is such that

$$G(x,z)z = g(y,\phi(\eta+z,y)) - g(y,\phi(\eta,y)),$$

with $g(\cdot) = [g_1(\cdot)^\top, g_2(\cdot)^\top]^\top$.

Then the system (6.19) in closed loop with (5.23) and $u = \phi(\hat{x})$ has a globally asymptotically stable equilibrium at the origin.

Proof. Consider the system (6.19) and the output feedback control law $u = \phi(\hat{\eta}, y) = \phi(\eta + z, y)$, where $z = \hat{\eta} - \eta$, and note that the closed-loop system is given in the (η, y) co-ordinates by

$$\begin{aligned}\dot{\eta} &= F_1(y)\gamma(Cy + H\eta) + \delta(y, H\eta) + g_2(y, \phi(x)) + G_1(x,z)z,\\ \dot{y} &= F_2(y)\gamma(Cy + H\eta) + g_2(y, \phi(x)) + G_1(x,z)z.\end{aligned} \qquad (6.29)$$

The claim then follows from boundedness of (6.28) and the fact that z is integrable, by differentiating $L(V(\hat{x}))$ along the trajectories of (6.29). □

Example 6.5. Consider the two-dimensional system

$$\begin{aligned}\dot{\eta} &= y\eta + u,\\ \dot{y} &= \eta + \eta^3 - y,\end{aligned} \qquad (6.30)$$

with input u and output y. From (5.23), a reduced-order observer for the system (6.30) is given by

$$\dot{\xi} = y\hat{\eta} + u - \frac{\partial \beta}{\partial y}\left(\hat{\eta} + \hat{\eta}^3 - y\right),$$
$$\hat{\eta} = \xi + \beta(y).$$

The dynamics of the observer error $z = \hat{\eta} - \eta$ are given by $\dot{z} = \Lambda(\eta, y, z)z$, where

$$\Lambda(\eta, y, z) = y - \frac{\partial \beta}{\partial y}\left(1 + 3\eta^2 + 3\eta z + z^2\right).$$

As in Example 6.4, we exploit the fact that $1 + 3\eta^2 + 3\eta z + z^2 \geq 1$ and select the function $\beta(y)$ so that

$$\frac{\partial \beta}{\partial y} = \rho + y^2 + a,$$

for some constant a, yielding

$$\beta(y) = \rho y + \frac{y^3}{3} + ay.$$

Selecting $a > 1/4$ yields

$$\Lambda(\eta, y, z) \leq -\rho - \left(y^2 - y + a\right) \leq -\rho.$$

Note now that a globally stabilising state feedback control law for the system (6.30) is given by

$$u = -y\eta - y\left(1 + \eta^2\right) - \eta,$$

which satisfies assumption (D1) of Theorem 6.4 with

$$V(x) = \frac{1}{2}y^2 + \frac{1}{2}\eta^2.$$

Moreover, the function $L(V) = \log(1 + V)$ is such that the term in (6.28) becomes

$$\left|\frac{(2y+1)\eta}{1 + \frac{1}{2}\left(y^2 + \eta^2\right)}\right|,$$

which is bounded for any η and y, hence Theorem 6.4 applies.

Note that, although this case appears to be simpler than the one in Example 6.4, we are not able to apply Theorem 6.3—at least not with the chosen scaling function $L(\cdot)$ and with a polynomial $\beta(\cdot)$—due to the unboundedness of the term in (6.20), which is caused by the quadratic term in y appearing in $\frac{\partial \beta}{\partial y}$. On the contrary, Theorem 6.4 yields a solution. ∎

Note 6.2. The results in Theorems 6.3 and 6.4 are based on two alternative factorisations of the system dynamics that include an integrable term and, similarly to [177, Theorem 1], they exploit the dependence of the observer convergence on the state. This leads to less conservative conditions (comparing, for instance, with the BIBS condition (B1) of Theorem 6.2). The contribution of Theorem 6.3 is to extend the certainty equivalence design of [177, Section 4] to a wider class of systems which may include several (possibly nonmonotonic) nonlinearities. Note that the certainty-equivalent output feedback designs derived from [177] rely on stabilisation of the state estimates, rather than the states themselves, therefore can be considered as *indirect* schemes according to the classification in [8], whereas the result in Theorem 6.4 is a *direct* scheme. ◁

6.5 Systems in Output Feedback Form

The design approach used so far consists of two steps. First, find a full-information control law that stabilises the system. Second, replace the unknown state η with an "estimate" $\hat{\eta} + \beta(y)$ and assign the dynamics of the "estimation error" $z = \hat{\eta} - \eta + \beta(y)$ so that the output feedback control law behaves asymptotically as the full-information control law, while keeping all trajectories bounded. As shown in Section 4.3, the converse approach can also be taken. Namely, starting from a stable estimation error dynamics, find an output feedback control law that robustifies the system with respect to this error. In this section the latter approach is applied to the class of systems in parametric output feedback form[8].

Consider the system

$$\dot{x}_1 = x_2 + \varphi_1(y)^\top \theta,$$
$$\vdots$$
$$\dot{x}_i = x_{i+1} + \varphi_i(y)^\top \theta, \qquad (6.31)$$
$$\vdots$$
$$\dot{x}_n = u + \varphi_n(y)^\top \theta,$$
$$y = x_1$$

with states $x_i \in \mathbb{R}$, $i = 1, \ldots, n$, where $u \in \mathbb{R}$ is the control input, $\theta \in \mathbb{R}^q$ is a vector of unknown parameters, and y is the measurable output. The adaptive output feedback tracking problem is to find a dynamic output feedback control law of the form

$$\dot{\hat{\eta}} = w(y, \hat{\eta}),$$
$$u = v(y, \hat{\eta}), \qquad (6.32)$$

such that all trajectories of the closed-loop system (6.31), (6.32) are bounded and

$$\lim_{t \to \infty} (y(t) - y^*) = 0, \qquad (6.33)$$

where $y^* = y^*(t)$ is a bounded $\mathcal{C}^n$ reference signal.

Proposition 6.1. *Consider the system (6.31) and suppose that there exists a mapping $\beta(\cdot)$ such that the zero equilibrium of the system*

$$\dot{z} = A(y)z, \qquad (6.34)$$

with $z \in \mathbb{R}^{n-1+q}$ and

$$A(y) = \begin{bmatrix} 0 & 0 & 0 & \cdots & 0 \\ \varphi_2(y)^\top & 0 & 1 & \cdots & 0 \\ \vdots & \vdots & & \ddots & \\ \varphi_{n-1}(y)^\top & 0 & 0 & \cdots & 1 \\ \varphi_n(y)^\top & 0 & 0 & \cdots & 0 \end{bmatrix} - \frac{\partial \beta}{\partial y} \begin{bmatrix} \varphi_1(y)^\top \mid 1 & 0 & \cdots & 0 \end{bmatrix},$$

is uniformly globally stable and, moreover, for any y,

$$\begin{bmatrix} \varphi_1(y)^\top \mid 1 & 0 & \cdots & 0 \end{bmatrix} z(t) \in \mathcal{L}_2. \qquad (6.35)$$

Then there exists a dynamic output feedback control law described by equations of the form (6.32) such that all trajectories of the closed-loop system (6.31), (6.32) are bounded and (6.33) holds.

---

[8] See the paper [135] and the monographs [137, 123].

6.5 Systems in Output Feedback Form

Proof. As in the state feedback case, the proof consists in constructing a control law u that ensures the closed-loop system is $\mathcal{L}_2$ stable with respect to the perturbation (6.35).

To begin with, define the vector of unmeasured states and unknown parameters
$$\eta = \begin{bmatrix} \theta & x_2 & \ldots & x_n \end{bmatrix}^\top$$
and the "observation error"
$$z = \hat{\eta} - \eta + \beta(y),$$
where $\beta(\cdot) = \begin{bmatrix} \beta_1(\cdot)^\top, \ldots, \beta_n(\cdot) \end{bmatrix}^\top$ is a $\mathcal{C}^1$ mapping to be defined and
$$\hat{\eta} = \begin{bmatrix} \hat{\theta} & \hat{x}_2 & \ldots & \hat{x}_n \end{bmatrix}^\top$$
is the observer state. Assigning the observer dynamics as
$$\dot{\hat{\theta}} = -\frac{\partial \beta_1}{\partial y}\left[\hat{x}_2 + \beta_2(y) + \varphi_1(y)^\top\left(\hat{\theta} + \beta_1(y)\right)\right],$$
$$\dot{\hat{x}}_2 = \hat{x}_3 + \beta_3(y) + \varphi_2(y)^\top\left(\hat{\theta} + \beta_1(y)\right)$$
$$-\frac{\partial \beta_2}{\partial y}\left[\hat{x}_2 + \beta_2(y) + \varphi_1(y)^\top\left(\hat{\theta} + \beta_1(y)\right)\right],$$
$$\vdots$$
$$\dot{\hat{x}}_n = u + \varphi_n(y)^\top\left(\hat{\theta} + \beta_1(y)\right) - \frac{\partial \beta_n}{\partial y}\left[\hat{x}_2 + \beta_2(y) + \varphi_1(y)^\top\left(\hat{\theta} + \beta_1(y)\right)\right],$$

yields the error system (6.34). Consider now the dynamics of $\tilde{x}_1 = y - y^*$, which are described by the equation
$$\dot{\tilde{x}}_1 = \hat{x}_2 + \beta_2(y) - z_2 + \varphi_1(y)^\top\left(\hat{\theta} + \beta_1(y) - z_1\right) - \dot{y}^*.$$

Define the error $\tilde{x}_2 = \hat{x}_2 - x_2^*$, where
$$x_2^* = \lambda_1(y, \hat{\theta}) - \beta_2(y) - \varphi_1(y)^\top\left(\hat{\theta} + \beta_1(y)\right) + \dot{y}^*,$$
for some function $\lambda_1(\cdot)$ yet to be specified. Note that the dynamics of $\tilde{x}_2$ are given by
$$\dot{\tilde{x}}_2 = \hat{x}_3 + \beta_3(y) + \varphi_2(y)^\top\left(\hat{\theta} + \beta_1(y)\right) - \frac{\partial x_2^*}{\partial \hat{\theta}}\dot{\hat{\theta}}$$
$$-\frac{\partial \beta_2}{\partial y}\left[\hat{x}_2 + \beta_2(y) + \varphi_1(y)^\top\left(\hat{\theta} + \beta_1(y)\right)\right]$$
$$-\frac{\partial x_2^*}{\partial y}\left[\hat{x}_2 + \beta_2(y) - z_2 + \varphi_1(y)^\top\left(\hat{\eta} + \beta_1(y) - z_1\right)\right].$$

Define the error $\tilde{x}_3 = \hat{x}_3 - x_3^*$, where

$$x_3^* = \lambda_2(y, \hat{x}_2, \hat{\theta}) - \beta_3(y) - \varphi_2(y)^\top (\hat{\theta} + \beta_1(y)) + \frac{\partial x_2^*}{\partial \hat{\theta}} \dot{\hat{\theta}}$$
$$+ \left(\frac{\partial \beta_2}{\partial y} + \frac{\partial x_2^*}{\partial y} \right) \left[\hat{x}_2 + \beta_2(y) + \varphi_1(y)^\top (\hat{\theta} + \beta_1(y)) \right],$$

for some function $\lambda_2(\cdot)$. Continuing with this procedure through the dynamics of $\tilde{x}_3, \ldots, \tilde{x}_n$ and defining functions $x_4^*, \ldots, x_n^*$ yields

$$\dot{\tilde{x}}_n = u + \varphi_n(y)^\top (\hat{\theta} + \beta_1(y)) - \frac{\partial \beta_n}{\partial y} \left[\hat{x}_2 + \beta_2(y) + \varphi_1(y)^\top (\hat{\theta} + \beta_1(y)) \right]$$
$$- \frac{\partial x_n^*}{\partial y} \left[\hat{x}_2 + \beta_2(y) - z_2 + \varphi_1(y)^\top (\hat{\theta} + \beta_1(y) - z_1) \right] - \sum_{i=2}^{n-1} \frac{\partial x_n^*}{\partial \hat{x}_i} \dot{\hat{x}}_i - \frac{\partial x_n^*}{\partial \hat{\theta}} \dot{\hat{\theta}}.$$

Finally, the control law u is selected as

$$u = \lambda_n(y, \hat{x}_2, \ldots, \hat{x}_n, \hat{\theta}) - \varphi_n(y)^\top (\hat{\theta} + \beta_1(y)) + \sum_{i=2}^{n-1} \frac{\partial x_n^*}{\partial \hat{x}_i} \dot{\hat{x}}_i + \frac{\partial x_n^*}{\partial \hat{\theta}} \dot{\hat{\theta}}$$
$$+ \left(\frac{\partial \beta_n}{\partial y} + \frac{\partial x_n^*}{\partial y} \right) \left[\hat{x}_2 + \beta_2(y) + \varphi_1(y)^\top (\hat{\theta} + \beta_1(y)) \right].$$

Note that the $\tilde{x}$-subsystem is described by the equations

$$\begin{aligned}
\dot{\tilde{x}}_1 &= \lambda_1(y, \hat{\theta}) + \tilde{x}_2 - (z_2 + \varphi_1(y)^\top z_1), \\
\dot{\tilde{x}}_2 &= \lambda_2(y, \hat{x}_2, \hat{\theta}) + \tilde{x}_3 + \frac{\partial x_2^*}{\partial y} (z_2 + \varphi_1(y)^\top z_1), \\
&\vdots \\
\dot{\tilde{x}}_n &= \lambda_n(y, \hat{x}_2, \ldots, \hat{x}_n, \hat{\theta}) + \frac{\partial x_n^*}{\partial y} (z_2 + \varphi_1(y)^\top z_1).
\end{aligned} \quad (6.36)$$

Consider now the function $W(\tilde{x}) = |\tilde{x}|^2$, whose time-derivative along the trajectories of (6.36) is given by

$$\dot{W} = 2\tilde{x}_1 \lambda_1(y, \hat{\theta}) + 2\tilde{x}_1 \tilde{x}_2 - 2\tilde{x}_1 (z_2 + \varphi_1(y)^\top z_1)$$
$$+ 2\tilde{x}_2 \lambda_2(y, \hat{x}_2, \hat{\theta}) + 2\tilde{x}_2 \tilde{x}_3 + 2\tilde{x}_2 \frac{\partial x_2^*}{\partial y} (z_2 + \varphi_1(y)^\top z_1)$$
$$+ \cdots + 2\tilde{x}_n \lambda_n(y, \hat{x}_2, \ldots, \hat{x}_n, \hat{\theta}) + 2\tilde{x}_n \frac{\partial x_n^*}{\partial y} (z_2 + \varphi_1(y)^\top z_1).$$

The proof is completed by noting that the functions $\lambda_i(\cdot)$ can be selected in such a way that, *for any* positive constants α and ε,

$$\dot{W} \leq -\alpha |\tilde{x}|^2 + \varepsilon (z_2 + \varphi_1(y)^\top z_1)^2.$$

Hence, from (6.35) and the fact that the zero equilibrium of the system (6.34) is stable, all states remain bounded. As a result, the signal in (6.35) converges to zero, hence the whole vector $\tilde{x}$ converges to zero and (6.33) holds, which proves the claim. □

6.6 Robust Output Feedback Stabilisation

In this section we use the design tools developed so far to deal with a class of systems with unstructured uncertainties. The globally stabilising output feedback control law that is obtained in this way can be applied to systems with unstable zero dynamics and with cross-terms between the output and the unmeasured states.

Consider the class of systems with unstructured uncertainties and non-trivial zero dynamics described by the equations

$$\begin{aligned}\dot{x}_0 &= f(x_0, y), \\ \dot{x}_1 &= x_2 + \Delta_1(x_0, y), \\ &\vdots \\ \dot{x}_i &= x_{i+1} + \Delta_i(x_0, y), \\ &\vdots \\ \dot{x}_n &= u + \Delta_n(x_0, y), \\ y &= x_1, \end{aligned} \quad (6.37)$$

where $(x_0, x_1, \ldots, x_n) \in \mathbb{R}^m \times \mathbb{R} \times \cdots \times \mathbb{R}$ is the state of the system, y is the measurable output, $u \in \mathbb{R}$ is the control input and the functions $\Delta_i(\cdot)$ describe the model uncertainties.

Note 6.3. A solution to the stabilisation problem for systems of the form (6.37) under suitable assumptions has been obtained in [83] by combining the tuning functions methodology of [123] with the nonlinear small-gain theorem [86] and the notion of input-to-state stability [203]. Note that a particular case of the form (6.37) is the well-known output feedback form [135, 137, 123], considered in the previous section, where the functions $\Delta_i(\cdot)$ are replaced by the *structured* uncertainties $\varphi_i(y)^\top x_0$, where x_0 is a vector of unknown parameters (*i.e.*, $\dot{x}_0 = 0$). Special instances of the system (6.37) have also been studied in [136, 176, 113]. In [136] the matrix $\frac{\partial f}{\partial x_0}$ is constant and Hurwitz, the perturbation functions $\Delta_i(\cdot), i = 1, \ldots, n-1$, depend only on the output y and $\Delta_n(\cdot)$ is linear in x_0. In [113], see also [182], the functions $\Delta_i(\cdot)$ are allowed to depend on the unmeasured states $x_2, \ldots, x_i$, but they must satisfy a *global Lipschitz* condition. In [176] a Lipschitz-like condition with an output-dependent upper bound is used instead.

A common hypothesis in the aforementioned methods is that the zero dynamics of the considered systems possess some strong stability property, *i.e.*, they have a globally asymptotically stable equilibrium or they are ISS. A method that relaxes this assumption has been proposed in [80] and has been shown to achieve *semiglobal practical* stability for systems that are possibly nonminimum-phase. Note that in [80] the functions $\Delta_i(\cdot)$ may depend also on the unmeasured states $x_2, \ldots, x_i$. Two global results for subclasses of nonlinear nonminimum-phase systems have been reported in [138] and [6], see also the survey paper [8]. ◁

We are interested in the case in which the zero equilibrium of the x_0-subsystem with $y = 0$ is not necessarily stable and therefore the output feedback scheme must incorporate an observer for the state x_0. In view of this fact, we consider a modified class of uncertain nonlinear systems described by equations of the form

$$\begin{aligned}
\dot{x}_0 &= F(y)x_0 + G(y) + \Delta_0(x_0, y), \\
\dot{x}_1 &= x_2 + \varphi_1(y)^\top x_0 + \Delta_1(x_0, y), \\
&\vdots \\
\dot{x}_i &= x_{i+1} + \varphi_i(y)^\top x_0 + \Delta_i(x_0, y), \\
&\vdots \\
\dot{x}_n &= u + \varphi_n(y)^\top x_0 + \Delta_n(x_0, y), \\
y &= x_1,
\end{aligned} \qquad (6.38)$$

with state $(x_0, x_1, \ldots, x_n) \in \mathbb{R}^m \times \mathbb{R} \times \cdots \times \mathbb{R}$, input $u \in \mathbb{R}$ and output $y \in \mathbb{R}$, where the functions $\Delta_i(\cdot)$ describe the model uncertainties. Assuming that the origin is an equilibrium for the system (6.38) with $u = 0$, i.e., $\Delta_i(0,0) = 0$ and $G(0) = 0$, and all functions are sufficiently smooth, the robust output feedback stabilisation problem is to find a dynamic output feedback control law described by equations of the form

$$\begin{aligned}
\dot{\hat{\eta}} &= w(y, \hat{\eta}), \\
u &= v(y, \hat{\eta}),
\end{aligned} \qquad (6.39)$$

with $\hat{\eta} \in \mathbb{R}^p$, such that the closed-loop system (6.38), (6.39) has a globally asymptotically stable equilibrium at the origin.

Note that the system (6.38) has relative degree n and its zero dynamics are given by $\dot{x}_0 = F(0)x_0 + \Delta_0(x_0, 0)$, hence they are not necessarily stable. Moreover, the functions $\Delta_i(\cdot)$ need not be bounded. However, the following conditions must hold.

Assumption 6.1. There exist positive-definite, locally quadratic and smooth functions $\rho_{i1}(\cdot)$ and $\rho_{i2}(\cdot)$, $i = 0, \ldots, n$, such that

$$|\Delta_i(x_0, y)|^2 \le \rho_{i1}(|x_0|) + \rho_{i2}(|y|). \qquad (6.40)$$

Assumption 6.2. There exists a smooth function $y^*(x_0)$ such that the system

$$\dot{x}_0 = F(y^*(x_0 + d_1) + d_2)x_0 + G(y^*(x_0 + d_1) + d_2) + \Delta_0(x_0, y^*(x_0 + d_1) + d_2)$$

is ISS with respect to d_1 and d_2, i.e., there exists a positive-definite and proper function $V_1(x_0)$ such that

$$\dot{V}_1 \le -\kappa_{11}(|x_0|) + \gamma_{11}(|d_1|) + \gamma_{12}(|d_2|),$$

where $\kappa_{11}(\cdot)$, $\gamma_{11}(\cdot)$ and $\gamma_{12}(\cdot)$ are smooth class-$\mathcal{K}_\infty$ functions.

Assumption 6.2 is a *robust stabilisability* condition on the zero dynamics. In the linear case, i.e., when the matrix F is constant and the vectors $G(y)$ and $\Delta_0(x_0, y)$ are linear functions, it is satisfied if the pair $(F + \frac{\partial \Delta_0}{\partial x_0}, G + \frac{\partial \Delta_0}{\partial y})$ is stabilisable or if the pair (F, G) is stabilisable and $\frac{\partial \Delta_0}{\partial x_0}$ and $\frac{\partial \Delta_0}{\partial y}$ are sufficiently small.

Note 6.4. In [83] it is assumed that the x_0-subsystem is ISS with respect to y, *i.e.*, Assumption 6.2 holds for $y^*(x_0) = 0$. It must be noted that in [83] the functions $\rho_{i1}(\cdot)$, $\rho_{i2}(\cdot)$ of Assumption 6.1 are multiplied by unknown coefficients which are estimated on-line using standard Lyapunov techniques. ◁

In the following sections a solution to the considered robust output feedback stabilisation problem is proposed based on a reduced-order observer and a combination of backstepping and small-gain ideas. In particular, it is shown that the closed-loop system can be described as the interconnection of ISS subsystems, whose gains can be tuned to satisfy the small-gain theorem.

6.6.1 Robust Observer Design

To begin with, a reduced-order observer is constructed for the unmeasured states x_0 and $x_2, \ldots, x_n$, following the approach in Chapter 5. To this end, let

$$\eta = \begin{bmatrix} x_0^\top, x_2, \ldots, x_n \end{bmatrix}^\top$$

and define the variable $z = \hat{\eta} - \eta + \beta(y)$ and the update law

$$\dot{\hat{\eta}} = A(y)(\hat{\eta} + \beta(y)) + \begin{bmatrix} G(y)^\top \mid 0 \cdots 0 \end{bmatrix}^\top, \tag{6.41}$$

where

$$A(y) = \begin{bmatrix} F(y) & 0 & 0 & \cdots & 0 \\ \varphi_2(y)^\top & 0 & 1 & \cdots & 0 \\ \vdots & \vdots & & \ddots & \\ \varphi_{n-1}(y)^\top & 0 & 0 & \cdots & 1 \\ \varphi_n(y)^\top & 0 & 0 & \cdots & 0 \end{bmatrix} - \frac{\partial \beta}{\partial y} \begin{bmatrix} \varphi_1(y)^\top \mid 1 & 0 & \cdots & 0 \end{bmatrix} \tag{6.42}$$

and $\beta(\cdot) = [\beta_0(\cdot)^\top, \beta_2(\cdot), \ldots, \beta_n(\cdot)]^\top$ is a $\mathcal{C}^1$ function to be defined. The observer error dynamics are described by the system

$$\dot{z} = A(y)z - \bar{\Delta}(x_0, y) + \Delta_1(x_0, y)\frac{\partial \beta}{\partial y}, \tag{6.43}$$

where $z = \begin{bmatrix} z_0^\top, z_2, \ldots, z_n \end{bmatrix}^\top$ and $\bar{\Delta}(\cdot) = \begin{bmatrix} \Delta_0(\cdot)^\top, \Delta_2(\cdot), \ldots, \Delta_n(\cdot) \end{bmatrix}^\top$. In addition to the observation error z, define the *output error* $\tilde{y} = y - y^*$, where

$$y^* = y^*(\hat{x}_0 + \beta_0(y)) = y^*(x_0 + z_0)$$

verifies Assumption 6.2.

Consider now the function $V_2(z) = z^\top P z$, where P is a constant, positive-definite matrix, and its time-derivative along the trajectories of (6.43), namely

$$\dot{V}_2 = z^\top \left(A(y)^\top P + P A(y) \right) z - 2\bar{\Delta}(x_0, y)^\top P z + 2\Delta_1(x_0, y)\frac{\partial \beta}{\partial y}^\top P z.$$

136   6 Robust Stabilisation via Output Feedback

Define the matrix
$$B(y) = I + \frac{\partial \beta}{\partial y} \frac{\partial \beta}{\partial y}^\top$$

and note that
$$\dot{V}_2 \leq z^\top \left[A(y)^\top P + PA(y) + \frac{PB(y)P}{\gamma(y)} \right] z + \gamma(y) \left(|\bar{\Delta}(x_0, y)|^2 + |\Delta_1(x_0, y)|^2 \right),$$

for any function $\gamma(y) > 0$. From Assumption 6.1 and the definition of $\tilde{y}$ it is possible to select functions $\gamma_{21}(\cdot)$, $\gamma_{22}(\cdot)$ and $\gamma_{23}(\cdot)$ such that

$$\dot{V}_2 \leq z^\top \left[A(y)^\top P + PA(y) + \frac{PB(y)P}{\gamma(y)} \right] z$$
$$+ \gamma_{21}(|x_0|) + \gamma_{22}(|\tilde{y}|) + \gamma(y)\gamma_{23}(|z_0|). \tag{6.44}$$

Consider now the following assumption.

Assumption 6.3. There exist functions $\beta(y)$, $\gamma(y)$, a positive-definite matrix P and a class-$\mathcal{K}_\infty$ function $\kappa_{21}(\cdot)$ such that, for any y,

$$z^\top \left[A(y)^\top P + PA(y) + \frac{PB(y)P}{\gamma(y)} \right] z + \gamma(y)\gamma_{23}(|z_0|) \leq -\kappa_{21}(|z|). \tag{6.45}$$

Assumption 6.3 is a *robust detectability* condition on the system (6.38) and can be regarded as a dual of Assumption 6.2. In fact, in the linear case, it is a necessary and sufficient condition for detectability when $\Delta_i(x_0, y) = 0$, for $i = 0, \ldots, n$ (see Section 6.6.6).

The main restriction in the condition (6.45) is the presence of the term $\gamma_{23}(|z_0|)$ which stems from the dependence of the perturbations on y^*. If the system is minimum-phase, then Assumption 6.2 is satisfied with $y^* = 0$, hence $\gamma_{23}(|z_0|) = 0$ in (6.44). Then the inequality (6.45) can be satisfied by selecting $\beta(y)$ so that, for some $P > 0$ and all y, the matrix $A(y)^\top P + PA(y)$ is negative-definite, and taking $\gamma(y)$ sufficiently large.

6.6.2 Stabilisation via a Small-gain Condition

Consider again the x_0-subsystem which is described by the equation

$$\dot{x}_0 = F(y^*(x_0+z_0)+\tilde{y})x_0 + G(y^*(x_0+z_0)+\tilde{y}) + \Delta_0(x_0, y^*(x_0+z_0)+\tilde{y}), \tag{6.46}$$

and note that, from Assumption 6.2, it follows that

$$\dot{V}_1 \leq -\kappa_{11}(|x_0|) + \gamma_{11}(|z_0|) + \gamma_{12}(|\tilde{y}|). \tag{6.47}$$

Moreover, Assumption 6.3 and condition (6.44) imply that the system

$$\dot{z} = A(y^*(x_0+z_0)+\tilde{y})z - \bar{\Delta}(x_0, y^*(x_0+z_0)+\tilde{y})$$
$$+ \Delta_1(x_0, y^*(x_0+z_0)+\tilde{y})\frac{\partial \beta}{\partial y} \tag{6.48}$$

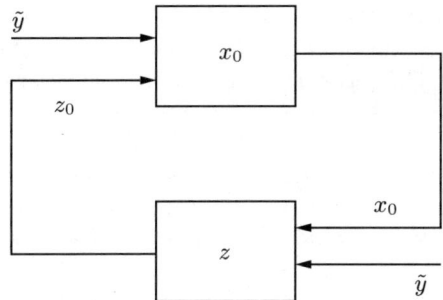

Fig. 6.2. Block diagram of the interconnected systems (6.46) and (6.48).

is ISS with respect to x_0 and $\tilde{y}$, i.e.,

$$\dot{V}_2 \leq -\kappa_{21}(|z|) + \gamma_{21}(|x_0|) + \gamma_{22}(|\tilde{y}|). \tag{6.49}$$

Thus each of the systems (6.46) and (6.48) can be rendered ISS by selecting the functions $\beta(y)$ and $y^*(x_0 + z_0)$ appropriately. In the following the stability properties of their interconnection (depicted in Figure 6.2) are studied by means of the Lyapunov formulation of the nonlinear small-gain theorem[9].

To this end, define class-$\mathcal{K}_\infty$ functions $\kappa_1(\cdot)$, $\kappa_2(\cdot)$, $\gamma_1(\cdot)$, $\gamma_2(\cdot)$ such that

$$\begin{aligned}\gamma_2^{-1} \circ \gamma_{21}(|x_0|) &\leq V_1(x_0) \leq \kappa_1^{-1} \circ \kappa_{11}(|x_0|), \\ \gamma_1^{-1} \circ \gamma_{11}(|z_0|) &\leq V_2(z) \leq \kappa_2^{-1} \circ \kappa_{21}(|z|),\end{aligned} \tag{6.50}$$

and note that conditions (6.47) and (6.49) can be written as

$$\begin{aligned}\dot{V}_1 &\leq -\kappa_1(V_1) + \gamma_1(V_2) + \gamma_{12}(|\tilde{y}|), \\ \dot{V}_2 &\leq -\kappa_2(V_2) + \gamma_2(V_1) + \gamma_{22}(|\tilde{y}|).\end{aligned}$$

The following theorem states the main result of this section.

Theorem 6.5. *Consider a system described by equations of the form (6.38) and such that Assumptions 6.1, 6.2 and 6.3 hold. Let $\kappa_1(\cdot)$, $\kappa_2(\cdot)$, $\gamma_1(\cdot)$ and $\gamma_2(\cdot)$ be class-$\mathcal{K}_\infty$ functions satisfying (6.50), with $V_1(x_0)$ as in Assumption 6.2 and $V_2(z) = z^\top P z$ as in Assumption 6.3, and suppose that there exist constants $0 < \varepsilon_1 < 1$ and $0 < \varepsilon_2 < 1$ such that*

$$\frac{1}{1-\varepsilon_1}\kappa_1^{-1} \circ \gamma_1 \circ \left(\frac{1}{1-\varepsilon_2}\kappa_2^{-1} \circ \gamma_2(r)\right) < r, \tag{6.51}$$

for all $r > 0$. Then the system (6.46), (6.48) with input $\tilde{y}$ is ISS. If, in addition, the ISS gain of this system is locally linear, then there exists a dynamic output feedback control law, described by equations of the form (6.39), such that the closed-loop system (6.38), (6.39) has a globally asymptotically stable equilibrium at the origin.

---

[9] See Theorem A.5 or [86, Theorem 3.1].

Theorem 6.5 states that it is possible to globally asymptotically stabilise the zero equilibrium of the system (6.38), where the functions $\Delta_i(\cdot)$ satisfy the growth condition (6.40), provided three subproblems are solvable. The first problem is the robust stabilisation of the zero equilibrium of the x_0-subsystem with input y (Assumption 6.2). The second problem is the input-to-state stabilisation of the observer dynamics with respect to x_0 (Assumption 6.3). The third problem is the stabilisation of the zero equilibrium of the interconnection of the two subsystems, which can be achieved by satisfying the small-gain condition (6.51)[10].

Proof. From condition (6.51) and Theorem A.5, the system (6.46), (6.48) with input $\tilde{y}$ is ISS. Since the gain of this system is locally linear, it suffices to prove that there exists a continuous control law $u(y, \hat{x}_2, \ldots, \hat{x}_n, \hat{x}_0)$ such that the gain of the system with state $(\tilde{y}, \hat{x}_2, \ldots, \hat{x}_n, \hat{x}_0)$, output $\tilde{y}$ and input (x_0, z) can be arbitrarily assigned. This can be achieved using a standard backstepping construction which can be described by the following recursive procedure.

Step 1. Consider the dynamics of $\tilde{x}_1 = \tilde{y}$, which are described by the equation

$$\dot{\tilde{x}}_1 = \hat{x}_2 + \beta_2(y) - z_2 + \varphi_1(y)^\top (\hat{x}_0 + \beta_0(y) - z_0)$$
$$+ \Delta_1(x_0, y) - \frac{\partial y^*}{\partial(x_0 + z_0)} \Big[F(y)(\hat{x}_0 + \beta_0(y)) + G(y) \Big]$$
$$+ \frac{\partial \beta_0}{\partial y} \Big(-z_2 - \varphi_1(y)^\top z_0 + \Delta_1(x_0, y) \Big).$$

Note that the term $\frac{\partial y^*}{\partial(x_0 + z_0)}$ is known. Consider $\hat{x}_2$ as a virtual control input and define the error $\tilde{x}_2 = \hat{x}_2 - x_2^*$, where

$$x_2^* = \lambda_1(y, \hat{x}_0) - \beta_2(y) - \varphi_1(y)^\top (\hat{x}_0 + \beta_0(y))$$
$$+ \frac{\partial y^*}{\partial(x_0 + z_0)} \left[F(y)(\hat{x}_0 + \beta_0(y)) + G(y) \right],$$

for some function $\lambda_1(\cdot)$ yet to be defined.

Step 2. The dynamics of $\tilde{x}_2$ are given by

$$\dot{\tilde{x}}_2 = \hat{x}_3 + \beta_3(y) + \varphi_2(y)^\top (\hat{x}_0 + \beta_0(y))$$
$$- \frac{\partial \beta_2}{\partial y} \Big[\hat{x}_2 + \beta_2(y) + \varphi_1(y)^\top (\hat{x}_0 + \beta_0(y)) \Big] - \frac{\partial x_2^*}{\partial \hat{x}_0} \dot{\hat{x}}_0$$
$$- \frac{\partial x_2^*}{\partial y} \Big[\hat{x}_2 + \beta_2(y) - z_2 + \varphi_1(y)^\top (\hat{x}_0 + \beta_0(y) - z_0) + \Delta_1(x_0, y) \Big].$$

Consider $\hat{x}_3$ as a virtual control input and define the error $\tilde{x}_3 = \hat{x}_3 - x_3^*$, where

$$x_3^* = \lambda_2(y, \hat{x}_2, \hat{x}_0) - \beta_3(y) - \varphi_2(y)^\top (\hat{x}_0 + \beta_0(y))$$
$$+ \left(\frac{\partial \beta_2}{\partial y} + \frac{\partial x_2^*}{\partial y} \right) \Big[\hat{x}_2 + \beta_2(y) + \varphi_1(y)^\top (\hat{x}_0 + \beta_0(y)) \Big] + \frac{\partial x_2^*}{\partial \hat{x}_0} \dot{\hat{x}}_0.$$

---

[10] This "reduction" idea is also the basis of the stabilisation result in [80], although therein an entirely different route is followed.

6.6 Robust Output Feedback Stabilisation

Continuing with this step-by-step design through the dynamics of $\tilde{x}_3, \ldots, \tilde{x}_n$, at the final step the control u appears.

Step n. The dynamics of $\tilde{x}_n$ are given by

$$\dot{\tilde{x}}_n = u + \varphi_n(y)^\top (\hat{x}_0 + \beta_0(y)) - \frac{\partial \beta_n}{\partial y}\left[\hat{x}_2 + \beta_2(y) + \varphi_1(y)^\top (\hat{x}_0 + \beta_0(y))\right]$$
$$- \frac{\partial x_n^*}{\partial y}\left[\hat{x}_2 + \beta_2(y) - z_2 + \varphi_1(y)^\top (\hat{x}_0 + \beta_0(y) - z_0) + \Delta_1(x_0, y)\right]$$
$$- \sum_{i=2}^{n-1} \frac{\partial x_n^*}{\partial \hat{x}_i}\dot{\hat{x}}_i - \frac{\partial x_n^*}{\partial \hat{x}_0}\dot{\hat{x}}_0.$$

Finally, the control law u is selected as

$$u = \lambda_n(y, \hat{x}_2, \ldots, \hat{x}_n, \hat{x}_0) - \varphi_n(y)^\top (\hat{x}_0 + \beta_0(y)) + \sum_{i=2}^{n-1}\frac{\partial x_n^*}{\partial \hat{x}_i}\dot{\hat{x}}_i + \frac{\partial x_n^*}{\partial \hat{x}_0}\dot{\hat{x}}_0$$
$$+ \left(\frac{\partial \beta_n}{\partial y} + \frac{\partial x_n^*}{\partial y}\right)\left[\hat{x}_2 + \beta_2(y) + \varphi_1(y)^\top (\hat{x}_0 + \beta_0(y))\right].$$
(6.52)

Note that the $\tilde{x}$-subsystem is described by the equations

$$\dot{\tilde{x}}_1 = \lambda_1(y, \hat{x}_0) + \tilde{x}_2 + \left(\frac{\partial y^*}{\partial(x_0 + z_0)}\frac{\partial \beta_0}{\partial y} - 1\right)\left(z_2 + \varphi_1(y)^\top z_0 - \Delta_1(x_0, y)\right),$$
$$\dot{\tilde{x}}_2 = \lambda_2(y, \hat{x}_2, \hat{x}_0) + \tilde{x}_3 + \frac{\partial x_2^*}{\partial y}\left(z_2 + \varphi_1(y)^\top z_0 - \Delta_1(x_0, y)\right),$$
$$\vdots$$
$$\dot{\tilde{x}}_n = \lambda_n(y, \hat{x}_2, \ldots, \hat{x}_n, \hat{x}_0) + \frac{\partial x_n^*}{\partial y}\left(z_2 + \varphi_1(y)^\top z_0 - \Delta_1(x_0, y)\right).$$
(6.53)

Consider now the function $W(\tilde{x}) = |\tilde{x}|^2$, whose time-derivative along the trajectories of (6.53) is given by

$$\dot{W} = 2\tilde{y}\lambda_1(y, \hat{x}_0) + 2\tilde{y}\tilde{x}_2 + 2\tilde{y}\left(\frac{\partial y^*}{\partial(x_0 + z_0)}\frac{\partial \beta_0}{\partial y} - 1\right)$$
$$\times \left(z_2 + \varphi_1(y)^\top z_0 - \Delta_1(x_0, y)\right) + 2\tilde{x}_2\lambda_2(y, \hat{x}_2, \hat{x}_0) + 2\tilde{x}_2\tilde{x}_3$$
$$+ 2\tilde{x}_2\frac{\partial x_2^*}{\partial y}\left(z_2 + \varphi_1(y)^\top z_0 - \Delta_1(x_0, y)\right) + \cdots$$
$$+ 2\tilde{x}_n\lambda_n(y, \hat{x}_2, \ldots, \hat{x}_n, \hat{x}_0) + 2\tilde{x}_n\frac{\partial x_n^*}{\partial y}\left(z_2 + \varphi_1(y)^\top z_0 - \Delta_1(x_0, y)\right).$$

The functions $\lambda_i(\cdot)$ can be selected in such a way that, for some positive constant ε and some smooth function $\alpha_2(\cdot)$ of class-$\mathcal{K}_\infty$,

$$\dot{W} \leq -\varepsilon W - \alpha_1(W) + \alpha_2(|(x_0^\top, z^\top)^\top|), \tag{6.54}$$

where $\alpha_1(\cdot)$ is *any* smooth function of class-$\mathcal{K}_\infty$. Note that the dependence of $\alpha_2(\cdot)$ on x_0 is a result of the perturbation $\Delta_1(\cdot)$. Using Assumption 6.1 both $\alpha_1(\cdot)$ and $\alpha_2(\cdot)$ can be made locally linear. Finally, as said previously, by hypotheses and by application of the gain assignment technique[11], an appropriate choice of $\alpha_1(\cdot)$ completes the proof of Theorem 6.5. □

Note 6.5. The design summarised in Theorem 6.5 partly extends the results of [83] by relaxing the hypothesis that the η-subsystem is ISS with respect to y and replacing it with an input-to-state stabilisability condition (see Assumption 6.2). It is worth noting that Theorem 6.5 is more general than some of the results in [135, 136, 123], although therein unknown parameters are also present.

Parametric uncertainties can be treated in the present framework either by incorporating the unknown parameters into the perturbation terms $\Delta_i(\cdot)$, or (in the linear parameterisation case) by including them in the vector η. While the former (similarly to [136]) requires only that the parameter vector belongs to a known bounded set, the latter implies that the origin may not be an equilibrium for the system (6.38) and so a somewhat different formulation is needed (see, *e.g.*, the approach in Section 6.5). ◁

6.6.3 Systems Without Zero Dynamics

In this section the applicability of Theorem 6.5 is discussed for a special class of systems described by equations of the form (6.38). Suppose that x_0 is an empty vector, *i.e.*, there are no zero dynamics, and that Assumption 6.1 holds. Note that, in this case, Assumption 6.2 is trivially satisfied. Then the matrix (6.42) is reduced to

$$A(y) = \begin{bmatrix} -\dfrac{\partial \beta_2}{\partial y} & 1 & \cdots & 0 \\ \vdots & & \ddots & \\ -\dfrac{\partial \beta_{n-1}}{\partial y} & 0 & \cdots & 1 \\ -\dfrac{\partial \beta_n}{\partial y} & 0 & \cdots & 0 \end{bmatrix}.$$

The above matrix can be rendered constant and Hurwitz by selecting

$$\beta_i(y) = k_i y,$$

for $i = 2, \ldots, n$, and choosing the constants k_i appropriately. As a result, Assumption 6.3 is trivially satisfied for any linear function $\kappa_{21}(\cdot)$ by taking $\gamma(y)$ to be a sufficiently large constant and condition (6.51) holds.

---

[11] See, *e.g.*, [83].

6.6.4 Systems with ISS Zero Dynamics

Suppose now that Assumptions 6.1 and 6.2 hold for $y^* = 0$, *i.e.*, the x_0-subsystem is ISS with respect to y. Then condition (6.47) reduces to

$$\dot{V}_1 \leq -\kappa_{11}(x_0) + \gamma_{12}(|y|),$$

i.e., $\gamma_{11}(|z_0|) = 0$, hence condition (6.51) holds. Finally, in Assumption 6.3, $\gamma_{23}(|z_0|) = 0$. Note that, in this case, it is possible to define new perturbation functions

$$\Delta_i'(x_0, y) = \varphi_i(y)^\top x_0 + \Delta_i(x_0, y),$$

for $i = 1, \ldots, n$, and select the functions $\beta_i(y) = k_i y$, for $i = 2, \ldots, n$, as before to yield a constant Hurwitz matrix A[12].

6.6.5 Unperturbed Systems

Finally, Assumption 6.3 and condition (6.51) can be relaxed in the case of an unperturbed system, *i.e.*, a system with $\Delta_i(\cdot) = 0$, for all i, as the following corollary shows.

Corollary 6.1. *Consider a system described by equations of the form (6.38) with $\Delta_i(\cdot) = 0$, $i = 0, 1, \ldots, n$, and such that Assumption 6.2 holds. Suppose that there exist functions $\beta_i(y)$, $i = 1, \ldots, n$ and a positive-definite matrix P such that*

$$z^\top \left(A(y)^\top P + P A(y)\right) z \leq -\kappa_{21}(|z|),$$

for any y, where $A(y)$ is given by (6.42). Then there exists a dynamic output feedback control law, described by equations of the form (6.39), such that the closed-loop system (6.38), (6.39) has a globally asymptotically stable equilibrium at the origin.

Proof. To begin with, note that Assumption 6.1 is trivially satisfied and Assumptions 6.2 and 6.3 hold by hypothesis. Consider now conditions (6.50) and note that the function $\gamma_{21}(\cdot)$ is zero, hence $\gamma_2(\cdot)$ can be arbitrarily selected and condition (6.51) holds. As a result, from Theorem 6.5, the system (6.46), (6.48) with input $\tilde{y}$ is ISS. A globally asymptotically stabilising control law u can be constructed by following the proof of Theorem 6.5 and noting that $\Delta_1(\cdot) = 0$ in (6.53), hence the functions $\lambda_i(\cdot)$ can be selected so that

$$\dot{W} \leq -\varepsilon W + \delta \kappa_{21}(|z|),$$

for any constant $\delta > 0$. As a result, the zero equilibrium of the $(\tilde{x}, z)$-subsystem is globally asymptotically stable with the Lyapunov function $W(\tilde{x}) + (1 + \delta) z^\top P z$, hence $\tilde{y}$ is bounded and asymptotically converges to zero. Since the system (6.46), (6.48) is ISS, this implies that x_0 and z converge to zero, which completes the proof. □

---
[12] In this way we recover the design proposed in [83].

6.6.6 Linear Nonminimum-phase Systems

Consider a linear system described by equations of the form[13]

$$\dot{x}_0 = F_0 x_0 + Gy + \Delta_0(x_0, y),$$
$$\dot{x}_1 = x_2 + F_1^\top x_0 + \Delta_1(x_0, y),$$
$$\vdots$$
$$\dot{x}_i = x_{i+1} + F_i^\top x_0 + \Delta_i(x_0, y), \qquad (6.55)$$
$$\vdots$$
$$\dot{x}_n = u + F_n^\top x_0 + \Delta_n(x_0, y),$$
$$y = x_1,$$

with $x_0 \in \mathbb{R}^m$, and suppose that Assumption 6.1 holds for some quadratic functions $\rho_{i1}(\cdot)$ and $\rho_{i2}(\cdot)$ and Assumption 6.2 holds for a linear function $y^*(x_0 + z_0)$. The system (6.55) can be written in matrix form as

$$\begin{bmatrix} \dot{\eta} \\ \dot{y} \end{bmatrix} = \begin{bmatrix} A_0 & \begin{array}{c} G \\ \vdots \\ 0 \end{array} \\ \hline C_0 & 0 \end{bmatrix} \begin{bmatrix} \eta \\ y \end{bmatrix} + \begin{bmatrix} 0 \\ \vdots \\ 1 \\ 0 \end{bmatrix} u + \Delta(x_0, y),$$

where $\eta = \begin{bmatrix} x_0^\top, x_2, \ldots, x_n \end{bmatrix}^\top$, $\Delta(\cdot) = \begin{bmatrix} \Delta_0(\cdot)^\top, \Delta_2(\cdot), \ldots, \Delta_n(\cdot), \Delta_1(\cdot) \end{bmatrix}^\top$,

$$A_0 = \begin{bmatrix} F_0 & 0 & 0 & \cdots & 0 \\ F_2^\top & 0 & 1 & \cdots & 0 \\ \vdots & \vdots & & \ddots & \\ F_{n-1}^\top & 0 & 0 & \cdots & 1 \\ F_n^\top & 0 & 0 & \cdots & 0 \end{bmatrix}, \qquad C_0 = \begin{bmatrix} F_1^\top & 1 & 0 & \cdots & 0 \end{bmatrix}.$$

Let
$$\beta(y) = Ky,$$

where K is a constant vector, and note that the matrix (6.42) can be written as
$$A = A_0 - KC_0.$$

As a result, Assumption 6.3 can be replaced by the following one.

Assumption 6.4. There exist a vector K, a positive-definite matrix P and a constant $\kappa_{21} > 0$ such that

$$z^\top \left[A^\top P + PA + \frac{P(I + KK^\top)P}{\gamma} \right] z + \gamma \gamma_{23} |z_0|^2 \leq -\kappa_{21} |z|^2.$$

---

[13] The form (6.55) can be obtained, for instance, from any transfer function of relative degree n and denominator of degree $n+m$, with known high-frequency gain and coefficients belonging to a known range.

If the system (6.55) is detectable for $\Delta = 0$, then the pair $\{A_0, C_0\}$ is also detectable, hence there exists a positive-definite matrix P such that the matrix $A^\top P + PA$ is negative-definite. Note that, due to the presence of γ_{23}, this does not imply (in general) that Assumption 6.4 holds. However, if the system (6.55) is also minimum-phase, then $\gamma_{23} = 0$ and Assumption 6.4 is always satisfied for sufficiently large γ.

Finally, conditions (6.47) and (6.49) reduce respectively to

$$\dot{V}_1 \leq -\kappa_{11}|x_0|^2 + \gamma_{11}|z_0|^2 + \gamma_{12}|\tilde{y}|^2,$$
$$\dot{V}_2 \leq -\kappa_{21}|z|^2 + \gamma_{21}|x_0|^2 + \gamma_{22}|\tilde{y}|^2.$$

Hence, the small-gain condition (6.51) reduces to

$$\frac{\gamma_{11}}{(1-\varepsilon_1)\kappa_{11}} \frac{\gamma_{21}}{(1-\varepsilon_2)\kappa_{21}} < 1.$$

Example 6.6. We demonstrate the applicability of Theorem 6.5 with a simple example, whose zero dynamics are linear and unstable. Consider the three-dimensional system

$$\begin{aligned}\dot{x}_0 &= x_0 + y + \delta(t)y, \\ \dot{x}_1 &= x_2 + \left(1 + y^2\right) x_0, \\ \dot{x}_2 &= u + \left(2 + y^2\right) x_0, \\ y &= x_1,\end{aligned} \qquad (6.56)$$

where $\delta(t)$ is an unknown disturbance such that $|\delta(t)| \leq p$, for all t, with $p \in [0, 1)$ a known constant. Hence, Assumption 6.1 is satisfied with $\rho_1(|x_0|) = 0$ and $\rho_2(|y|) = p^2 y^2$.

Assumption 6.2 is also satisfied by selecting

$$y^*(x_0) = -k_0 x_0,$$

for some positive constant k_0. In fact, the time-derivative of the function $V_1(x_0) = \frac{1}{2}x_0^2$ along the trajectories of the system

$$\dot{x}_0 = x_0 + (1 + \delta(t))\left(y^*(x_0 + z_0) + \tilde{y}\right) \qquad (6.57)$$

is

$$\dot{V}_1 = -\left(k_0\left(1 + \delta(t)\right) - 1\right)x_0^2 - k_0\left(1 + \delta(t)\right)z_0 x_0 + \left(1 + \delta(t)\right)\tilde{y} x_0,$$

which implies

$$\dot{V}_1 \leq -\left(k_0\left(1 - p\right) - 1 - \gamma_1\right)x_0^2 + \frac{k_0^2(1+p)^2}{2\gamma_1}z_0^2 + \frac{(1+p)^2}{2\gamma_1}\tilde{y}^2, \qquad (6.58)$$

for some $\gamma_1 > 0$. Hence, for $k_0 > (1 + \gamma_1)/(1 - p)$, the system (6.57) is ISS with respect to z_0 and $\tilde{y}$. The error dynamics (6.43) are given by the system

$$\begin{bmatrix} \dot{z}_0 \\ \dot{z}_2 \end{bmatrix} = \begin{bmatrix} 1 - \dfrac{\partial \beta_0}{\partial y}(1+y^2) & -\dfrac{\partial \beta_0}{\partial y} \\ 2+y^2 - \dfrac{\partial \beta_2}{\partial y}(1+y^2) & -\dfrac{\partial \beta_2}{\partial y} \end{bmatrix} \begin{bmatrix} z_0 \\ z_2 \end{bmatrix} - \begin{bmatrix} \delta(t)y \\ 0 \end{bmatrix}.$$

Consider the Lyapunov function $V_2(z) = z^\top P z$ with $P = \operatorname{diag}(1, a)$, where $a > 1$ is a constant. Assigning the functions $\beta_0(y)$ and $\beta_2(y)$ so that

$$\frac{\partial \beta_0}{\partial y} = \frac{a}{1+y^2}, \qquad \frac{\partial \beta_2}{\partial y} = \frac{2+y^2}{1+y^2} - \frac{1}{(1+y^2)^2},$$

namely

$$\beta_0(y) = a \arctan(y), \qquad \beta_2(y) = y + \frac{1}{2}\arctan(y) - \frac{y}{2(1+y^2)},$$

yields

$$\dot{V}_2 \leq -c_1 z_0^2 - c_2 z_2^2 + \gamma_2 p^2 y^2 + \frac{1}{4\gamma_2} z_0^2,$$

for some $\gamma_2 > 0$, where $c_2 = 1 + c_1 = a$. Noting, from Corollary A.3, that

$$y^2 = (\tilde{y} - k_0 (x_0 + z_0))^2$$
$$\leq (1+d)\tilde{y}^2 + k_0^2 \left(1 + \frac{1}{d}\right)^2 x_0^2 + k_0^2 \left(1 + \frac{1}{d}\right)(1+d) z_0^2,$$

with $d > 0$, yields

$$\dot{V}_2 \leq -\left(c_1 - \frac{1}{4\gamma_2} - \gamma_2 p^2 k_0^2 \left(1 + \frac{1}{d}\right)(1+d)\right) z_0^2 - c_2 z_2^2$$
$$+ \gamma_2 p^2 k_0^2 \left(1 + \frac{1}{d}\right)^2 x_0^2 + \gamma_2 p^2 (1+d) \tilde{y}^2.$$

Hence, for sufficiently large c_1, Assumption 6.3 is satisfied. The design is completed by choosing all the constants in the foregoing inequalities to satisfy the small-gain condition. Clearly, such a selection is possible since the constants c_1 and c_2 can be chosen arbitrarily large.

Figure 6.3 shows the response of the closed-loop system with the initial conditions $x_0(0) = -1$, $x_1(0) = x_2(0) = 0$ for various disturbances $\delta(t)$. ∎

Note 6.6. Applying the change of co-ordinates $\xi_1 = x_1$, $\xi_2 = x_2 + (1 + x_1^2) x_0$, the system (6.56) can be transformed into the system

$$\dot{x}_0 = x_0 + \xi_1 + \delta(t)\xi_1,$$
$$\dot{\xi}_1 = \xi_2,$$
$$\dot{\xi}_2 = u + (3 + 2\xi_1^2 + 2\xi_1 \xi_2) x_0 + (1 + \xi_1^2) \xi_1 (1 + \delta(t)),$$
$$y = \xi_1,$$

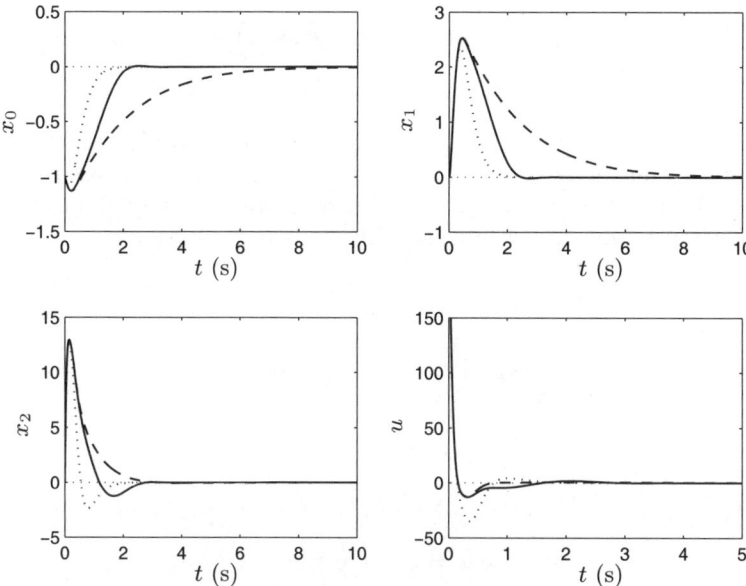

Fig. 6.3. Response of the controlled system (6.56) for various disturbances. Dotted line: $\delta(t) = 0$. Dashed line: $\delta(t) = -0.4$. Solid line: $\delta(t) = -0.4\cos(t)$.

for which the result in [80] is applicable. However, its application hinges upon the hypothesis (see Assumption 2 in [80]) that a robust global output feedback stabiliser is available for the auxiliary system

$$\dot{x}_0 = x_0 + \xi_1 + \delta(t)\xi_1,$$
$$\dot{\xi}_1 = u_a,$$
$$y_a = \left(3 + 2\xi_1^2 + 2\xi_1 u_a\right) x_0 + \left(1 + \xi_1^2\right) \xi_1 \left(1 + \delta(t)\right),$$

with input u_a and output y_a. Although it may be possible in this case to find such a stabiliser, it is certainly not a trivial task.   ◁

6.7 Example: Translational Oscillator/Rotational Actuator

In this section the proposed approach is used to design a globally stabilising output feedback controller for a translational oscillator with a rotational actuator (TORA), which has been considered as a benchmark nonlinear system[14].

---

[14] An introduction to this problem can be found in [32]. Output feedback controllers requiring measurement of the rotational and translational positions but not

The TORA system, depicted in Figure 6.4, is described by the equations

$$(M+m)\ddot{x}_d + ml\left(\ddot{\theta}\cos\theta - \dot{\theta}^2\sin\theta\right) = -kx_d,$$
$$(J+ml^2)\ddot{\theta} + m\ddot{x}_d l\cos\theta = \tau,$$

where θ is the angle of rotation, x_d is the translational displacement and τ is the control torque. The positive constants k, l, J, M and m denote the spring stiffness, the radius of rotation, the moment of inertia, the mass of the cart and the eccentric mass, respectively. Define the co-ordinates[15]

$$x_{01} = x_d + \frac{ml}{M+m}\sin\theta,$$
$$x_{02} = \dot{x}_d + \frac{ml}{M+m}\dot{\theta}\cos\theta,$$
$$x_1 = \theta,$$
$$x_2 = \dot{\theta}\psi(\theta),$$

with $\psi(\theta) = \sqrt{(J+ml^2)(M+m) - m^2l^2\cos^2\theta}$, and the control input

$$u = (M+m)\tau.$$

Note that $\psi(\theta) > \sqrt{J(M+m)} > 0$, hence the above transformation is well-defined. In these co-ordinates the system is described by a set of equations of the form (6.38), namely

$$\dot{x}_0 = \begin{bmatrix} 0 & 1 \\ -\epsilon_2 & 0 \end{bmatrix} x_0 + \begin{bmatrix} 0 \\ \epsilon_3 \sin y \end{bmatrix},$$
$$\dot{x}_1 = \frac{1}{\psi(y)} x_2,$$
$$\dot{x}_2 = \frac{1}{\psi(y)} u + \frac{k\epsilon_1 \cos y}{\psi(y)} x_{01} - \frac{\epsilon_2 \epsilon_1^2 \sin y \cos y}{\psi(y)},$$
$$y = x_1,$$
(6.59)

where $\epsilon_1 = ml$, $\epsilon_2 = k/(M+m)$ and $\epsilon_3 = kml/(M+m)^2$. It is assumed that only the output y is available for measurement, thus the system is only weakly minimum-phase. The control objective is to stabilise the zero equilibrium, so that both the translational displacement and the rotation angle converge to zero.

Following the construction of Section 6.6.1, we define the variables

$$z_0 = \hat{x}_0 - x_0 + \beta_0(y),$$
$$z_2 = \hat{x}_2 - x_2 + \beta_2(y),$$

---

of the velocities have been proposed in [33, 84, 89], while controllers using measurements of the rotational position alone have appeared in [33, 54].

[15] The co-ordinates transformation used here follows [32] and [84], but avoids the normalisations and time scaling used therein.

6.7 Example: Translational Oscillator/Rotational Actuator

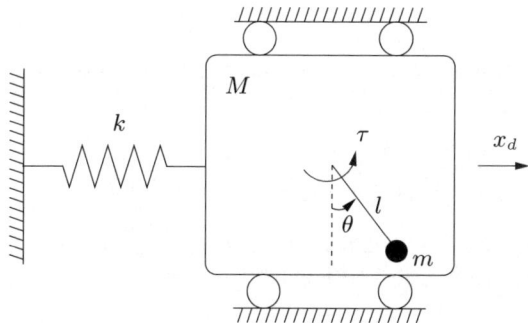

Fig. 6.4. A translational oscillator with a rotational actuator (TORA).

and the update laws

$$\dot{\hat{x}}_0 = \begin{bmatrix} 0 & 1 \\ -\epsilon_2 & 0 \end{bmatrix} (\hat{x}_0 + \beta_0(y)) + \begin{bmatrix} 0 \\ \epsilon_3 \sin y \end{bmatrix} - \frac{\partial \beta_0}{\partial y} \frac{1}{\psi(y)} (\hat{x}_2 + \beta_2),$$

$$\dot{\hat{x}}_2 = \frac{1}{\psi(y)} u + \begin{bmatrix} \frac{k\epsilon_1 \cos y}{\psi(y)} & 0 \end{bmatrix} (\hat{x}_0 + \beta_0(y)) - \frac{\epsilon_2 \epsilon_1^2 \sin y \cos y}{\psi(y)}$$
$$- \frac{\partial \beta_2}{\partial y} \frac{1}{\psi(y)} (\hat{x}_2 + \beta_2(y)).$$

Selecting the functions

$$\beta_0(y) = \begin{bmatrix} k_1 k \epsilon_1 \sin y \\ 0 \end{bmatrix}, \qquad \beta_2(y) = k_2 y,$$

where $k_1 > 0$ and $k_2 > 0$, yields the error dynamics

$$\dot{z} = \begin{bmatrix} 0 & 1 & -\dfrac{k_1 k \epsilon_1 \cos y}{\psi(y)} \\ -\epsilon_2 & 0 & 0 \\ \hline \dfrac{k\epsilon_1 \cos y}{\psi(y)} & 0 & -\dfrac{k_2}{\psi(y)} \end{bmatrix} z. \tag{6.60}$$

Note that the Lyapunov function $V_2(z) = z^\top P z$ with $P = \frac{1}{2}\mathrm{diag}\,(1, 1/\epsilon_2, k_1)$ is such that

$$\dot{V}_2(z) = -\frac{k_1 k_2}{\psi(y)} z_2^2 \leq -\frac{k_1 k_2}{\sqrt{(J+ml^2)(M+m)}} z_2^2,$$

hence $z(t) \in \mathcal{L}_\infty$ and $z_2(t) \in \mathcal{L}_2$. It follows, from boundedness of $\dot{z}$ and Corollary A.1, that $\lim_{t \to \infty} z_2(t) = 0$. Although this property is weaker than Assumption 6.3, it is sufficient to construct a globally asymptotically stabilising control law. Towards this end, consider the x_0-subsystem and the function

$$y^* = y^*(x_0 + z_0) = -\arctan(\begin{bmatrix} 0, & k_0 \end{bmatrix}(x_0 + z_0)),$$

with $k_0 > 0$ constant, and note that the function $V_1(x_0) = \frac{1}{2\epsilon_3}\left(\epsilon_2 x_{01}^2 + x_{02}^2\right)$ is such that

$$\dot{V}_1(x_0) = -\frac{k_0 x_{02}(x_{02} + z_{02})}{\sqrt{1 + k_0^2(x_{02} + z_{02})^2}}\cos\tilde{y} + \frac{x_{02}}{\sqrt{1 + k_0^2(x_{02} + z_{02})^2}}\sin\tilde{y}.$$

Using the identity $\cos\tilde{y} = 1 - 2\sin^2(\tilde{y}/2)$ and the inequalities $2ab \le da^2 + b^2/d$, with $d > 0$, and $|\sin(x)/x| \le 1$ yields

$$\dot{V}_1(x_0) \le -\left(1 - \frac{1}{\gamma_{11}} - \frac{1}{\gamma_{12}}\right)\frac{k_0 x_{02}^2}{\sqrt{1 + k_0^2(x_{02} + z_{02})^2}} + \frac{\gamma_{11} k_0 z_{02}^2}{4\sqrt{1 + k_0^2(x_{02} + z_{02})^2}}$$
$$+ \frac{\gamma_{12} k_0}{2}(x_{02} + z_{02})^2 \tilde{y}^2 + \frac{\gamma_{12}}{2k_0}\tilde{y}^2,$$

for some constants $\gamma_{11} > 1$, $\gamma_{12} > 1$. Consider now the output error $\tilde{y} = y - y^*$. Defining the control law as in (6.52) yields the system

$$\begin{aligned}\dot{\tilde{x}}_1 &= \frac{1}{\psi(y)}\left(\lambda_1(y, \hat{x}_0) + \tilde{x}_2 - z_2\right), \\ \dot{\tilde{x}}_2 &= \frac{1}{\psi(y)}\left(\lambda_2(y, \hat{x}_2, \hat{x}_0) + \frac{\partial x_2^*}{\partial y}z_2\right).\end{aligned} \quad (6.61)$$

Hence, the selection

$$\lambda_1(y, \hat{x}_0) = -a_1\psi(y)\tilde{y} - \varepsilon\tilde{y} - \delta\psi(y)\hat{x}_{02}^2\tilde{y},$$
$$\lambda_2(y, \hat{x}_2, \hat{x}_0) = -a_2\psi(y)\tilde{x}_2 - \tilde{y} - \varepsilon\left(\frac{\partial x_2^*}{\partial y}\right)^2\tilde{x}_2,$$

where a_1, a_2, ε and δ are positive constants, is such that $W(\tilde{x}) = \frac{1}{2}\left(\tilde{x}_1^2 + \tilde{x}_2^2\right)$ is an ISS Lyapunov function for the $\tilde{x}$-subsystem. In particular, the time-derivative of $W(\tilde{x})$ along the trajectories of (6.61) is such that

$$\dot{W} \le -a_1\tilde{x}_1^2 - a_2\tilde{x}_2^2 - \delta\hat{x}_{02}^2\tilde{y}^2 + \frac{1}{2\varepsilon\psi(y)}z_2^2.$$

As a result, since z_2 goes to zero asymptotically, so does $\tilde{x}_1$. This, from (6.60), implies that the entire vector z converges to zero. By combining $W(\tilde{x})$ with $V_1(x_0)$, it follows that the $(x_0, \tilde{x})$-subsystem with input z and output x_{02} is input-to-output stable (IOS). Since it is also zero-state detectable, x_0 converges to zero. Hence, the origin is globally asymptotically stable.

The closed-loop system has been simulated using the parameters[16] $J = 0.0002175$ kg m^2, $M = 1.3608$ kg, $m = 0.096$ kg, $l = 0.0592$ m and $k =$

---

[16]The simulations have been carried out using the same parameters and initial conditions as in the paper [54], which also provides a globally asymptotically stabilising controller requiring only measurement of the rotational position.

6.7 Example: Translational Oscillator/Rotational Actuator

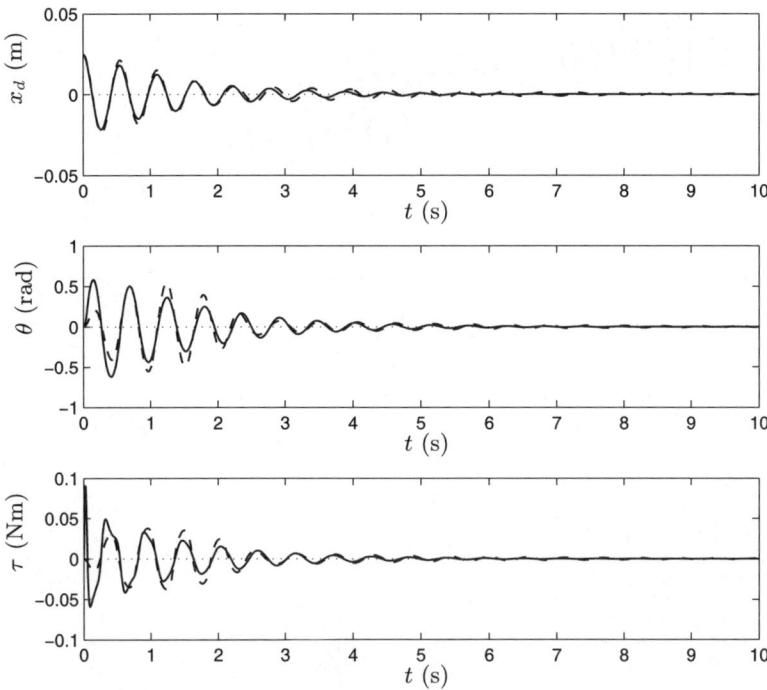

Fig. 6.5. Response of the TORA system. Dashed line: Passivity-based controller. Solid line: Proposed controller.

186.3 N/m and the initial state $x_0(0) = [0.025, 0]^\top$, $x_1(0) = x_2(0) = 0$. The controller parameters have been set to $k_1 = k_2 = 2$, $\varepsilon = 0.01$, $\delta = 0.001$, $a_1 = a_2 = 1$, and $k_0 = 3$.

Figure 6.5 shows the response of the closed-loop system in the original co-ordinates x_d and θ, together with the control torque τ, for the proposed controller and for a *passivity-based controller*[17]. The convergence rate of x_d and θ can be seen in Figure 6.6, where their norm has been plotted in logarithmic scale. Observe that the response is considerably faster than the passivity-based control scheme, while the control effort remains within the physical constraints $|\tau| \leq 0.1$ Nm. The performance can be further improved (at the expense of the control effort) by increasing the parameter k_0. Finally, the observer errors z_0 and z_2 are shown in Figure 6.7.

---

[17] See [54] for details.

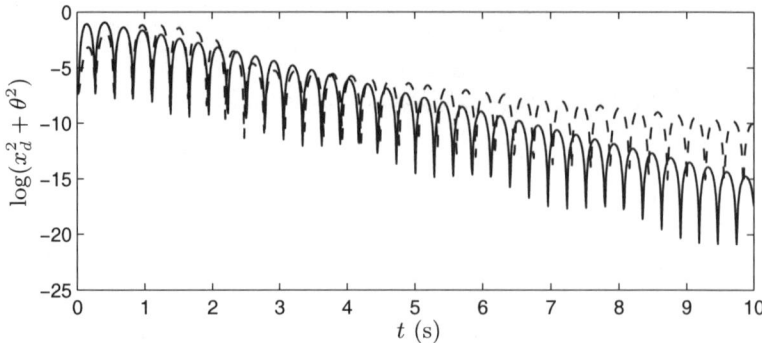

Fig. 6.6. Convergence rate of the TORA system. Dashed line: Passivity-based controller. Solid line: Proposed controller.

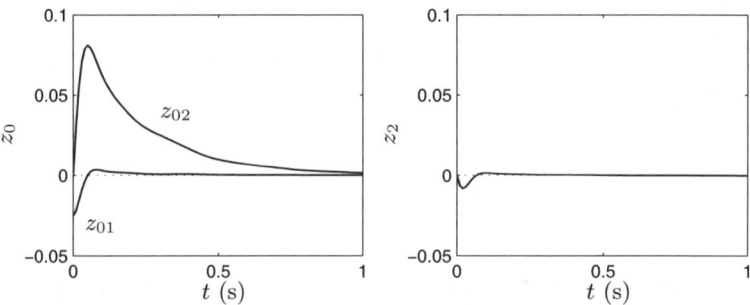

Fig. 6.7. Time histories of the observer errors z_0 and z_2.

7

Nonlinear PI Control of Uncertain Systems

A novel approach to stabilisation and trajectory tracking for nonlinear systems with unknown parameters and uncertain disturbances is developed in this chapter. We depart from the classical adaptive control approach consisting of a parameterised feedback law and an identifier, which tries to minimise a tracking (or prediction) error. Instead, we propose a simple nonlinear *proportional-integral* (PI) structure that generates a stable error equation—similar to the off-the-manifold dynamics obtained in the I&I approach—with a perturbation function that exhibits at least one root. Trajectories are forced to converge to this root by suitably adjusting the proportional and integral gains of the nonlinear PI controller.

We consider two basic problems: (i) nonlinear systems with matched uncertainties, *i.e.*, the uncertain terms are in the image of the input matrix, and (ii) linear systems with unknown control direction, *i.e.*, the control signal is multiplied by a gain of unknown sign. It is shown that, without knowing the system parameters and with only basic information on the uncertainties, it is possible to achieve global regulation and tracking.

Although most of the results are derived assuming full state measurement, we also present an observer-based solution for a chain of integrators with unknown control direction.

7.1 Introduction

It is well-known that (linear) PI controllers, if suitably tuned, provide satisfactory solutions to many practical applications without requiring a detailed description of the system dynamics. In the presence of strong nonlinear effects, however, their performance deteriorates and it is necessary to re-tune the controller appealing to gain scheduling or adaptive procedures. The design of these tuning procedures is complicated when only a very coarse description of the uncertain nonlinearities is available, some prototypical examples being the presence of friction and eccentricity in mechanical systems and the lack of knowledge about the reaction functions in chemical and biological processes.

In the adaptive control approach it is assumed that we can fit a model (*e.g.*, a polynomial or an experimental data-fitting curve or a neural network) to the uncertain function. Then, a parameter identifier, which tries to minimise a tracking (or prediction) error, is implemented. Two drawbacks of adaptive methods are that they usually lead to highly complex designs, as also illustrated in this book, and significant prior knowledge is required for a successful operation[1].

In this chapter we adopt a different perspective to the problem, abandoning its parametric formulation. Namely, instead of trying to fit a parameterised function to the uncertainty, we aim at generating an error equation with a perturbation term that can be driven to zero. More precisely, choosing the functions that define the nonlinear PI controller, we *shape* the perturbation function so as to exhibit at least one root, towards which we force the trajectories to converge. This root search happens in a one-dimensional space, in contrast with adaptive control, where the search takes place in a high-dimensional parameter space.

Note 7.1. The approach proposed in this chapter is relevant for the problem of feedback stabilisation of linear time-invariant systems with reduced prior knowledge. As thoroughly studied in [131], besides the model reference and pole placement adaptive control paradigms, there are two different methods to solve this problem: adaptive high-gain feedback, á la Willems and Byrnes [227], [75], or dense (open-loop) searches in parameter space, as proposed by Martensson [140]. While the first approach is restricted to minimum-phase systems and suffers from high noise sensitivity, the convergence rate of the second method is inadequate for practical applications. ◁

To illustrate the approach we present four motivating problems for which nonlinear PI control provides simple robust solutions.

Example 7.1 (Eccentricity compensation). Consider the one-degree-of-freedom rotational system

$$\ddot{y} = a\cos(by + c) + u, \qquad (7.1)$$

where the angular position y and velocity $\dot{y}$ are measurable and a, b, c are unknown positive parameters, and the problem of designing a controller that ensures $\dot{y}(t)$ asymptotically tracks any arbitrary bounded reference $\dot{y}^*(t)$ with known bounded first-order derivative[2].

Note that the system (7.1) is a special case of the more general class of systems described by the equation $\ddot{y} = \varphi(y) + u$, with $\varphi(\cdot)$ any bounded and (possibly) periodic function. ∎

---

[1]See also [199] for an alternative treatment of this class of problems.

[2]A solution to this problem, using an ingenious adaptive control technique, has been proposed in [38].

Example 7.2 (Friction compensation). Friction is an ubiquitous phenomenon in mechanical systems that is difficult to model and often requires to be compensated.

Consider the one-degree-of-freedom mechanical system[3]

$$\ddot{y} = -F(\dot{y}, t) + u, \qquad (7.2)$$

where the position y and speed $\dot{y}$ are measurable and the friction force $F(\cdot)$ is an unknown continuous function satisfying the bound

$$|F(\dot{y}(t), t)| \leq \bar{F}(1 + |\dot{y}(t)|),$$

for all $t \geq 0$, with $\bar{F}$ a known positive parameter, and the problem of designing a controller that ensures $y(t)$ asymptotically tracks any arbitrary bounded reference $y^*(t)$ with known bounded first- and second-order derivatives. ∎

Example 7.3 (Neural network function approximation). Consider the n-dimensional system

$$\begin{aligned}
\dot{x}_1 &= x_2, \\
\dot{x}_2 &= x_3, \\
&\vdots \\
\dot{x}_n &= \sum_{i=1}^{n} a_i x_i + \sum_{i=1}^{N} \frac{\delta_i}{1 + \alpha_i e^{-\beta_i x_i}} + u,
\end{aligned} \qquad (7.3)$$

where N is a known positive integer and all parameters a_i, $i = 1, \ldots, n$, $\alpha_i > 0$, $\beta_i > 0$ and $\delta_i > 0$, $i = 1, \ldots, N$, are unknown, and the problem of designing a state feedback controller that ensures $x(t)$ asymptotically tracks some desired bounded reference $x^*(t)$. ∎

Example 7.4 (Chain of integrators with unknown control direction). Consider the n-dimensional linear time-invariant system

$$\begin{aligned}
\dot{x}_1 &= x_2, \\
\dot{x}_2 &= x_3, \\
&\vdots \\
\dot{x}_n &= bu,
\end{aligned} \qquad (7.4)$$

where x_1 and x_n are measurable and b is an unknown parameter, and the problem of designing a controller that drives $x(t)$ to zero for all initial conditions $x(0) \in \mathbb{R}^n$. ∎

We show in this chapter that all these problems can be solved with nonlinear PI controllers, which are defined as follows.

---

[3]The basic (normalised) model (7.2) covers a variety of friction models, which have been reported, e.g., in [156].

7 Nonlinear PI Control of Uncertain Systems

Definition 7.1. *Given a set of measurable signals $y \in \mathbb{R}^n$, references $y^* = y^*(t) \in \mathbb{R}^n$ and a scalar manipulated variable $u \in \mathbb{R}$, define three mappings $\beta_P : \mathbb{R}^n \times \mathbb{R}^n \to \mathbb{R}^q$, $w_I : \mathbb{R}^n \times \mathbb{R}^n \times \mathbb{R}^q \to \mathbb{R}^q$, $v : \mathbb{R}^q \times \mathbb{R}^n \times \mathbb{R}^n \to \mathbb{R}$. The triple $\{\beta_P, w_I, v\}$ defines a nonlinear PI controller via the qth-order dynamical system*

$$\dot{\beta}_I = w_I(y, y^*, \beta_I),$$
$$u = v(\beta_P(y, y^*) + \beta_I, y, y^*). \tag{7.5}$$

The classical linear PI scheme for single-input single-output systems is recovered from the above definition by setting $n = q = 1$ and choosing the linear functions

$$\beta_P(y, y^*) = k_P(y - y^*),$$
$$w_I(y, y^*) = k_I(y - y^*),$$
$$u = -\beta_P(y, y^*) - \beta_I,$$

where the constants k_P and k_I are the proportional and integral gains, respectively.

Note that, according to the definition of $\dot{\beta}_I$ in (7.5), it is possible to include a stabilising factor in the integral action. Finally, note that the form (7.5) is similar to the adaptive I&I control structure used in Chapters 3 and 4, with β_I playing the role of the "parameter estimate" $\hat{\theta}$, $\beta_P(\cdot)$ corresponding to the mapping $\beta(\cdot)$, and $\beta_P(y, y^*) + \beta_I$ corresponding to the off-the-manifold co-ordinates.

7.2 Control Design

To illustrate the rationale of the nonlinear PI control design procedure we first consider the problem of regulating to zero the state of a first-order system subject to a bounded uncertainty. The result is then extended to tracking, high-dimensional systems (with unknown parameters) and unbounded uncertainty. We also show how it is possible to address in a simple unified way the problems described in Section 7.1.

7.2.1 Bounded Uncertainty

Consider the simple scalar system

$$\dot{y} = \varphi(y) + u, \tag{7.6}$$

where $\varphi(y)$ is an unknown continuous function such that $\varphi_m < \varphi(y) < \varphi_M$, for some constants $\varphi_m < \varphi_M$. First, select the desired closed-loop dynamics to be given by $\dot{y} = -\lambda y$, with $\lambda > 0$. Defining the signal

$$z = \beta_P(y) + \beta_I \tag{7.7}$$

and closing the loop with the nonlinear PI controller (7.5) yields the system dynamics in *perturbed* form as

$$\dot{y} = -\lambda y + [\varphi(y) + v(z, y) + \lambda y]. \tag{7.8}$$

The control objective is then to drive the term in brackets asymptotically to zero. Towards this end, we consider the *root searching* problem of finding functions $\beta_P(\cdot)$, $w_I(\cdot)$ and $v(\cdot)$ such that, for each y, there exists (at least one) $\bar{z}_y$ that solves the algebraic equation[4] $\varphi(y) + v(\bar{z}_y, y) + \lambda y = 0$, and z converges asymptotically towards $\bar{z}_y$.

Consider now the dynamics of z, which are described by the equation

$$\dot{z} = w_I(y, \beta_I) + \frac{\partial \beta_P}{\partial y}[\varphi(y) + v(z, y)].$$

In order to force the *root* of the disturbance term, $\bar{z}_y$, to be an equilibrium of the z dynamics, we fix the integral parameter of the nonlinear PI controller as

$$w_I(y, \beta_I) = \frac{\partial \beta_P}{\partial y}\lambda y,$$

which yields

$$\dot{z} = \frac{\partial \beta_P}{\partial y}[\varphi(y) + v(z, y) + \lambda y] \triangleq f_y(z). \tag{7.9}$$

To complete the description of the nonlinear PI controller we must define the functions $\beta_P(y)$ and $v(z, y)$ to force convergence of the trajectories of the closed-loop system (7.8), (7.9) to the equilibrium $(y, z) = (0, \bar{z}_y)$.

We now prove that the problem is solved if we can ensure that y is bounded and that there exists $M = M(y) > 0$ such that

$$z f_y(z) \leq 0, \tag{7.10}$$

for all $|z| > M$. It is straightforward to show that a function $v(z, y)$ that ensures (7.10) holds is given by

$$v(z, y) = -\lambda y - \varphi_m - \frac{\varphi_M - \varphi_m}{1 + e^{-z/y}},$$

provided $\frac{\partial \beta_P}{\partial y}$ has the same sign as y.

Note now that this selection of $v(z, y)$ is such that $y(t) \in \mathcal{L}_\infty$, hence, by (7.9) and (7.10), $z(t) \in \mathcal{L}_\infty$. Consider now the function $W(y, z) = \beta_P(y) - z$, whose time-derivative is given by[5] $\dot{W} = -\lambda \frac{\partial \beta_P}{\partial y} y$. This simple calculation motivates the choice

---

[4] The sub-index notation $(\cdot)_y$ is used to underscore the fact that, in general, the root depends on y.

[5] From (7.7) we see that $\dot{W} = -\beta_I$. Hence, the form of $\dot{W}$ is independent of the plant dynamics, and it stems only from the choice of w_I.

$$\beta_P(y) = \frac{1}{2}y^2, \qquad (7.11)$$

which is consistent with the fact that $\frac{\partial \beta_P}{\partial y}$ should have the same sign as y. In this way, we have $W(y,z) = \frac{1}{2}y^2 - z$ and $\dot{W} = -\lambda y^2 \leq 0$. Since we have shown above that z is bounded, we conclude that $W(y,z)$ is bounded from below and $y(t) \in \mathcal{L}_2 \cap \mathcal{L}_\infty$. To establish convergence of y to zero it suffices to prove that y is uniformly continuous—for instance, showing that $\dot{y}$ is also bounded. Recalling that $\varphi(y)$ is bounded (by assumption), this follows from (7.8) and the fact that $v(z,y) + \lambda y$ is also bounded.

It is clear from the proof that we have some freedom in the choice of the nonlinear PI structure. In particular, the only requirement on the function $v(\cdot)$ is that it is bounded and (7.10) holds for all $|z| \leq M(y)$, for some function $M(y) > 0$. Also, $\beta_P(\cdot)$ can be any function such that $\frac{\partial \beta_P}{\partial y} y \geq \delta y^2$, for some $\delta > 0$. These degrees of freedom can be used to tailor the root searching in specific applications where additional information on $\varphi(y)$ is available.

We have thus established the following proposition.

Proposition 7.1. *Consider the scalar system (7.6) with $\varphi(y)$ such that $\varphi_m < \varphi(y) < \varphi_M$, for all y and for some constants $\varphi_m < \varphi_M$, in closed loop with the nonlinear PI controller*

$$u = -\lambda y - \varphi_m - \frac{\varphi_M - \varphi_m}{1 + e^{-(\beta_P(y) + \beta_I)/y}},$$

$$\beta_P(y) = \frac{1}{2}y^2,$$

$$w_I(y) = \lambda y^2.$$

Then for all $\lambda > 0$ and all initial conditions $(y(0), \beta_I(0))$, $\lim_{t \to \infty} y(t) = 0$ with all signals bounded.

It must be noted that the above nonlinear PI control law is discontinuous in y. If $\varphi(0) \neq 0$, and without the knowledge of $\varphi(0)$, it is clear that *any* controller that ensures convergence of y for the whole class of admissible $\varphi(y)$ should be discontinuous. This obstruction appears because there does not exist a function $v(\cdot)$, continuous in y, such that condition (7.10) holds. Indeed, as y changes sign and $\varphi(y)$ remains of the same sign, the only way to enforce the inequality $zy[\varphi(y) + v(z,y) + \lambda y] \leq 0$ is that $v(\cdot)$ goes through a jump. (Selecting another function $\beta_P(y)$ will not help because $\frac{\partial \beta_P}{\partial y}$ should have the same sign as y to ensure $\dot{W} \leq 0$.) If $\varphi(0) = 0$, a smooth controller can be designed provided further information on $\varphi(\cdot)$ is available, for instance, if a bound on its derivative at zero is known.

7.2.2 Comparison with Adaptive Control

To put the nonlinear PI approach in perspective it is interesting to see how this simple example can be tackled with the classical parameter adaptive viewpoint[6].

First, we assume a (nonlinear) parameterisation for the uncertainty, say $\varphi(y) = \psi(y,\theta)$, where the function $\psi(y,\theta)$ is known, but the constant parameters $\theta \in \mathbb{R}^r$ are *unknown*. Then, we consider a *certainty-equivalent* controller

$$u = -\lambda y - \psi(y,\hat{\theta}),$$
$$\dot{\xi} = \alpha(\xi, y),$$
$$\hat{\theta} = \beta(\xi, y),$$

where $\alpha(\cdot)$ and $\beta(\cdot)$ are continuous functions to be defined, and $\hat{\theta}$ plays the role of an estimate for θ. In the indirect approach a *bona fide* parameter estimator that would, hopefully, drive $\hat{\theta}$ towards θ, is built with some filters and approximation considerations for the calculation of a gradient descent (in the parameter space). On the other hand, in the direct approach we try to find instead a Lyapunov function candidate whose time-derivative along the trajectories of the system should be made nonpositive. Solutions to this problem have been presented in Chapter 3 for the restricted class of linearly parameterised functions, *i.e.*, $\psi(y,\theta) = \psi_0(y)\theta$, with $\psi_0(y)$ known.

To pursue the comparison we review the direct adaptive control solution, noting that similar conclusions can be drawn for indirect schemes. Let $\psi(y,\theta) = \psi_0(y)\theta$ and consider the Lyapunov function candidate

$$V(y,\hat{\theta}) = \frac{1}{2}\left(y^2 + |\hat{\theta} - \theta|^2\right).$$

Picking $\dot{\hat{\theta}} = -\psi_0(y)y$ yields $\dot{V} = -\lambda y^2$, which proves boundedness of y and $\hat{\theta}$ and convergence of y to zero.

On the other hand, in nonlinear PI control, similarly to the I&I approach, a *cascade* connection between the plant dynamics (7.8) and the z dynamics (7.9) is established. Although the latter plays a role similar to the term induced by the parameter estimation error of adaptive control, in nonlinear PI we do not aim at its cancellation (or domination). Instead, we exploit the particular features of the nonlinear PI structure to drive it to zero.

7.2.3 Extensions

The result in Proposition 7.1 can be easily extended to the problem of tracking a bounded reference signal $y^* = y^*(t)$, as the following proposition shows.

---

[6]We refer the reader to [148] for a unifying overview of the field, covering the issues of parameterisations of linear systems and underscoring the importance of *implicit tuning*, a class of tuning algorithms to which nonlinear PI belongs.

Proposition 7.2. *Consider the first-order system (7.6) with $\varphi(y)$ such that $\varphi_m < \varphi(y) < \varphi_M$, for all y and for some constants $\varphi_m < \varphi_M$, in closed loop with the nonlinear PI controller*

$$u = \dot{y}^* - \lambda \tilde{y} - \varphi_m - \frac{\varphi_M - \varphi_m}{1 + e^{-(\beta_P(y) + \beta_I)/\tilde{y}}},$$

$$\beta_P(y) = \frac{1}{2}\tilde{y}^2,$$

$$w_I(y) = \lambda \tilde{y}^2,$$

where $\tilde{y} = y - y^*$ is the output tracking error. Then, for all initial conditions $(y(0), \beta_I(0))$ and all bounded references $y^*(t)$ with bounded derivatives, $\lim_{t \to \infty} \tilde{y}(t) = 0$, with all signals bounded.

Proof. Let $z = \beta_P(y) + \beta_I$ and consider the dynamics[7]

$$\dot{\tilde{y}} = -\lambda \tilde{y} + \left(\varphi(y) - \varphi_m - \frac{\varphi_M - \varphi_m}{1 + e^{-z/\tilde{y}}} \right),$$

$$\dot{z} = \tilde{y}\left(\varphi(y) - \varphi_m - \frac{\varphi_M - \varphi_m}{1 + e^{-z/\tilde{y}}} \right).$$

Since $\tilde{y}$ is bounded, condition (7.10) is satisfied for all $|z| > M(y)$, for some $M(y) > 0$, and we have, as before, that z is bounded. Mimicking again the proof of Proposition 7.1, we consider the function $W(\tilde{y}, z) = \frac{1}{2}\tilde{y}^2 - z$, with $\dot{W} = -\lambda \tilde{y}^2$. Given that z is bounded, $W(\tilde{y}, z)$ is bounded from below and we can conclude that $\tilde{y}$ is square-integrable and bounded. Convergence of $\tilde{y}(t)$ to zero follows immediately from boundedness of $\dot{\tilde{y}}$ and Corollary A.1. □

The extension of Proposition 7.1 to nth-order systems with unknown parameters[8] is readily obtained as follows.

Proposition 7.3. *Consider the n-dimensional single-input LTI system*

$$\dot{x} = Ax + B[u + \varphi(x)],$$
$$y = C^\top x,$$

and suppose that $C^\top B$ and a bound on $|C^\top A|$ are known, and that $C^\top B \neq 0$. Moreover, suppose that $y(t) \in \mathcal{L}_\infty$ implies $x(t) \in \mathcal{L}_\infty$, $y(t) \in \mathcal{L}_2$ implies $x(t) \in \mathcal{L}_2$, and the uncertain function $\varphi(x)$ is such that $|\varphi(x)| \leq \varphi_M$, for all x and for some constant $\varphi_M > 0$. Then the system in closed loop with a nonlinear PI controller with parameters

---

[7] We have recovered again the structure of Section 7.2.1, where the $\tilde{y}$ dynamics consists of a stable part plus a perturbation and the z dynamics have an equilibrium at the "roots" of the disturbance term. This situation is repeated in all the cases considered in the chapter.

[8] This problem has been studied, for instance, in [11, 153, 60].

$$u = \frac{1}{C^\top B}(1+|x|)\left(-y + N - \frac{2N}{1+e^{-(\beta_P(y)+\beta_I)/y}}\right),$$

$$\beta_P(y) = \frac{1}{2}y^2,$$

$$w_I(x) = y^2(1+|x|),$$

where $N \geq |C^\top A| + |C^\top B|\varphi_M$, is such that, for all initial conditions $(x(0), \beta_I(0))$, $\lim_{t \to \infty} x(t) = 0$ with all signals bounded.

Proof. Let $z = \beta_P(y) + \beta_I$ and note that

$$\dot{y} = -(1+|x|)y + (1+|x|)\left(N - \frac{2N}{1+e^{-z/y}} + \frac{C^\top Ax + C^\top B\varphi(x)}{1+|x|}\right),$$

$$\dot{z} = y(1+|x|)\left(N - \frac{2N}{1+e^{-z/y}} + \frac{C^\top Ax + C^\top B\varphi(x)}{1+|x|}\right).$$

Given that

$$0 \leq N + \frac{C^\top Ax + C^\top B\varphi(x)}{1+|x|} \leq 2N,$$

we have that $y(t) \in \mathcal{L}_\infty$. Moreover, the function $V(z) = z^2$ satisfies $\dot{V} \leq 0$ for $|z| > M(y)$, for some $M(y) > 0$, hence z is bounded. Proceeding as in the proof of Proposition 7.1, we consider the function $W(y,z) = \frac{1}{2}y^2 - z$, which is such that $\dot{W} = -\lambda(1+|x|)y^2$. Since z is bounded, $W(y,z)$ is bounded from below and $y(t) \in \mathcal{L}_2 \cap \mathcal{L}_\infty$. As a result, x is bounded and square-integrable, by assumption. Convergence of $y(t)$ and $x(t)$ to zero follows immediately from boundedness of $\dot{y}$ and $\dot{x}$, respectively. □

The assumption $C^\top B \neq 0$ can be replaced by the assumption that $C^\top A^i B \neq 0$ and we know $C^\top A^i B$ and a bound on $|C^\top A^i|$, and, moreover, $C^\top B = C^\top AB = C^\top A^2 B = \ldots = C^\top A^{i-1}B = 0$, for some i. In this case, we can repeat the proof of Proposition 7.3 taking instead of y, $C^\top(I + \alpha_1 A + \ldots + \alpha_{i-1}A^{(i-1)})x$, for some suitably chosen coefficients α_i.

Finally, note that with some simple calculations it is possible to extend the result of Proposition 7.3 to force the output $y(t)$ to track signals generated by $\dot{x}^* = A^*x^* + Br$, $y^* = C^{*\top}x^*$, for arbitrary known bounded $r(t)$ with bounded derivatives.

7.2.4 Unbounded Uncertainty

The technique developed in Section 7.2.1 can still be applied if the uncertainty is unbounded, but we know $N > 0$ and $k \geq 1$ such that, for all y,

$$|\varphi(y)| \leq N(1 + |y| + |y|^k). \tag{7.12}$$

7 Nonlinear PI Control of Uncertain Systems

Proposition 7.4. *Consider the system (7.6), where $\varphi(y)$ verifies (7.12) for some constants $N > 0$ and $k \geq 1$, in closed loop with the nonlinear PI controller*

$$u = N(1 + |y| + |y|^k)\left(-y + 1 - \frac{2}{1 + e^{-(\beta_P(y) + \beta_I)/y}}\right),$$

$$\beta_P(y) = \frac{1}{2}y^2,$$

$$w_I(y) = -y^2 N(1 + |y| + |y|^k).$$

Then for all initial conditions $(y(0), \beta_I(0))$, $\lim_{t \to \infty} y(t) = 0$ with all signals bounded.

Proof. The closed-loop dynamics are given by

$$\dot{y} = -N(1 + |y| + |y|^k)y$$
$$+ N(1 + |y| + |y|^k)\left(1 - \frac{2}{1 + e^{-z/y}} + \frac{\varphi(y)}{N(1 + |y| + |y|^k)}\right),$$

$$\dot{z} = yN(1 + |y| + |y|^k)\left(1 - \frac{2}{1 + e^{-z/y}} + \frac{\varphi(y)}{N(1 + |y| + |y|^k)}\right),$$

where we have used again the new co-ordinate $z = \beta_P(y) + \beta_I$ and the definitions of $\beta_P(\cdot)$ and $w_I(\cdot)$. The proof can be completed by repeating the steps of the proof of Proposition 7.3. □

Using the construction of Proposition 7.4 it is possible to extend the results of Proposition 7.2 (tracking) and Proposition 7.3 (high-order systems) to handle possibly unbounded functions $\varphi(y)$ that satisfy (7.12).

We now illustrate how the techniques developed above apply to the solution of some of the motivating examples of Section 7.1.

Example 7.5 (Eccentricity compensation). In this case the uncertain term is bounded and the system is second-order, hence Proposition 7.3 applies. Note, however, that only velocity tracking is required and this can be achieved with the nonlinear PI controller

$$u = \ddot{y}^* - (1 + |x|)\left(\dot{\tilde{y}} + N - \frac{2N}{1 + e^{-(\beta_P(y) + \beta_I)/y}}\right),$$

$$\beta_P(y) = \dot{\tilde{y}},$$

$$w_I(x) = \dot{\tilde{y}}(1 + |x|),$$

where $\dot{\tilde{y}} = \dot{y} - \dot{y}^*$, $|x| = |(y, \dot{y})|$, $N \geq a$, and $\dot{y}^* = \dot{y}^*(t)$ is the velocity reference to be tracked. Finally, it is clear that we do not assume any particular form for the disturbance $\varphi(\cdot)$, as the proposed controller yields asymptotic tracking for all bounded functions[9]. ∎

---

[9] This is in contrast with the result of [38], which requires the exact knowledge of the form (but not of the parameters) of the nonlinearity $\varphi(\cdot)$.

Example 7.6 (Friction compensation). This problem is conceptually similar to the previous one, the only difference being that the nonlinearity, representing the friction force, is linearly bounded, *i.e.*, $|F(\dot{y}, t)| \leq \bar{F}(1 + |\dot{y}|)$. Hence, a combination of the results in Proposition 7.3 and Proposition 7.4 has to be used. ∎

Example 7.7 (Neural network function approximation). This problem can be treated as the previous one, provided that the whole term

$$\sum_{i=1}^{n} a_i x_i + \sum_{i=1}^{N} \frac{\delta_i}{1 + \alpha_i e^{-\beta_i x_i}}$$

is regarded as an unknown perturbation. Note finally that, because of the simple structure of the system, it is a trivial task to find an output map satisfying the hypothesis of Proposition 7.3. ∎

7.3 Unknown Control Direction

In this section we show how the nonlinear PI framework can provide a solution to the problem of stabilising systems with unknown control direction (*i.e.*, unknown high-frequency gain in the case of linear systems). It is shown that the nonlinear PI design yields an alternative to the Nussbaum gain stabilisers that—in contrast with the latter and due to the presence of the proportional term in the PI—is shown to be robust in the presence of (fast) unmodelled dynamics.

Note 7.2. The problem of stabilising systems with unknown control direction has been extensively studied in the adaptive control literature, see [131, 76] and the references therein. A major breakthrough in this problem was the introduction of the so-called Nussbaum gains, which—similarly to the "root-searching" functions developed in the present chapter—continuously change sign until they latch to the right sign of the high-frequency gain. Although of significant theoretical interest, these schemes are impractical due to their poor robustness, as revealed in [64]. ◁

As before, we start with the simple problems of regulation and tracking for a scalar system. Finally, we present a partial state feedback stabiliser for a chain of integrators.

7.3.1 State Feedback

Proposition 7.5. *Consider the system*

$$\dot{y} = ay + bu, \tag{7.13}$$

with $y \in \mathbb{R}$ and $u \in \mathbb{R}$, where a and b are unknown parameters, in closed loop with the nonlinear PI controller

162   7 Nonlinear PI Control of Uncertain Systems

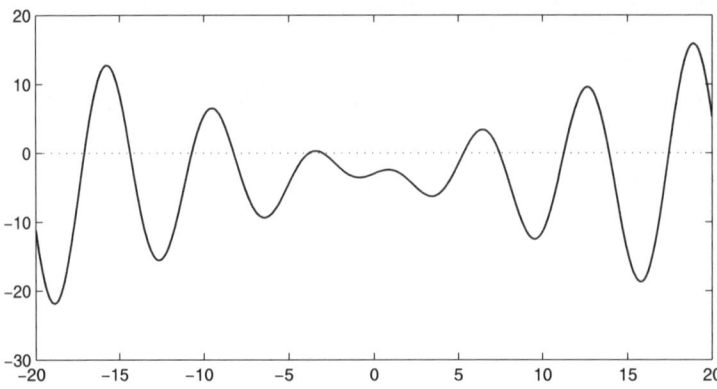

Fig. 7.1. Sketch of the function $a + bz\cos(z) + \lambda$.

$$u = (\beta_P(y) + \beta_I)\cos(\beta_P(y) + \beta_I)y,$$
$$\beta_P(y) = \frac{1}{2}y^2, \quad (7.14)$$
$$w_I(y) = \lambda y^2,$$

with $\lambda > 0$. Then for all initial conditions $(y(0), \beta_I(0))$, $\lim_{t\to\infty} y(t) = 0$ with all signals bounded.

Proof. Replacing (7.14) in (7.13) and defining $z = \beta_P(y) + \beta_I$ yields the closed-loop dynamics

$$\dot{y} = -\lambda y + [a + bz\cos(z) + \lambda]y,$$
$$\dot{z} = y^2[a + bz\cos(z) + \lambda],$$

where we have added and subtracted λy in the first equation to underscore the perturbed structure.

Notice that $\dot{z} = y^2 f(z)$, with $f(z)$ a function that has an infinite number of roots, which define bounded intervals that are invariant for the z dynamics, ensuring that $z(t)$ is bounded (see Figure 7.1). Moreover, this property holds true independently of the behaviour of the "time-scaling" factor y^2. The proof can be completed following the steps of the proof of Proposition 7.4 by considering the function $W(y, z) = \frac{1}{2}y^2 - z$, whose time-derivative satisfies $\dot{W} = -\lambda y^2 \leq 0$, and proving uniform continuity of y. □

Proposition 7.6. *Consider the system*

$$\dot{y} = bu, \quad (7.15)$$

with $y \in \mathbb{R}$ and $u \in \mathbb{R}$, where b is an unknown parameter, in closed loop with the nonlinear PI controller

$$u = -(\beta_P(y, y^*) + \beta_I)\cos(\beta_P(y, y^*) + \beta_I)\left(\tilde{y} - \frac{1}{\lambda}\dot{y}^*\right),$$

$$\beta_P(y, y^*) = \frac{1}{2}\tilde{y}^2 - \frac{1}{\lambda}\dot{y}^*\tilde{y}, \tag{7.16}$$

$$w_I(y, y^*) = \frac{1}{\lambda}\ddot{y}^*\tilde{y} + \left(\tilde{y} - \frac{1}{\lambda}\dot{y}^*\right)\dot{y}^* + \lambda\left(\tilde{y} - \frac{1}{\lambda}\dot{y}^*\right)^2,$$

with $\lambda > 0$. Then for all initial conditions $(y(0), \beta_I(0))$ and all bounded references $y^* = y^*(t)$ with bounded derivatives, $\lim_{t\to\infty} \tilde{y}(t) = 0$ with all signals bounded.

Proof. The closed-loop dynamics are given by

$$\dot{\tilde{y}} = -\lambda\tilde{y} + (\tilde{y} - \frac{1}{\lambda}\dot{y}^*)[-bz\cos(z) + \lambda], \tag{7.17}$$

$$\dot{z} = (\tilde{y} - \frac{1}{\lambda}\dot{y}^*)^2[-bz\cos(z) + \lambda]. \tag{7.18}$$

Arguing as in the proof of Proposition 7.5 we have that $z(t)$ is bounded. Define now $s = \tilde{y} - \frac{1}{\lambda}\dot{y}^*$ and write the $\tilde{y}$ dynamics in terms of s as

$$\dot{s} = -bz\cos(z)s - \dot{y}^* - \frac{1}{\lambda}\ddot{y}^*. \tag{7.19}$$

Consider the function $W(s, z) = \frac{1}{2}s^2 - z$, whose time-derivative satisfies

$$\dot{W} = -\lambda s^2 - s(\dot{y}^* + \frac{1}{\lambda}\ddot{y}^*) \leq -\frac{\lambda}{2}s^2 + N,$$

where $N \geq \frac{1}{2\lambda}(\dot{y}^* + \frac{1}{\lambda}\ddot{y}^*)^2$. From boundedness of z we have that there exists a positive constant c_1 such that $W \leq \frac{1}{2}s^2 + c_1$ yielding the differential inequality $\dot{W} \leq -\lambda W + N + c_1$, from which we conclude that $W(t)$, and consequently s and $\tilde{y}$, are bounded.

We now establish convergence of $\tilde{y}$. First, introducing the time scaling $\frac{d\tau}{dt} = s^2$ and looking at the z dynamics in the τ-time scale, we see that the trajectories of (7.18) converge to a constant z_∞ such that $-bz_\infty\cos(z_\infty) + \lambda = 0$ if $\lim_{t\to\infty} \int_0^t s^2(\tau)d\tau = \infty$. Hence, if $s(t) \notin \mathcal{L}_2$, we conclude from (7.17) and boundedness of $\tilde{y}$ and $\dot{y}^*$ that $\tilde{y}$ converges to zero. On the other hand, if $s(t) \in \mathcal{L}_2$, and recalling that the term in square brackets in (7.17) is bounded, we have that $\tilde{y}$ is the output of a strictly proper, exponentially stable, linear system with square-integrable input, hence $\tilde{y}(t)$ converges to zero[10]. □

The following lemma allows to define a class of functions that could be used instead of $z\cos(z)$ in the PI function $v(\cdot)$ in Propositions 7.5 and 7.6.

---

[10] See Theorem 4.9 of [49].

164   7 Nonlinear PI Control of Uncertain Systems

Lemma 7.1. *Consider the scalar non-autonomous system* $\dot{z} = \alpha(z) + m(t)$ *with initial condition* $z_0 \in \mathbb{R}$. *Assume there exist two numbers* $z^+(z_0) > z_0$ *and* $z^-(z_0) < z_0$ *such that*

$$\sup_t(m(t)) + \alpha(z^+(z_0)) < 0,$$

$$\inf_t(m(t)) + \alpha(z^-(z_0)) > 0.$$

Then $z(t)$ *is bounded.*

Note 7.3. The simplicity of the result in Proposition 7.5 should be contrasted with the complexity of the analysis of Nussbaum gain controllers for the same problem [131]. Moreover, the nonlinear PI control design can be applied to plants of the form $\dot{y} = ay + \varphi(y) + bu$, with $\varphi(0) = 0$ and $\varphi(\cdot)$ a bounded and differentiable function. Unfortunately, except for the case $a = 0$ given in Proposition 7.6, it is not possible to extend this result to tracking.   ◁

7.3.2 Observer-based Design

We present now the following output feedback result.

Proposition 7.7. *Consider the n-dimensional single-input single-output, LTI system*

$$\begin{aligned}\dot{x} &= Ax + Bbu,\\ y &= C^\top x,\end{aligned} \quad (7.20)$$

where $A \in \mathbb{R}^{n \times n}$, $B \in \mathbb{R}^n$, $C \in \mathbb{R}^n$, *and* $b \in \mathbb{R}$ *are unknown,* $B^\top B \in \mathbb{R}$ *is known and the only signals available for measurement are* y, $B^\top x$ *and* $B^\top Ax$. *Assume that the system (7.20) is stabilisable and detectable and, moreover, there exist a known matrix* $F \in \mathbb{R}^{n \times n}$ *and known vectors* $L, K \in \mathbb{R}^n$ *such that the* $2n \times 2n$ *matrix*

$$\begin{bmatrix} A & BK^\top \\ LC^\top & F \end{bmatrix} \quad (7.21)$$

is Hurwitz[11]. *Then the observer-based nonlinear PI controller with parameters*

$$\begin{aligned}u &= (\beta_P(x,\xi) + \beta_I)\cos(\beta_P(x,\xi) + \beta_I)K^\top \xi,\\ \beta_P(x,\xi) &= x^\top BK^\top \xi,\\ w_I(x,\xi) &= -\xi^\top K B^\top(Ax + BK^\top \xi) - x^\top BK^\top(F\xi + Ly),\end{aligned} \quad (7.22)$$

and ξ *generated as*

$$\dot{\xi} = F\xi + Ly, \quad (7.23)$$

is such that, for all initial conditions $(x(0), \beta_I(0), \xi(0))$, $\lim_{t \to \infty} x(t) = 0$ *with all signals bounded.*

---

[11] This is equivalent to saying that the system (7.20) with $b = 1$ is dynamic output feedback stabilisable.

7.3 Unknown Control Direction

Proof. Replacing (7.22) in the system dynamics yields the perturbed closed-loop system

$$\begin{bmatrix} \dot{x} \\ \dot{\xi} \end{bmatrix} = \begin{bmatrix} A & BK^\top \\ LC^\top & F \end{bmatrix} \begin{bmatrix} x \\ \xi \end{bmatrix} + \begin{bmatrix} BK^\top \xi \left(bz \cos(z) - 1 \right) \\ 0 \end{bmatrix}, \quad (7.24)$$

where $z = \beta_P(x, \xi) + \beta_I$. Evaluating the time-derivative of z and using (7.22) yields

$$\dot{z} = \xi^\top K B^\top B K^\top \xi \left(bz \cos(z) - 1 \right), \quad (7.25)$$

from which we conclude that z is bounded. To complete the proof we proceed similarly to the convergence proof of Proposition 7.6. Namely, we consider first the case $BK^\top \xi(t) \notin \mathcal{L}_2$, which is sufficient for convergence to zero of the disturbance term $\epsilon(t) = bz(t)\cos(z(t)) - 1$. Now, from (7.24), boundedness of z, and the fact that the matrix (7.21) is Hurwitz, a simple Lyapunov argument allows to establish the existence of a matrix $P = P^\top > 0$ such that the Lyapunov function $V(x,\xi) = \frac{1}{2}[x^\top, \xi^\top] P [x^\top, \xi^\top]^\top$ satisfies

$$\dot{V} \leq -\alpha \left(1 - \epsilon(t) \right) V,$$

for some $\alpha > 0$ and vanishing $\epsilon(t)$. As a result, x converges to zero. On the other hand, if $BK^\top \xi(t) \in \mathcal{L}_2$, we also conclude that x converges to zero, as the output of an exponentially stable linear system with square-integrable input[12]. □

Example 7.8 (Chain of integrators with unknown control direction). Proposition 7.7 directly provides a solution to the problem introduced in Example 7.4. In fact, in the case of a chain of integrators we have

$$A = \begin{bmatrix} 0 & 1 & \cdots & 0 \\ \vdots & & \ddots & \\ 0 & 0 & \cdots & 1 \\ 0 & 0 & \cdots & 0 \end{bmatrix}, \quad B = \begin{bmatrix} 0 \\ \vdots \\ 0 \\ 1 \end{bmatrix}, \quad C = \begin{bmatrix} 1 \\ \vdots \\ 0 \\ 0 \end{bmatrix}.$$

Hence, $B^\top x = x_n$ and $C^\top x = x_1$, while $B^\top A = 0$. Finally, as A and B are fully known, it is trivial to find F, L and K such that the matrix (7.21) is Hurwitz. ∎

7.3.3 Robustness

It is interesting to compare the nonlinear PI derived in Proposition 7.5 with the well-known Nussbaum gain controller, which (in its basic form) is given by[13]

---

[12] This is established, as in the proof of Proposition 7.6, using (7.24), boundedness of z and Theorem 4.9 of [49].
[13] See [131] and [64].

$$u = \hat{\theta}^2 \cos(\hat{\theta})y,$$
$$\dot{\hat{\theta}} = \lambda y^2.$$

Comparing with (7.5) and (7.14) we see that they differ on the presence of a proportional term $\beta_P(y)$ in the nonlinear PI controller and the use of a quadratic term $\hat{\theta}^2$ in the Nussbaum scheme, instead of a linear term in the nonlinear PI. Although the latter difference is not essential, the proportional adaptation term effectively enhances the robustness of the nonlinear PI.

To illustrate this point consider the effect of unmodelled dynamics and place the controller (7.5), (7.14) in closed loop with an arbitrary nth-order linear system $\dot{x} = Ax + Bu$, $y = C^\top x$. After some simple calculations we obtain the closed-loop dynamics

$$\begin{aligned} \dot{x} &= Ax + Bz\cos(z), \\ \dot{z} &= x^\top C^\top \left[\lambda x + Ax + Bz\cos(z)\right]. \end{aligned} \qquad (7.26)$$

Note that the function $V(y,z) = \frac{1}{2}y^2 - z$ is always such that $\dot{V} = -\lambda y^2 \leq 0$. Hence, as before, the central issue is the boundedness of z.

If we assume that the plant has relative degree one and it is minimum-phase, then it admits a representation of the form

$$\begin{aligned} \dot{\eta} &= F\eta + Gy, \\ \dot{y} &= H^\top \eta + ay + bu, \end{aligned}$$

for some a, b, H and G, and a Hurwitz matrix F. The z dynamics then become

$$\dot{z} = y^2[a + bz\cos(z) + \lambda] + yH^\top \eta.$$

Using the stability of F and doing some simple bounding yields

$$\dot{z} \leq \sup_{t \geq 0} |y(t)|^2 \left[|a + bz\cos(z) + \lambda| + \kappa\right] + \epsilon(t),$$

where $\kappa > 0$ depends on the plant parameters and $\epsilon(t)$ is an exponentially decaying term that depends on the initial conditions. From the expression above we conclude that z is bounded, hence establishing the robustness of the nonlinear PI controller in the presence of relative-degree-one, minimum-phase, unmodelled dynamics.

The robustifying effect of the proportional term can also be highlighted adopting a passive systems viewpoint. To this end, note that the system (7.26) can be represented by the block diagram of Figure 7.2. If z is bounded, the feedback map $u_1 \mapsto y_1$ described by

$$\begin{aligned} \dot{z} &= u_1, \\ y_1 &= z\cos(z) \end{aligned}$$

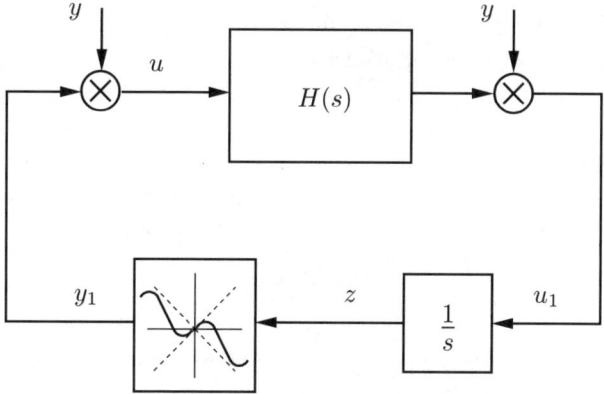

Fig. 7.2. Block diagram of the system (7.26).

is passive, with storage function $z\sin(z) + \cos(z)$. On the other hand, the forward path is passive if the transfer function

$$H(s) = C^\top \left[(\lambda I + A)(sI - A)^{-1} + I\right] B$$

is positive real. Consider now the case when the plant is a simple integrator perturbed by a fast parasitic first-order unmodelled dynamics, that is,

$$y(s) = \frac{\mu}{s(s+\mu)} u(s),$$

where $\mu > 0$. In this case $H(s) = \mu \frac{s+\lambda}{s(s+\mu)}$, which is positive real for all $\lambda < \mu$[14]. If we remove the proportional term, as in the Nussbaum gain controller, the zero disappears and we get $H(s) = \frac{\lambda\mu}{s(s+\mu)}$ which is clearly not positive real. This provides an alternative explanation of the poor robustness properties of Nussbaum gain controllers[15].

7.4 Example: Visual Servoing

In this section we revisit the visual servoing problem of Section 3.6 in order to compare the adaptive I&I and the nonlinear PI control approaches.

---

[14] This fact by itself is not conclusive for stability because passivity of the feedback operator relies on boundedness of z. However, using standard centre manifold arguments, it is possible to show that the equilibria of the closed-loop system, which are of the form $(0, 0, \bar{z})$, are stable. Moreover, this equilibrium manifold is locally attractive.

[15] For this particular case, it has been shown in [64], via simulations and some analysis, that the Nussbaum gain controller is unstable for $\mu > 1$.

168   7 Nonlinear PI Control of Uncertain Systems

Recall that the dynamic model of the visual servoing system, depicted in Figure 3.3, is given by the equations

$$\dot{x} = ae^{J\theta}u, \tag{7.27}$$

where $x \in \mathbb{R}^2$ are the image-space co-ordinates of the robot tip, $\theta \in (-\frac{\pi}{2}, \frac{\pi}{2})$ and $a > 0$ are *unknown* parameters representing the camera orientation and scale factor, respectively, $u \in \mathbb{R}^2$ are the transformed joint velocities which act as control inputs, and

$$J = \begin{bmatrix} 0 & -1 \\ 1 & 0 \end{bmatrix}, \quad e^{J\theta} = \begin{bmatrix} \cos(\theta) & -\sin(\theta) \\ \sin(\theta) & \cos(\theta) \end{bmatrix}.$$

Note that, in contrast with the adaptive control design of Section 3.6, in this section it is not assumed that a lower bound on the scale factor a is known. However, it is presumed that the camera orientation parameter θ is strictly between the critical values $\pm \pi/2$. This is necessary in order to implement the root searching procedure for the perturbed error dynamics.

As in Proposition 3.2, it is convenient to define a two-dimensional error reference system

$$\dot{w} = Rw - [\tan(\theta)J]\dot{x}^*, \tag{7.28}$$

where $R = -\epsilon[I + \tan(\theta)J]$ is a Hurwitz matrix and $\epsilon > 0$ is a tuning parameter. In the proposition that follows we present a nonlinear PI controller that ensures the tracking error asymptotically tracks $w(t)$. To understand the implications of this property consider the function $V_w = \frac{1}{2}|w|^2$, whose time-derivative along the trajectories of (7.28) satisfies

$$\dot{V}_w = -\epsilon|w|^2 - \tan(\theta)w^\top J\dot{x}^* \leq -\epsilon V_w + \frac{1}{2\epsilon}\tan^2(\theta)|\dot{x}^*|^2,$$

which implies that, for all values of ϵ and all initial conditions $w(0)$, w converges to a residual set, the size of which is $\mathcal{O}\{|\dot{x}^*(t)|\}$. Hence, as for the adaptive controller, good regulation is achieved for slowly-varying references. Note, moreover, that we can shrink the residual set by increasing ϵ.

Proposition 7.8. *Consider the system (7.27) and a bounded reference trajectory $x^* = x^*(t)$, with bounded first- and second-order derivatives $\dot{x}^*, \ddot{x}^*$. Then the nonlinear PI controller*

$$\begin{aligned} u &= -z\cos(z)s, \\ \dot{\beta}_I &= s^\top \dot{x}^* + \tfrac{1}{\epsilon}\tilde{x}^\top \ddot{x}^* + \epsilon|s|^2, \end{aligned} \tag{7.29}$$

with

$$z = \beta_I \frac{1}{2}|\tilde{x}|^2 - \frac{1}{\epsilon}\tilde{x}^\top \dot{x}^* \tag{7.30}$$

and $s = \tilde{x} - \frac{1}{\epsilon}\dot{x}^*$, is such that all trajectories of the closed-loop system (7.27), (7.29) are bounded and the tracking error either satisfies

$$\lim_{t \to \infty} |\tilde{x}(t) - w(t)| = 0, \tag{7.31}$$

with $w(t)$ the solution of (7.28), or

$$\lim_{t \to \infty} |s(t)| = 0, \quad \lim_{t \to \infty} \tilde{x}(t) = 0. \tag{7.32}$$

In particular, if $\lim_{t \to \infty} |\dot{x}^*(t)| = 0$, then $\lim_{t \to \infty} |\tilde{x}(t)| = 0$.

Proof. Substituting the controller (7.29) into (7.27) and using the definition of s yields

$$\dot{\tilde{x}} = -ae^{J\theta} z \cos(z)\tilde{x} + \left(\frac{a}{\epsilon} e^{J\theta} z \cos(z) - I\right) \dot{x}^*. \tag{7.33}$$

Differentiating (7.30), using (7.29) and (7.33), yields

$$\dot{z} = -|s|^2 \left(a \cos(\theta) z \cos(z) - \epsilon\right), \tag{7.34}$$

where we have used the fact that $s^\top a e^{J\theta} s = a \cos(\theta) |s|^2$. Note that the dynamics of the closed-loop system are described by the system (7.33), (7.34)[16].

Similarly to the I&I controller, it can be directly concluded from (7.34) and the assumption $\cos(\theta) \neq 0$ that z is bounded. We prove now that $\tilde{x}$ is also bounded. To this end, rewrite the $\tilde{x}$ dynamics in terms of s as

$$\dot{s} = -ae^{J\theta} z \cos(z) s - \left(\dot{x}^* + \frac{1}{\epsilon} \ddot{x}^*\right) \tag{7.35}$$

and consider the function $W(s, z) = \frac{1}{2}|s|^2 - z$, whose derivative along the trajectories of (7.34), (7.35) satisfies

$$\dot{W} = -\epsilon |s|^2 - s^\top \left(\dot{x}^* + \frac{1}{\epsilon} \ddot{x}^*\right) \leq -\frac{\epsilon}{2}|s|^2 + d_2,$$

where $d_2 \geq \frac{1}{2\epsilon}|\dot{x}^* + \frac{1}{\epsilon}\ddot{x}^*|^2$, and we have used the inequality $|s||\dot{x}^* + \frac{1}{\epsilon}\ddot{x}^*| \leq \frac{\epsilon}{2}|s|^2 + \frac{1}{2\epsilon}|\dot{x}^* + \frac{1}{\epsilon}\ddot{x}^*|^2$, which holds for arbitrary $\epsilon > 0$. From the fact that z is bounded it follows, as in the proof of Proposition 3.2, that $\dot{W} \leq -\epsilon W + \epsilon d_2 + c_2$, for some $c_2 > 0$, from which we conclude $W(t) \in \mathcal{L}_\infty$, and consequently $s(t) \in \mathcal{L}_\infty$ and $\tilde{x}(t) \in \mathcal{L}_\infty$.

Finally, the proof of the convergence results (7.31) and (7.32) exactly mimics the one given in Proposition 3.2. In particular, when $s(t) \notin \mathcal{L}_2$, it follows from (7.34) that z converges to some root z_∞ of $(a \cos(\theta)) z_\infty \cos(z_\infty) - \epsilon = 0$, hence the dynamics (7.33) can be written in the form

$$\dot{\tilde{x}} = [R + B_2(t)]\tilde{x} - [\tan(\theta)J + B_2(t)]\dot{x}^*,$$

for some bounded matrix $B_2(t)$, with $\lim_{t \to \infty} |B_2(t)| = 0$. The convergence result (7.31) then follows subtracting (7.28) from the equation above. On the other hand, when $s(t) \in \mathcal{L}_2$, it follows from (7.35) and the boundedness of s, $\dot{x}^*$ and $\ddot{x}^*$, that $\dot{s}$ is also bounded. Hence, by Lemma A.2, we conclude that $\lim_{t \to \infty} |s(t)| = 0$, which, from (7.33), implies that $\lim_{t \to \infty} |\tilde{x}(t)| = 0$. □

170   7 Nonlinear PI Control of Uncertain Systems

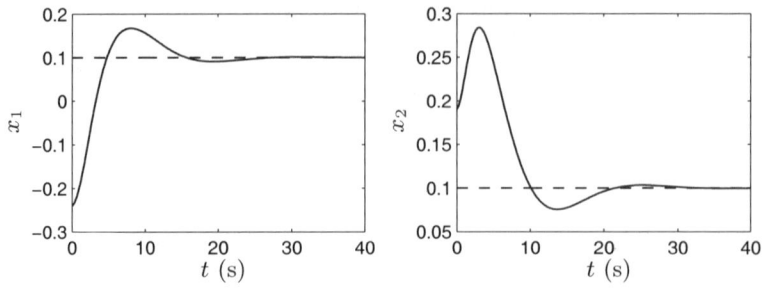

Fig. 7.3. Time histories of the normalised image output signals $x_1(t)$ and $x_2(t)$ for $x_1^* = x_2^* = 0.1$ and $\epsilon = 3$.

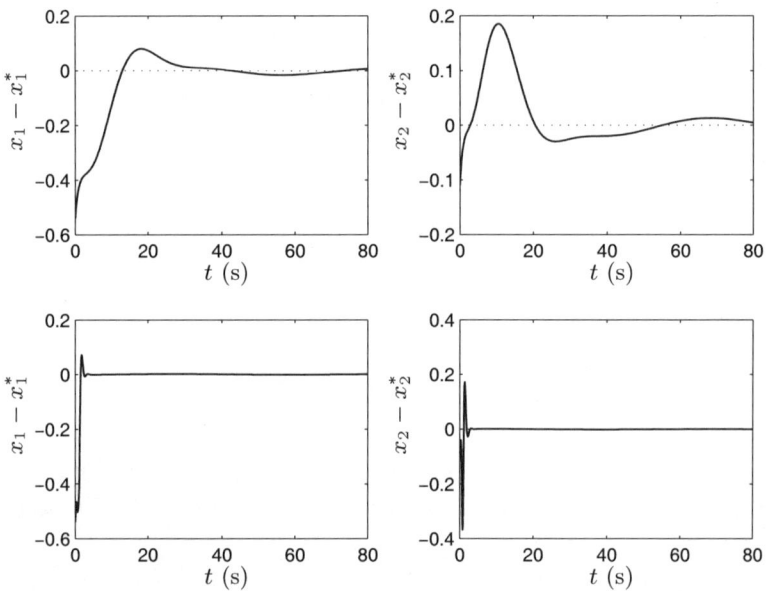

Fig. 7.4. Time histories of the tracking errors $x_1(t) - x_1^*(t)$ and $x_2(t) - x_2^*(t)$ for $\epsilon = 0.5$ (top graphs) and $\epsilon = 20$ (bottom graphs).

The nonlinear PI controller (7.29) has been tested through simulations using the same setup as for the adaptive I&I controller, which is described in Section 3.6.

For the set-point control case (Figure 7.3), with $x_1^* = x_2^* = 0.1$, the convergence is observed in 25 seconds with $\epsilon = 3$, which is still within 4% of

---
[16] Compare with (3.35), (3.36).

the final value, but significantly slower than the adaptive controller. However, the control velocities for the nonlinear PI controller are much smaller, not exceeding 0.17 rad/s. The convergence rate can be improved by increasing the parameter ϵ. However, this leads to a surge in the control effort.

The tracking case has also been tested with the same reference trajectory as for the adaptive controller, with $w_r = 0.07$, and for two different values of ϵ, namely $\epsilon = 0.5$ and $\epsilon = 20$ (Figure 7.4). It is clear from these graphs that, by increasing the parameter ϵ, the tracking performance improves. The price to be paid is the increased control effort. Nonetheless, it is lower (in absolute value) than 5 rad/s for $\epsilon = 20$, whereas for $\epsilon = 0.5$ it is lower than 0.17 rad/s.

8
Electrical Systems

This chapter presents four case studies which illustrate the applicability of the I&I technique to practical control design problems for electrical systems and in particular to the control of power electronic systems. First, we describe an application of the methodology in Section 3.1 to the design of a controller for a *thyristor-controlled series capacitor* (TCSC), which is a power electronic device that is used to regulate the power flow in an AC transmission line. Then, the results of Chapters 3, 5 and 6 are applied to the design of nonlinear adaptive controllers for single-phase power converters and an experimental study of their properties is carried out.

8.1 Power Flow Control Using TCSC

Series capacitive compensation in AC transmission systems can yield several benefits, such as increased power transfer capability and enhanced transient stability. Thyristor-controlled series capacitors (TCSCs) are beginning to find applications as adjustable series capacitive compensators, since they provide a continuously variable capacitance by controlling the firing angle delay of a thyristor controlled reactor connected in parallel with a fixed capacitor[1]. Besides regulating the power flow, TCSCs have a potential to provide other benefits, such as damping power swing oscillations, mitigating subsynchronous resonance and reducing fault currents. Hence, effective firing control strategies are required to exploit all advantages that a TCSC installation might offer.

In this section we present an application of the I&I technique for the design of a controller that guarantees stable regulation and improves (speeds-up) the response of a TCSC used to control the power flow in a distribution line. The controller is derived using a modification of the standard *dynamic phasor* model[2] and achieves global asymptotic regulation of the voltage.

---

[1] See [68] for more details.
[2] This model has been proposed for the TCSC circuit in [141].

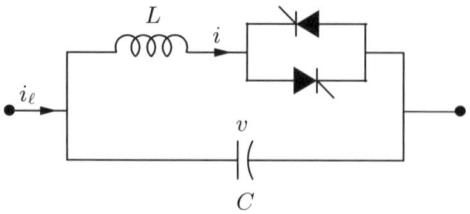

Fig. 8.1. Thyristor-controlled series capacitor (TCSC) circuit.

8.1.1 Problem Formulation

Consider the single-phase TCSC circuit shown in Figure 8.1, which is used to control the power flow in a transmission line. A two-dimensional (averaged) model for the TCSC circuit in terms of the phasor quantities is obtained by assuming that the dynamics of the current i are much faster than the dynamics of the voltage v and thus can be neglected. This model is described by the equations

$$C\dot{V} = I_\ell - Jw_s C_{\text{eq}}(\beta) V, \qquad (8.1)$$

where C is the capacitance, w_s is the line fundamental frequency, and the vectors V and I_ℓ are the fundamental Fourier coefficients, or 1-phasors, for the voltage v across the capacitor and the line current i_ℓ, respectively. In this model we have used the matrix

$$J = -J^\top = \begin{bmatrix} 0 & -1 \\ 1 & 0 \end{bmatrix},$$

instead of the imaginary unit $\jmath$, to express all phasor quantities as real vectors, i.e., $V = [V_d, V_q]^\top \in \mathbb{R}^2$ and $I_\ell = [I_{\ell,d}, I_{\ell,q}]^\top \in \mathbb{R}^2$ (with real and imaginary part of the corresponding complex phasor as the entries); $C_{\text{eq}}(\beta)$ is the equivalent capacitance of the circuit in terms of the angle β which is equivalent to half of the prevailing conduction angle, and the term $w_s C_{\text{eq}}(\beta)$ is the effective (also called apparent) *quasi-steady-state* admittance of the TCSC and represents the control input.

The equivalent capacitance $C_{\text{eq}}(\beta)$ can be approximated by the expression

$$C_{\text{eq}}(\beta) = \left[\frac{1}{C} - \frac{4}{\pi} \left(\frac{S}{2C} \left(\beta + \frac{1}{2} \sin(2\beta) \right) \right. \right.$$
$$\left. \left. + w_s^2 L S^2 \cos^2(\beta) \left(\tan(\beta) - \eta \tan(\eta\beta) \right) \right) \right]^{-1}, \qquad (8.2)$$

where $\eta = \sqrt{1/(w_s^2 LC)} \neq 1$ and $S = \eta^2/(\eta^2 - 1)$. In addition, we introduce the variables

$$\beta = \beta_0 + 2\phi, \qquad \phi = \arctan(-V_q/V_d), \qquad (8.3)$$

which capture the effect of the capacitor voltage on the conduction angle of the thyristors. We assume that the line current has the form $i_\ell(t) = I_{\ell,q} \sin(w_s t)$, with $I_{\ell,q}$ a positive constant, thus $I_\ell = [0, -I_{\ell,q}]^\top$.

The control objective consists in finding an angle β_0 such that the voltage vector $V = [V_d, V_q]^\top$ asymptotically converges towards some predefined constant reference $V^* = [V_d^*, 0]^\top$. Note that, for simulation purposes, once the angle β_0 has been computed following the control design, we include the effect of the angle ϕ as described in equation (8.3), before substituting into equation (8.2).

8.1.2 Modified Model of the TCSC

In this section we propose an approximation of (8.2), which leads to the addition of a nonlinear term into the model (8.1), and we show that this term corresponds to an additional damping. The benefit from introducing this modification is that (in the known parameters case) it leads to a very simple (feedforward) control law.

This approximation is based on the expression

$$C_{\text{eq}}(\beta) = C_{\text{eq}}(\beta_0 + 2\phi) \cong C_{\text{eq}}(\beta_0) + 2\phi \left.\frac{\partial C_{\text{eq}}}{\partial \beta}\right|_{\beta_0} \triangleq C_{\text{eq}}(\beta_0) + 2\phi f(\beta_0).$$

Note that, in the capacitive region, $f(\beta_0) < 0$, for all β. Moreover, for $|\phi| \ll 1$ we have $\phi \cong -V_q/V_d$. As a result, the (voltage subsystem) equations (8.1) can be approximated by

$$C\dot{V} \cong I_\ell - J w_s C_{\text{eq}}(\beta_0) V + 2 J w_s \frac{V_q}{V_d} f(\beta_0) V. \qquad (8.4)$$

The last term can be rewritten as

$$2 J w_s \frac{V_q}{V_d} f(\beta_0) V = 2 w_s f(\beta_0) \begin{bmatrix} -(V_q/V_d)^2 & 0 \\ 0 & 1 \end{bmatrix} \begin{bmatrix} V_d \\ V_q \end{bmatrix} \triangleq \mathcal{R}(V, \beta_0) V.$$

From the foregoing equations and the structure of the matrix $\mathcal{R}(V, \beta_0)$, it is apparent that the last term in (8.4) is adding damping to the second row but not to the first row (recall that $f(\beta_0) < 0$). For this reason, the control design should be carried out in a way that the effect of this nonlinear term in the dynamics of V_d is either cancelled or dominated.

To facilitate the control design process, we neglect the contribution of the control signal β_0 in the term $\mathcal{R}(V, \beta_0) V$ and we only consider its contribution at the desired equilibrium point, i.e.,

$$\mathcal{R}(V, \beta_0) V \cong - \begin{bmatrix} 0 & 0 \\ 0 & 2 w_s |f(\bar{\beta}_0)| \end{bmatrix},$$

where the constant $\bar{\beta}_0$ is the equilibrium value of β_0. Defining the positive constant $R = 2 w_s |f(\bar{\beta}_0)|$ and taking the effective admittance as the control input[3], i.e.,

---

[3] Notice that the effective admittance u is restricted to take positive values if the TCSC circuit is required to behave as a capacitor.

176   8 Electrical Systems

$$u = w_s C_{\text{eq}}(\beta_0),$$

yields the system
$$\begin{aligned} C\dot{V}_d &= uV_q, \\ C\dot{V}_q &= -I_{\ell,q} - uV_d - RV_q. \end{aligned} \tag{8.5}$$

Note that the uncertainty on the parameter R is not significantly affecting the performance, as R is only adding damping to the system, hence we introduce the adaptation only to cope with the uncertainty on the parameter $I_{\ell,q}$.

8.1.3 Controller Design

An adaptive I&I controller for system (8.5), with $C = 1$ and $V_d^* = -1$, has been designed in Example 3.2 and has been shown to render the desired equilibrium $(V_d^*, 0)$ *locally* asymptotically stable. In this section we make this result global by enforcing the constraint $u > 0$ with a saturation function.

To begin with, note that the known-parameters (feedforward) control law $u = -I_{\ell,q}/V_d^*$ achieves the desired objective with the Lyapunov function

$$W(V) = \frac{1}{2}C(V_d - V_d^*)^2 + \frac{1}{2}CV_q^2.$$

Before deriving an adaptive controller using the I&I methodology, we first try the classical (direct) approach, as in Example 3.1, and propose the control law $u = -\hat{I}_{\ell,q}/V_d^*$, where $\hat{I}_{\ell,q}$ is an estimate for $I_{\ell,q}$. To obtain the adaptive law, we consider the function

$$W(V, \hat{I}_{\ell,q}) = \frac{1}{2}C(V_d - V_d^*)^2 + \frac{1}{2}CV_q^2 + \frac{1}{2}(\hat{I}_{\ell,q} - I_{\ell,q}),$$

whose time-derivative is given by

$$\begin{aligned} \dot{W} &= uV_q(V_d - V_d^*) - I_{\ell,q}V_q - uV_dV_q - RV_q^2 + \dot{\hat{I}}_{\ell,q}(\hat{I}_{\ell,q} - I_{\ell,q}) \\ &= (\hat{I}_{\ell,q} - I_{\ell,q})V_q - RV_q^2 + \dot{\hat{I}}_{\ell,q}(\hat{I}_{\ell,q} - I_{\ell,q}). \end{aligned}$$

Hence selecting $\dot{\hat{I}}_{\ell,q} = -V_q$ yields $\dot{W} = -RV_q^2 \leq 0$. Unfortunately, this condition does not imply convergence to the desired set-point, since the closed-loop system has a manifold of equilibria described by $\hat{I}_{\ell,q}V_d = I_{\ell,q}V_d^*$. It can also be shown that the indirect approach does not work either because of the lack of detectability.

A solution to the considered adaptive stabilisation problem can be obtained using the tools developed in Section 3.1. The result is summarised in the following statement.

Proposition 8.1. *Consider the system (8.5) and the adaptive I&I controller*

$$C\dot{\hat{I}}_{\ell,q} = -\frac{\partial \beta}{\partial V_d}uV_q + \frac{\partial \beta}{\partial V_q}\left(\hat{I}_{\ell,q} + \beta(V_d, V_q) + uV_d\right),$$

$$u = -\frac{1}{V_d^*}\text{sat}_{[\epsilon_1, \epsilon_2]}(\hat{I}_{\ell,q} + \beta(V_d, V_q)),$$

(8.6)

where ϵ_1, ϵ_2 are constants such that $0 < \epsilon_1 \leq I_{\ell,q} \leq \epsilon_2$, and

$$\beta(V_d, V_q) = k_d(V_d - V_d^*) - k_q V_q.$$

Then for any $k_d > 0$, $k_q > 0$, the closed-loop system (8.5), (8.6) has a globally asymptotically stable equilibrium at $(V, \hat{I}_{\ell,q}) = (V^*, I_{\ell,q})$.

Proof. Following the definition of the control law it is natural to define the off-the-manifold co-ordinate z as

$$z = \hat{I}_{\ell,q} - I_{\ell,q} + \beta(V_d, V_q).$$

Using (8.5), and considering that $I_{\ell,q}$ is constant, yields the off-the-manifold dynamics

$$C\dot{z} = C\dot{\hat{I}}_{\ell,q} + \frac{\partial \beta}{\partial V_d}uV_q - \frac{\partial \beta}{\partial V_q}\left(I_{\ell,q} + uV_d + RV_q\right).$$

Substituting the adaptive law from (8.6) yields

$$C\dot{z} = \frac{\partial \beta}{\partial V_q}(z - RV_q).$$

It is clear from the above equation that the condition $\frac{\partial \beta}{\partial V_q}V_q < 0$ is required to guarantee stability of the zero equilibrium of the off-the-manifold dynamics when $V_q = 0$. Substituting the expression for the function $\beta(\cdot)$, where $k_d > 0$, $k_q > 0$ are design parameters, yields the closed-loop dynamics

$$C\dot{V}_d = -\frac{V_q}{V_d^*}(I_{\ell,q} + z + \delta(z)),$$

$$C\dot{V}_q = \frac{V_d}{V_d^*}(I_{\ell,q} + z + \delta(z)) - I_{\ell,q} - RV_q,$$

(8.7)

$$C\dot{z} = -k_q z + k_q RV_q,$$

where we have defined the function

$$\delta(z) = \text{sat}_{[\epsilon_1, \epsilon_2]}(I_{\ell,q} + z) - (I_{\ell,q} + z).$$

Notice that, due to the presence of the term RV_q, in this case we do not obtain a cascade system as in Theorem 3.1. There is also an additional perturbation $\delta(z)$ due to the saturation. However, we can still carry out the stability analysis with the Lyapunov function

$$W(V,z) = \frac{1}{2}C(V_d - V_d^*)^2 + \frac{1}{2}CV_q^2 + \frac{1}{2Rk_q}z^2,$$

whose time-derivative along the trajectories of (8.7) is given by

$$\dot{W} = V_q \left(I_{\ell,q} + z + \delta(z) \right) - V_q I_{\ell,q} - RV_q^2 - \frac{1}{R}z^2 + V_q z$$

$$= -RV_q^2 + 2V_q \left(z + \frac{\delta(z)}{2} \right) - \frac{1}{R}z^2$$

$$= -R \left(V_q - \frac{1}{R}\left(z + \frac{\delta(z)}{2} \right) \right)^2 + \frac{1}{R}\delta(z)\left(z + \frac{\delta(z)}{4} \right).$$

It can readily be seen that the term $\delta(z)$ is zero inside the saturation limits and has the opposite sign of $z + \delta(z)/4$ everywhere else, hence

$$\dot{W} \leq -R \left(V_q - \frac{1}{R}\left(z + \frac{\delta(z)}{2} \right) \right)^2 \leq 0.$$

This establishes Lyapunov stability of the equilibrium $(V, z) = (V^*, 0)$. Further, by LaSalle's invariance principle (see Theorem A.2), all trajectories converge to the largest invariant set in $E = \{(V, z) \mid RV_q = z + \delta(z)/2\}$. But from (8.7) and the fact that $I_{\ell,q} + z + \delta(z) \geq \epsilon_1 > 0$ the largest invariant set in E is $\{(V, z) \mid V_q = 0, z = 0, V_d = V_d^*\}$, which is the desired equilibrium. Moreover, since z converges to zero, $\hat{I}_{\ell,q}$ converges to $I_{\ell,q}$. □

We remark that a more complex expression for $\beta(\cdot)$ can be proposed, but we have preferred the one above for its easy implementation and to simplify the presentation. Note finally that, in steady state and with the above selection of $\beta(\cdot)$, the control signal (8.6) converges towards $u^* = -I_{\ell,q}/V_d^*$, which is its desired expected value.

8.1.4 Simulation Results

For the purpose of simulation we use the model (8.1) with the following parameters[4]: $L = 6.8$ mH, $C = 176.80$ μF, at a constant frequency $f = 60$ Hz. The intervals where the apparent admittance u is restricted are given by

$$0.0151 \leq u \leq 0.0662 \quad \text{(capacitive region)}$$
$$-0.2469 \leq u \leq -0.0182 \quad \text{(inductive region)}$$

The line current has been set to $i_\ell = 600\sin(w_s t)$ A. We present simulations only for the capacitive operation. The test consists in fixing a reference for the apparent admittance $u^* = 0.06$ Ω^{-1} at the beginning, and then after time $t = 0.2$ s, switching to a new reference $u^* = 0.03$ Ω^{-1}. This corresponds to a step change in the desired voltage reference V_d^* from -10 kV to -20 kV.

---
[4] These parameters correspond to the Kayenta substation [68].

8.2 Partial State Feedback Control of the DC–DC Ćuk Converter

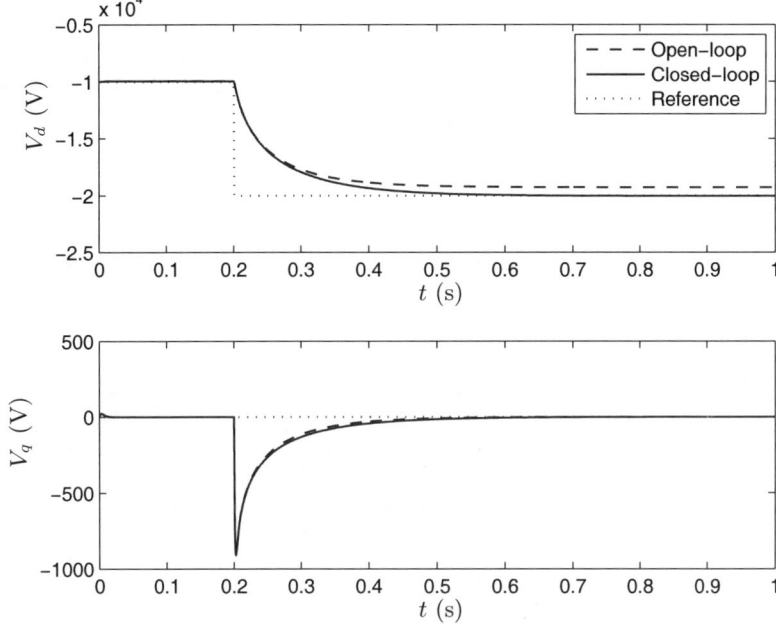

Fig. 8.2. Comparison between responses of phasor voltages V_d and V_q with the I&I controller (solid line) and in open loop (dashed line) with references V_d^* and V_q^* (dotted line).

Figure 8.2 presents a comparison between the response of the voltages V_d and V_q obtained with the proposed I&I controller (in solid line) and the response obtained in open loop (dashed line). In the open-loop case the control is simply taken as $u = u^*$. We observe that, in the open-loop operation, and due to inaccuracy in the generation of the table $u \mapsto \beta$, a considerable steady-state error exists between the voltage V_d and its reference V_d^* (dotted line). In contrast, with the I&I controller, the error is close to zero.

Figure 8.3 shows the time-domain signal v together with the phasor quantity V_d and its reference V_d^* (dashed line). The control signal $u = w_s C_{\text{eq}}(\beta)$ that results from the application of the proposed I&I controller and the parameter estimate $\hat{I}_{\ell,q}$ are also displayed.

8.2 Partial State Feedback Control of the DC–DC Ćuk Converter

The DC–DC Ćuk converter is one of the most widely studied power converters. Its goal is to invert the polarity of the input voltage and to step-up or

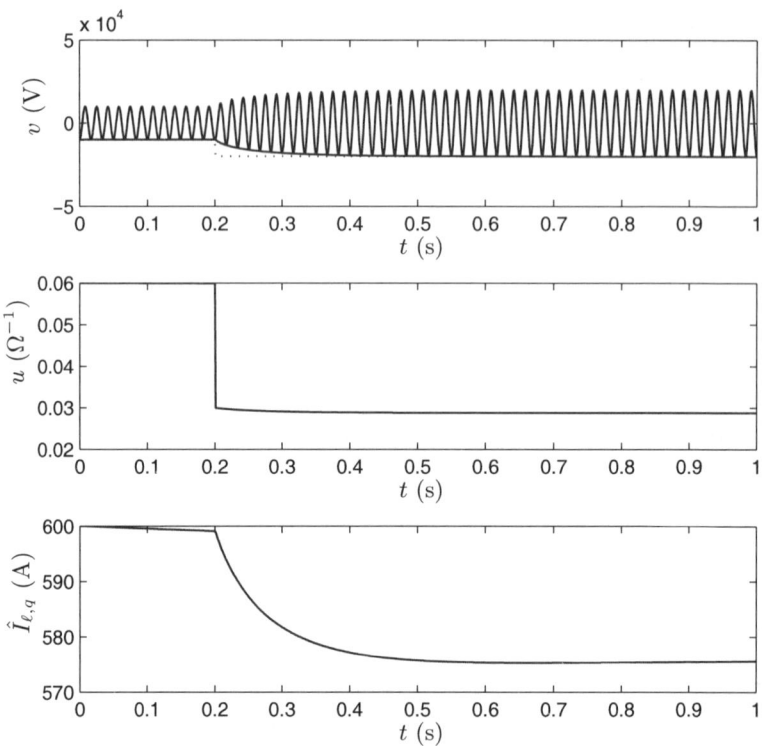

Fig. 8.3. Transient response of time-domain voltage v, control signal $u = w_s C_{\text{eq}}(\beta)$ and estimate $\hat{I}_{\ell,q}$.

step-down its absolute value. Its main application is in regulated DC power supplies, where an output voltage with negative polarity (with respect to the common terminal of the input voltage) is desired. A significant advantage of this converter over other inverting topologies, such as the buck-boost and flyback, is the use of inductors in the input and output loops (see Figure 8.4), which induces very little input and output current ripple.

The output voltage regulation problem for the DC–DC Ćuk converter, operating in continuous conduction mode, has attracted significant attention. Besides its practical relevance, this system is an interesting theoretical case study because it is a switched device whose averaged dynamics can be described by a fourth-order bilinear model with saturated input and uncertain parameters, namely the load resistance and the input voltage. Moreover, with respect to the available measured states, the system is nonminimum-phase.

8.2 Partial State Feedback Control of the DC–DC Ćuk Converter

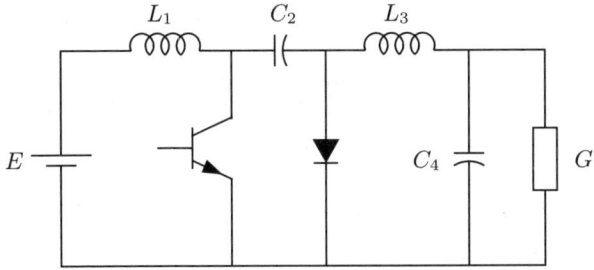

Fig. 8.4. DC–DC Ćuk converter circuit.

In this section the problem of output voltage regulation by means of partial state and parameter information for the DC–DC Ćuk converter (depicted in Figure 8.4) is addressed and solved based on a certainty equivalence point of view. In particular, it is shown that the combination of a full-information controller, derived using simple energy considerations[5], with a stable state and parameter estimator, derived using the I&I technique, yields a globally asymptotically stable closed-loop system. (Note that, since the design of the estimator is independent from the design of the stabilising controller, the estimator can be used in conjunction with any full-information control law.)

It is assumed that only the input capacitor voltage and the output inductor current are measured. These are then used to construct, under suitable assumptions, asymptotic estimates for the input inductor current, the output voltage, the load resistance and the input voltage (assumed constant). The estimated states and parameters are then used to replace unmeasured quantities in the full-information control law.

Note 8.1. The problem of output voltage regulation for the DC–DC Ćuk converter has been addressed and solved using various techniques. In particular, classical linear design tools have been used in [107], whereas the applicability of advanced nonlinear methods, such as feedback linearisation [190], passivation [163, 193], sliding-mode control [198] and H_∞ design [124], has also been investigated. ◁

8.2.1 Problem Formulation

The averaged model of the DC–DC Ćuk converter shown in Figure 8.4 is given by the equations
$$\begin{aligned} L_1 \dot{i}_1 &= -(1-u)v_2 + E, \\ C_2 \dot{v}_2 &= (1-u)i_1 + u i_3, \\ L_3 \dot{i}_3 &= -u v_2 - v_4, \\ C_4 \dot{v}_4 &= i_3 - G v_4, \end{aligned} \qquad (8.8)$$

---

[5] See [166] for an introductory description of energy-based control design.

where i_1 and i_3 describe the currents in the inductances L_1 and L_3, respectively; v_2 and v_4 are the voltages across the capacitors C_2 and C_4, respectively. (L_1, C_2, L_3 and C_4 are also used, with a slight abuse of notation, for the values of the capacitances and of the inductances.) Finally, G denotes the load admittance, E the input voltage and $u \in (0, 1)$ is a continuous control signal, which represents the duty ratio of the transistor switch.

It must be noted that, because of physical considerations, the state vector $x = [i_1, v_2, i_3, v_4]^\top$ is constrained in the set

$$\mathcal{D} = \mathbb{R}_{>0} \times \mathbb{R}_{>0} \times \mathbb{R}_{<0} \times \mathbb{R}_{<0}. \tag{8.9}$$

As a result, in what follows we implicitly assume that $x(t) \in \mathcal{D}$ for all t. The control goal is to regulate the voltage across the load (*i.e.*, the capacitor voltage v_4) to a constant value V_d.

Note that, by setting u to a constant value $\bar{u}$, the equilibria $(\bar{i}_1, \bar{v}_2, \bar{i}_3, \bar{v}_4)$ of system (8.8) verify the following relations

$$\bar{i}_1 = -\frac{\bar{u}}{1-\bar{u}} \bar{i}_3, \quad \bar{v}_2 = \frac{E}{1-\bar{u}}, \quad \bar{i}_3 = G\bar{v}_4, \quad \bar{v}_4 = -\frac{\bar{u}}{1-\bar{u}} E.$$

As a result, setting $\bar{v}_4 = -V_d$ yields the control input

$$u^* = \frac{V_d}{V_d + E}$$

and the desired operating equilibrium

$$i_1^* = \frac{GV_d^2}{E}, \quad v_2^* = V_d + E, \quad i_3^* = -GV_d, \quad v_4^* = -V_d.$$

In what follows we show that the considered control goal is achievable with a bounded control signal $u(t) \in (0, 1)$ and with partial state and parameter information, namely we assume that the only measured states are v_2 and i_3 and the parameters E and G are unknown.

8.2.2 Full-information Controller

In this section we present a full-information controller yielding global asymptotic stability of the desired equilibrium.

Proposition 8.2. *Consider the DC–DC Ćuk converter model (8.8) in closed loop with the full-information controller*

$$u_{\mathrm{FI}} = u^* + \lambda \frac{GV_d v_2 + E(i_3 - i_1)}{1 + (GV_d v_2 + E(i_3 - i_1))^2}, \tag{8.10}$$

where $\lambda = \lambda(x)$ is any nonnegative function of x such that

$$0 \le \lambda < 2\min(u^*, 1 - u^*).$$

8.2 Partial State Feedback Control of the DC–DC Ćuk Converter

Then, for any $V_d > 0$, the equilibrium $x^* = [i_1^*, v_2^*, i_3^*, -V_d]^\top$ is globally asymptotically stable. Moreover, the function

$$H_d(\tilde{x}) = \frac{1}{2}L_1 \tilde{i}_1^2 + \frac{1}{2}C_2 \tilde{v}_2^2 + \frac{1}{2}L_3 \tilde{i}_3^2 + \frac{1}{2}C_4 \tilde{v}_4^2, \qquad (8.11)$$

where $\tilde{x} = x - x^*$, is a Lyapunov function for the closed-loop system. Finally, u_{FI} is such that $u_{\text{FI}}(t) \in (0,1)$ for all t.

Proof. To begin with, note that, in terms of the *error variable* $\tilde{x}$, the dynamics (8.8) are described by the equations

$$\begin{aligned}
L_1 \dot{\tilde{i}}_1 &= -(1-u)\tilde{v}_2 + (V_d + E)\tilde{u}, \\
C_2 \dot{\tilde{v}}_2 &= (1-u)\tilde{i}_1 + u\tilde{i}_3 - GV_d \left(1 + \frac{V_d}{E}\right)\tilde{u}, \\
L_3 \dot{\tilde{i}}_3 &= -u\tilde{v}_2 - \tilde{v}_4 - (V_d + E)\tilde{u}, \\
C_4 \dot{\tilde{v}}_4 &= \tilde{i}_3 - G\tilde{v}_4,
\end{aligned} \qquad (8.12)$$

where $\tilde{u} = u - u^*$. Note now that the function $H_d(\tilde{x})$ is positive-definite, radially unbounded and has a unique minimum for $\tilde{x} = 0$. Moreover,

$$\dot{H}_d = -\frac{V_d + E}{E}\left(GV_d v_2 + E(i_3 - i_1)\right)\tilde{u} - G\tilde{v}_4^2. \qquad (8.13)$$

Replacing the control law (8.10) in (8.13), i.e., setting $u = u_{\text{FI}}$ and hence $\tilde{u} = u_{\text{FI}} - u^*$, yields

$$\dot{H}_d = -G\tilde{v}_4^2 - \lambda \frac{V_d + E}{E} \frac{(GV_d v_2 + E(i_3 - i_1))^2}{1 + (GV_d v_2 + E(i_3 - i_1))^2} \leq 0.$$

This implies that the closed-loop system has a globally stable equilibrium at $\tilde{x} = 0$ and that all its trajectories converge to the largest invariant set contained in

$$\mathcal{S} \triangleq \left\{x \in \mathcal{D} \mid \lambda\left(GV_d v_2 + E(i_3 - i_1)\right)^2 + G\tilde{v}_4^2 = 0\right\}.$$

Since, $u = u^*$ and $\tilde{v}_4 = 0$ on the set $\mathcal{S}$, it follows that the largest invariant set is the equilibrium point x^*. The proof is completed by noting that the control signal is the sum of u^* and the signal

$$\lambda \frac{GV_d v_2 + E(i_3 - i_1)}{1 + (GV_d v_2 + E(i_3 - i_1))^2}$$

which is bounded in modulo by $\lambda/2$. $\qquad \square$

As a simple consequence of Proposition 8.2 we give the following result.

Corollary 8.1. *Consider the DC–DC Ćuk converter model (8.8) in closed loop with the full-information controller*

$$u = u^* = \frac{V_d}{V_d + E}. \tag{8.14}$$

Then, for any $V_d > 0$, the equilibrium $x^* = [i_1^*, v_2^*, i_3^*, -V_d]^\top$ is globally asymptotically stable.

The control law in Corollary 8.1 is indeed very simple, as it requires knowledge of E only. However, the resulting performance may be unacceptable in practice.

8.2.3 I&I Adaptive Observer

In this section we use the methodology of Chapter 5 to construct a nonlinear observer that yields, under suitable assumptions, asymptotically converging estimates for the unmeasured states and the unknown parameters.

Proposition 8.3. *Consider the DC–DC Ćuk converter model (8.8) and the system*

$$\dot{\hat{E}} = -L_1\Gamma_1\big[(1-\hat{u})(\hat{i}_1 + C_2\Gamma_2 v_2) + \hat{u}i_3\big], \tag{8.15}$$

$$\dot{\hat{i}}_1 = \frac{1}{L_1}\big[-(1-\hat{u})v_2 + (\hat{E} + L_1 C_2 \Gamma_1 v_2)\big]$$
$$\qquad - \Gamma_2\big[(1-\hat{u})(\hat{i}_1 + C_2\Gamma_2 v_2) + \hat{u}i_3\big], \tag{8.16}$$

$$\dot{\hat{G}} = \frac{1}{L_3}(\hat{v}_4 - L_3\Gamma_3 i_3)\big[\hat{u}v_2 + (\hat{v}_4 - L_3\Gamma_3 i_3)\big] - i_3\dot{\hat{v}}_4, \tag{8.17}$$

$$\dot{\hat{v}}_4 = \frac{1}{C_4}\big[i_3 - (\hat{G} + \hat{v}_4 i_3 - \tfrac{1}{2}L_3\Gamma_3 i_3^2)(\hat{v}_4 - L_3\Gamma_3 i_3)\big]$$
$$\qquad + \Gamma_3\big[-\hat{u}v_2 - (\hat{v}_4 - L_3\Gamma_3 i_3)\big], \tag{8.18}$$

where $\hat{u}$ is any known signal such that $\hat{u}(t) \in (0, 1-\epsilon)$, for some $\epsilon \in (0,1)$ and for all t. Then there exist positive constants Γ_1, Γ_2 such that, for any $\Gamma_3 > 0$ and any initial condition $(\hat{E}(0), \hat{i}_1(0), \hat{G}(0), \hat{v}_2(0))$,

$$\begin{aligned}\lim_{t\to\infty}\big[\hat{E} + L_1 C_2 \Gamma_1 v_2 - E\big] &= 0,\\ \lim_{t\to\infty}\big[\hat{i}_1 + C_2\Gamma_2 v_2 - i_1\big] &= 0,\end{aligned} \tag{8.19}$$

exponentially, and

$$\begin{aligned}(\hat{G} + \hat{v}_4 i_3 - \tfrac{1}{2}L_3\Gamma_3 i_3^2 - G) &\in \mathcal{L}_\infty,\\ (\hat{v}_4 - L_3\Gamma_3 i_3 - v_4) &\in \mathcal{L}_\infty.\end{aligned} \tag{8.20}$$

Moreover, if $v_4(t) \in \mathcal{L}_\infty$ and $v_4(t) < -\sigma$, for some $\sigma > 0$ and for all t, then

8.2 Partial State Feedback Control of the DC–DC Ćuk Converter

$$\lim_{t \to \infty} \left[\hat{G} + \hat{v}_4 i_3 - \tfrac{1}{2} L_3 \Gamma_3 i_3^2 - G \right] = 0,$$
$$\lim_{t \to \infty} \left[\hat{v}_4 - L_3 \Gamma_3 i_3 - v_4 \right] = 0,$$
(8.21)

exponentially.

Proof. To begin with, define the estimation errors

$$\begin{aligned}
z_1 &= \hat{E} + L_1 C_2 \Gamma_1 v_2 - E, \\
z_2 &= \hat{i}_1 + C_2 \Gamma_2 v_2 - i_1, \\
z_3 &= \hat{G} + \hat{v}_4 i_3 - \tfrac{1}{2} L_3 \Gamma_3 i_3^2 - G, \\
z_4 &= \hat{v}_4 - L_3 \Gamma_3 i_3 - v_4.
\end{aligned}$$
(8.22)

Straightforward computations show that the estimation error dynamics are such that

$$\begin{aligned}
\dot{z}_1 &= -L_1 \Gamma_1 (1 - \hat{u}) z_2, \\
\dot{z}_2 &= \frac{1}{L_1} z_1 - \Gamma_2 (1 - \hat{u}) z_2, \\
\dot{z}_3 &= \frac{1}{L_3} (\hat{v}_4 - L_3 \Gamma_3 i_3) z_4, \\
\dot{z}_4 &= -\frac{1}{C_4} (\hat{v}_4 - L_3 \Gamma_3 i_3) z_3 - \left(\frac{G}{C_4} + \Gamma_3 \right) z_4.
\end{aligned}$$
(8.23)

Equations (8.23) can be regarded as two decoupled subsystems, one with state (z_1, z_2) and one with state (z_3, z_4). Therefore, to establish the claims, we consider these two subsystems separately.

First, rewrite the (z_1, z_2) subsystem as the feedback interconnection of the LTI system

$$\begin{bmatrix} \dot{z}_1 \\ \dot{z}_2 \end{bmatrix} = \begin{bmatrix} 0 & -L_1 \Gamma_1 \bar{d} \\ \frac{1}{L_1} & -\Gamma_2 \bar{d} \end{bmatrix} \begin{bmatrix} z_1 \\ z_2 \end{bmatrix} + \begin{bmatrix} L_1 \Gamma_1 \bar{d} \\ \Gamma_2 \bar{d} \end{bmatrix} w,$$
$$\bar{y} = \begin{bmatrix} 0 & 1 \end{bmatrix} \begin{bmatrix} z_1 \\ z_2 \end{bmatrix}$$
(8.24)

with the static nonlinearity

$$w = -\frac{\delta d}{\bar{d}} \bar{y},$$

where $\bar{d} = \tfrac{1}{2}(1+\epsilon)$, $\delta d = d - \bar{d}$, and $d = 1 - \hat{u}$. We now study this interconnected system using the small gain theorem. To this end, note that since $\Gamma_1 > 0$ and $\Gamma_2 > 0$, the unforced system (8.24) is (globally) asymptotically stable. Moreover, its H_∞ norm is given by

$$\gamma(\mu) = \frac{1}{\sqrt{1 + 2\mu^2} \sqrt{1 + \frac{2}{\mu} - 2\mu^2} - 2\mu},$$

where $\mu = \Gamma_1/\left(\Gamma_2^2 \bar{d}\right)$. Note now that $|\delta d/\bar{d}| < (1-\epsilon)/(1+\epsilon)$, therefore the closed-loop system is (exponentially) stable provided

$$\gamma(\mu)\frac{1-\epsilon}{1+\epsilon} < 1. \qquad (8.25)$$

Hence, to complete the proof of the claim we need to show that for any $\epsilon \in (0,1)$ it is possible to select $\Gamma_1 > 0$ and $\Gamma_2 > 0$ such that (8.25) holds. Note that $\gamma(\mu) \geq 1$, for any $\mu \geq 0$, and that (8.25) is equivalent to

$$\epsilon > \frac{\gamma(\mu)-1}{\gamma(\mu)+1}.$$

Observe now that the range of the function on the right-hand side is the open set $(0,1)$. Therefore, for any $\epsilon \in (0,1)$, there is a $\mu > 0$, and hence $\Gamma_1 > 0$ and $\Gamma_2 > 0$, such that condition (8.25) holds[6].

Consider now the (z_3, z_4) subsystem and the Lyapunov function

$$V_{34}(z_3, z_4) = \frac{1}{2}L_3 z_3^2 + \frac{1}{2}C_4 z_4^2.$$

A simple calculation shows that

$$\dot{V}_{34} = -C_4\left(\frac{G}{C_4} + \Gamma_3\right)z_4^2 \leq 0$$

and this proves that condition (8.20) holds.

Finally, if $v_4(t) \in \mathcal{L}_\infty$ and $v_4(t) < -\sigma$ for some $\sigma > 0$ and for all t, we conclude (invoking standard results on stability of skew-symmetric systems and the notion of persistence of excitation[7]) that the origin of the (z_3, z_4) subsystem is (globally) exponentially stable. □

8.2.4 Partial State Feedback Controller

In this section we present two asymptotically stabilising partial information dynamic control laws for the DC–DC Ćuk converter. These are obtained combining the full-information controllers (8.14) and (8.10) with the adaptive observer of Proposition 8.3.

---

[6] An alternative way of proving that the (z_1, z_2) subsystem has a globally exponentially stable equilibrium at the origin is via the partial change of co-ordinates $\bar{z}_1 = z_1 - \frac{L_1 \Gamma_1}{\Gamma_2} z_2$ and the Lyapunov function $V_{12}(\bar{z}_1, z_2) = \frac{1}{2}\bar{z}_1^2 + \frac{1}{2}\left(\frac{L_1 \Gamma_1}{\Gamma_2}\right)^2 z_2^2$, whose time-derivative is negative-definite provided that the constants $\Gamma_1 > 0$ and $\Gamma_2 > 0$ are selected so that $\Gamma_1/\Gamma_2^2 < \epsilon$. See Section 8.3 for details.

[7] See, e.g., [191, 152]. Note that the conditions $v_4(t) \in \mathcal{L}_\infty$ and $v_4 < -\sigma$ can be relaxed as detailed in [169].

8.2 Partial State Feedback Control of the DC–DC Ćuk Converter

Proposition 8.4. *Let $\epsilon \in (0,1)$. Consider the DC–DC Ćuk converter described by equations (8.8) in closed loop with the dynamic controller*

$$u = \mathrm{sat}_{[0,1-\epsilon]}\left(\frac{V_d}{V_d + \bar{E}}\right), \qquad (8.26)$$

where[8] $\bar{E} = \hat{E} + L_1 C_2 \Gamma_1 v_2$, $\bar{i}_1 = \hat{i}_1 + C_2 \Gamma_2 v_2$, $\bar{G} = \hat{G} + \hat{v}_4 i_3 - \frac{1}{2} L_3 \Gamma_3 i_3^2$, *and $\hat{E}$, $\hat{i}_1$ and $\hat{G}$ are computed from (8.15) and (8.18) with $\hat{u} = u$. Then, for any V_d such that*

$$0 < V_d < \frac{\epsilon - 1}{\epsilon} E, \qquad (8.27)$$

all trajectories of the closed-loop system are bounded and are such that $\lim_{t \to \infty} x(t) = x^*$.

Proof. To begin with, note that the closed-loop system is described in the x and z co-ordinates by

$$\dot{x} = A(u^*)x + bE + \varphi(z_1)Dx \qquad (8.28)$$

and (8.23), where $A(u^*)$, b and D are constant matrices and

$$\varphi(z_1) = \mathrm{sat}_{[0,1-\epsilon]}\left(\frac{V_d}{V_d + E + z_1}\right) - u^*.$$

Note now that system (8.28) with $\varphi(z_1) = 0$ has a globally exponentially stable equilibrium at the origin. Moreover, as $\varphi(0) = 0$, $|\varphi(z_1)| < \kappa_0$ for some $0 < \kappa_0 < 1$, and z_1 is exponentially converging to zero, it follows[9] that x converges to x^*, hence the claim. □

It must be noted that the control law in Proposition 8.4 requires only partial knowledge of the circuit parameters, as it does not use explicitly the parameters L_3 and C_4.

Proposition 8.5. *Let $\epsilon \in (0,1)$. Consider the DC–DC Ćuk converter described by equations (8.8) in closed loop with the dynamic controller*

$$u = \mathrm{sat}_{[0,1-\epsilon]}\left(\frac{\bar{G}V_d v_2 + \bar{E}(i_3 - \bar{i}_1)}{1 + (\bar{G}V_d v_2 + \bar{E}(i_3 - \bar{i}_1))^2} + \frac{V_d}{V_d + \bar{E}}\right), \qquad (8.29)$$

where $\bar{E} = \hat{E} + L_1 C_2 \Gamma_1 v_2$, $\bar{i}_1 = \hat{i}_1 + C_2 \Gamma_2 v_2$, $\bar{G} = \hat{G} + \hat{v}_4 i_3 - \frac{1}{2} L_3 \Gamma_3 i_3^2$, and $\hat{E}$, $\hat{i}_1$ and $\hat{G}$ are computed from (8.15) and (8.18) with $\hat{u} = u$. Assume that $v_4(t) \in \mathcal{L}_\infty$ and $v_4(t) < -\sigma$ for some $\sigma > 0$ and for all t. Then, for any V_d such that condition (8.27) holds, all trajectories of the closed-loop system are bounded and are such that $\lim_{t \to \infty} x(t) = x^*$.

---

[8] To take into account that $E \geq 0$ we could define $\bar{E} = \max(0, \hat{E} + L_1 C_2 \Gamma_1 v_2)$. In that case the forthcoming discussion requires minor modifications. We could also use a smooth saturation function as in Section 10.3.
[9] See [112].

Proof. To begin with, note that under the stated assumptions the state and parameters estimation errors converge exponentially to zero. Note now that the closed-loop system can be described by the equations (8.12) perturbed by an additive term which is bounded by

$$\kappa |x||u - u_{\mathrm{FI}}|,$$

for some positive constant κ, and it is such that

$$\lim_{t \to \infty} |u(t) - u_{\mathrm{FI}}(t)| = 0,$$

exponentially. Consider now the Lyapunov function $H_d(\tilde{x})$ in (8.11) and note that

$$\dot{H}_d \leq \kappa \left| \frac{\partial H_d}{\partial x} \right| ||x|| |u - u_{\mathrm{FI}}|,$$

hence

$$\dot{H}_d \leq \tilde{\kappa} |\tilde{x}| ||x|| |u - u_{\mathrm{FI}}| \leq \hat{\kappa}(H_d + \rho)|u - u_{\mathrm{FI}}|,$$

for some positive constants $\tilde{\kappa}$, $\hat{\kappa}$ and ρ. The above equation implies that, along the trajectories of the system in closed loop with the partial information controller, H_d is bounded. As a result $|x|$ is bounded and, by standard properties of cascaded systems, x converges to x^*. This completes the proof. □

8.2.5 Simulation Results

Numerical simulations have been carried out to assess the performance of the proposed controllers. The values of the converter parameters have been selected as[10] $L_1 = 10$ mH, $C_2 = 22.0$ μF, $L_3 = 10$ mH and $C_4 = 22.9$ μF. Moreover, it is assumed that the nominal values for the load admittance and the input voltage are $G_N = 0.0447$ S and $E_N = 12$ V, respectively. The initial conditions for all simulations are set to $x(0) = \begin{bmatrix} 0.5 & 10 & -1 & -12 \end{bmatrix}^\top$.

In all simulations we assume that the initial set point for the output voltage is $V_d = -5$ V, and then this is changed at $t = 0.05$ s to $V_d = -35$ V. Moreover, to assert the robustness of the controllers we consider two step changes of the load admittance and the input voltage: at $t = 0.1$ the load admittance is decreased to $G = 0.022$ S; and at $t = 0.15$ the input voltage is decreased to $E = 10$ V.

Figures 8.5 and 8.6 show the performance of the full-information controllers (8.14) and (8.10) (with $\lambda = 1$), respectively. As can be observed, the output voltage initially converges to the desired value. However, both controllers are sensitive to input voltage variation. Note that the additional term in the control law (8.10) increases the damping in the system, but it also makes the controller sensitive to load variations.

To test the controllers of Propositions 8.4 and 8.5 we study the same control goal as above. The parameters of the adaptive observer are set to

---
[10] See [124].

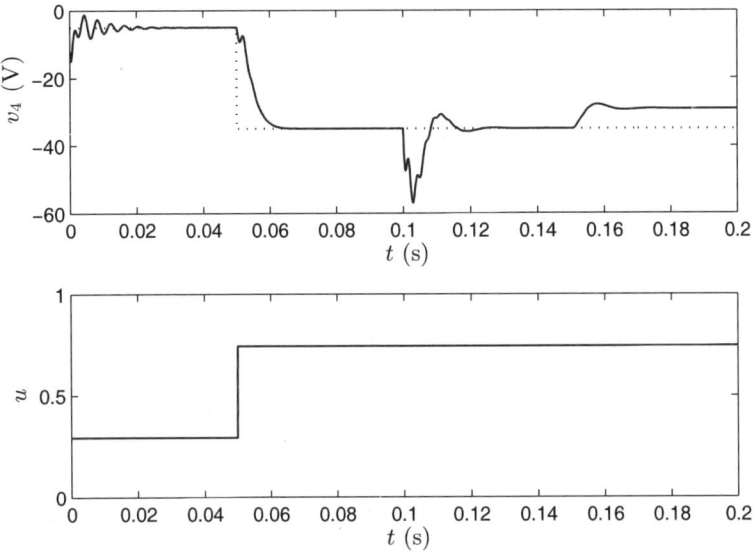

Fig. 8.5. Output voltage v_4 (top) and control input (8.14) (bottom).

$\epsilon = 0.05$, $\mu = 0.045$, $\Gamma_1 = 3 \times 10^5$, $\Gamma_2 = 2 \times 10^3$ and $\Gamma_3 = 9 \times 10^4$. The performance of the partial information controller (8.26) is shown in Figure 8.7. Observe that the output voltage now converges to the desired set-point despite the uncertainty in the input voltage and load admittance and the partial state feedback.

Finally, numerical simulations for the partial information controller (8.29), with $\lambda = 1$, are shown in Figures 8.8 and 8.9. These display time histories of the output voltage and the control signal and of the estimation errors.

8.3 Output Feedback Control of the DC–DC Boost Converter

In this section the output feedback stabilisation tools developed in Chapter 6, and in particular the result in Theorem 6.2, are applied to the design of an output voltage regulator for the DC–DC boost converter. These devices are widely used in power supplies whenever it is required to step-up the DC voltage.

The DC–DC boost converter (similarly to the Ćuk converter) poses an interesting control problem as it is described by a second-order bilinear model with a single bounded input and is nonminimum-phase with respect to the output to be regulated. In order to overcome the minimum-phase obstacle, a common practice is to control the output voltage *indirectly* by regulating

190   8 Electrical Systems

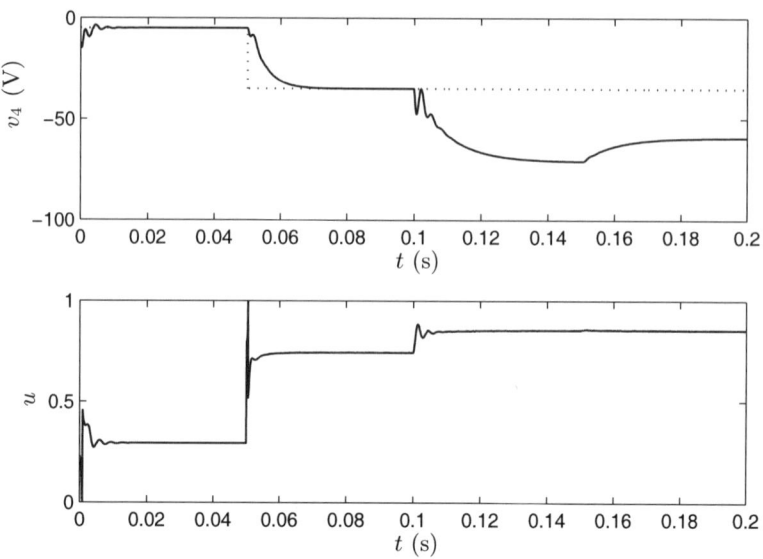

Fig. 8.6. Output voltage v_4 (top) and control input (8.10) with $\lambda = 1$ (bottom).

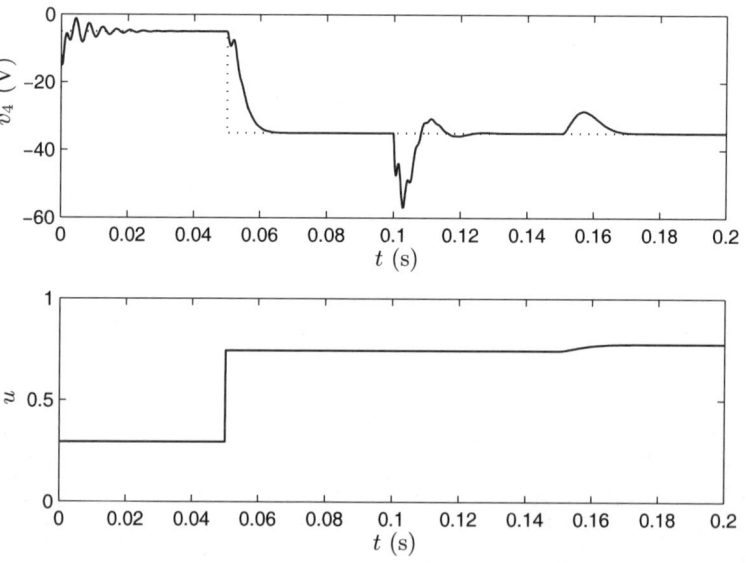

Fig. 8.7. Output voltage v_4 (top) and control input (8.26) (bottom).

8.3 Output Feedback Control of the DC–DC Boost Converter

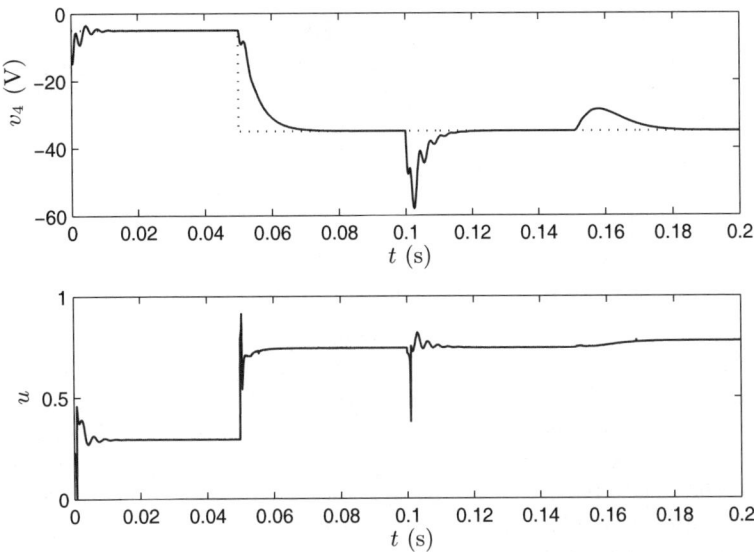

Fig. 8.8. Output voltage v_4 (top) and control input (8.29) with $\lambda = 1$ (bottom).

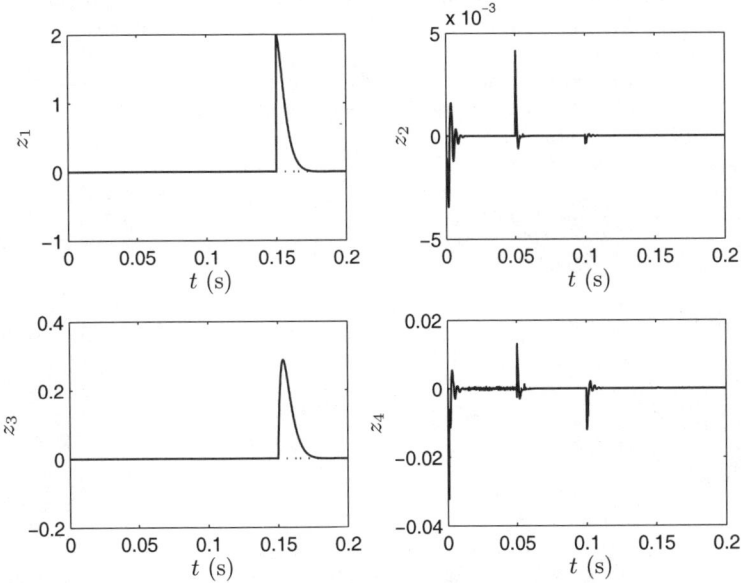

Fig. 8.9. Estimation errors for the input voltage (z_1), input inductor current (z_2), load admittance (z_3) and output capacitor voltage (z_4).

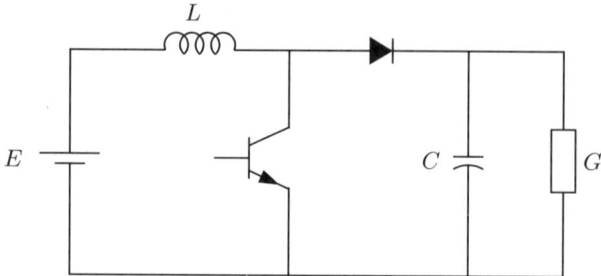

Fig. 8.10. DC–DC boost converter circuit.

the input current. The drawback of this approach is that it requires a current sensor and leads to control schemes that are sensitive to variations in the circuit parameters and in particular of the load. On the other hand, *direct* control schemes do not require input current measurements and are robust to load variations, but rely strongly on measuring the input voltage.

The adaptive output feedback controller developed in this section does not require measurement of the input current nor of the input voltage and is shown to be robust to load variations. Moreover, it allows to shape the response of the closed-loop system by introducing a damping term, which is similar to the one introduced in Section 8.2 for the Ćuk converter.

Note 8.2. The problem of regulating DC–DC converters by PWM control schemes has been treated in [107] and, in the framework of nonlinear control theory, in [163] and related references, see also [55] for an experimental comparative study and [187] for a design example. ◁

8.3.1 Problem Formulation

We consider a DC–DC boost converter with the topology shown in Figure 8.10. The averaged model is given by the equations

$$\begin{aligned} L\dot{x}_1 &= -ux_2 + E, \\ C\dot{x}_2 &= -Gx_2 + ux_1, \\ y &= x_2, \end{aligned} \tag{8.30}$$

where x_1 is the input (inductor) current, x_2 is the output (capacitor) voltage, u is the duty ratio of the transistor switch and L, C, G and E are positive constants representing the inductance, capacitance, load conductance and input voltage, respectively.

It is assumed that only the state $y = x_2$ is available for feedback, the constant E is unknown, and $u \in [\epsilon, 1]$ with $0 < \epsilon < 1$. The control problem is to regulate the output y (*i.e.*, the voltage across the capacitor) to a positive value $y^* = V_d$. Note that the corresponding equilibrium is given by

$$x_1^* = \frac{GV_d^2}{E}, \qquad x_2^* = V_d, \qquad u^* = \frac{E}{V_d}.$$

From the last equation we see that the control constraint $u \in [\epsilon, 1]$ imposes $V_d \geq E$.

Note that, to simplify the design, we have neglected all parasitic resistances in the circuit, including the internal resistance of the source. As is shown in the sequel, this leads to a significant mismatch between simulated and experimental results. However, the performance of the proposed adaptive output feedback controller is not affected, since the adaptation is able to compensate for this modelling error.

Before proceeding to the control design, we show that the converter model with output y is nonminimum-phase. Define the variables $\bar{u} = u - u^*$ and $\bar{y} = y - y^*$ and consider the linearisation of (8.30) around the equilibrium (x_1^*, x_2^*). The resulting transfer function from the input $\bar{u}$ to the output $\bar{y}$ is given by

$$\frac{\bar{y}(s)}{\bar{u}(s)} = \frac{b_1 s - b_2}{s^2 + a_1 s + a_2} \tag{8.31}$$

with $a_1 = G/C$, $a_2 = E^2/(LCV_d^2)$, $b_1 = GV_d^2/(CE)$, and $b_2 = E/(LC)$, from where we conclude that the linearised system has a zero on the right half-plane, *i.e.*, it is nonminimum-phase.

8.3.2 Full-information Controller

A full-information control law that depends on the unmeasured state x_1 and the unknown input voltage E is given in the following proposition.

Proposition 8.6. *Consider the DC–DC boost converter model (8.30) in closed loop with the full-information controller*[11]

$$u_{\mathrm{FI}} = u^* + \lambda \frac{E(x_1 - x_1^*) - GV_d(x_2 - x_2^*)}{1 + (E(x_1 - x_1^*) - GV_d(x_2 - x_2^*))^2}, \tag{8.32}$$

where $\lambda = \lambda(x)$ is any nonnegative function of x such that

$$0 \leq \lambda < 2\min(u^*, 1 - u^*).$$

Then, for any $V_d > E$, the equilibrium $x^* = \begin{bmatrix} GV_d^2/E, V_d \end{bmatrix}^\top$ is globally asymptotically stable. Moreover, u_{FI} is such that $u_{\mathrm{FI}}(t) \in (0,1)$ for all t.

---

[11] This controller, without the normalisation, has been proposed in [189].

Proof. Consider the proper Lyapunov function

$$H_d(\tilde{x}) = \frac{1}{2}L\tilde{x}_1^2 + \frac{1}{2}C\tilde{x}_2^2,$$

where $\tilde{x} = x - x^*$, and note that its time-derivative along the trajectories of (8.30) is given by

$$\dot{H}_d = \tilde{x}_1\left(-u\tilde{x}_2 - \tilde{u}V_d\right) + \tilde{x}_2\left(u\tilde{x}_1 + \tilde{u}\frac{GV_d^2}{E} - G\tilde{x}_2\right)$$

$$= -G\tilde{x}_2^2 - \frac{V_d}{E}\left(E\tilde{x}_1 - GV_d\tilde{x}_2\right)\tilde{u},$$

where $\tilde{u} = u - u^*$. Substituting $\tilde{u}$ with the last term in (8.32) yields $\dot{H}_d < 0$ for all $\tilde{x} \neq 0$, which implies that the equilibrium $\tilde{x} = 0$ is globally asymptotically stable. The proof is completed by noting that the signal $\tilde{u}$ is bounded in modulo by $\lambda/2$. □

8.3.3 Output Feedback Controller

Following the methodology proposed in Section 6.3, the system (8.30) can be rewritten in the form (6.14), namely

$$\begin{bmatrix} \dot{\eta}_1 \\ \dot{\eta}_2 \end{bmatrix} = \begin{bmatrix} 0 & 0 \\ \frac{1}{L} & 0 \end{bmatrix} \begin{bmatrix} \eta_1 \\ \eta_2 \end{bmatrix} + \begin{bmatrix} 0 \\ -\frac{1}{L}uy \end{bmatrix} \quad (8.33)$$

$$\dot{y} = -\frac{G}{C}y + \frac{1}{C}u\eta_2,$$

where $\eta_1 \triangleq E$ and $\eta_2 \triangleq x_1$. Note that when $u = 0$ the states η_1 and η_2 (the input voltage and current) become unobservable from the output y (the output voltage), which justifies the assumption $u \geq \epsilon > 0$. (Physically, $u = 0$ implies that the transistor switch is on, thus cutting off the load from the source.)

A solution to the output voltage regulation problem by means of output feedback is given by the following proposition.

Proposition 8.7. *Consider the DC–DC boost converter model (8.33) and the constant reference voltage y^*. Then, for any $0 < \epsilon < 1$ and any y^* such that*

$$\epsilon \leq \frac{\eta_1}{y^*} \leq 1, \quad (8.34)$$

there exists a dynamic output feedback control law described by equations of the form (6.2) such that all trajectories of the closed-loop system are bounded and

$$\lim_{t \to \infty} y(t) = y^*.$$

8.3 Output Feedback Control of the DC–DC Boost Converter

Proof. To begin with, note that the constraint (8.34) is a consequence of the control constraint $u \in [\epsilon, 1]$, i.e., if $u \in [\epsilon, 1]$ then all achievable equilibria of the system (8.33) are such that (8.34) holds. Consider now the full-information control law[12]

$$u^* = \text{sat}_{[\epsilon,1]}(\frac{\eta_1}{y^*} + \lambda \frac{E(x_1 - x_1^*) - GV_d(x_2 - x_2^*)}{1 + (E(x_1 - x_1^*) - GV_d(x_2 - x_2^*))^2}), \quad (8.35)$$

which has been shown in Proposition 8.6 to yield asymptotic regulation. Consider now the Lyapunov function

$$V(\eta_2, y) = \frac{1}{2}L\eta_2^2 + \frac{1}{2}Cy^2 - a\eta_2 y,$$

where a is a constant to be determined, and note that its time-derivative along the trajectories of (8.33) satisfies

$$\dot{V} = \eta_1\eta_2 - Gy^2 - \frac{au}{C}\eta_2^2 + \frac{aG}{C}\eta_2 y - \frac{a}{L}\eta_1 y + \frac{au}{L}y^2$$

$$\leq -\left(\frac{au}{C} - \delta_1\right)\eta_2^2 - \left(G - \frac{au}{L} - \delta_2\right)y^2 + \frac{\eta_1^2}{4\delta_1} + \frac{a^2\eta_1^2}{4\delta_2} + \frac{aG}{C}\eta_2 y$$

$$\leq -\left(\frac{a\epsilon}{C} - \delta_1\right)\eta_2^2 - \left(G - \frac{a}{L} - \delta_2\right)y^2 + \frac{\eta_1^2}{4\delta_1} + \frac{a^2\eta_1^2}{4\delta_2} + \frac{a\epsilon}{2C}\eta_2^2 + \frac{aG^2}{2C\epsilon}y^2$$

$$\leq -\left(\frac{a\epsilon}{2C} - \delta_1\right)\eta_2^2 - \left(G - \frac{a}{L} - \frac{aG^2}{2C\epsilon} - \delta_2\right)y^2 + \frac{\eta_1^2}{4\delta_1} + \frac{a^2\eta_1^2}{4\delta_2},$$

which implies that all trajectories of the system (8.33) are bounded, provided a is selected as

$$0 < a < \min(\sqrt{LC}, \frac{2GLC\epsilon}{2C\epsilon + G^2L}).$$

Moreover, this holds for any $u \in [\epsilon, 1]$. Hence, assumption (B1) of Theorem 6.2 holds. (Apparently, the fact that u is bounded away from zero is crucial not only for observability, as mentioned before, but also for boundedness of trajectories. This is justified by the fact that when $u(t) = 0$ for all t the inductor becomes short-circuited, therefore its current increases linearly. However, if there is a small resistance in the input circuit, which is always the case in practice, this situation is avoided and trajectories are bounded for any u. This can easily be shown by using the total electrical energy $\frac{1}{2}L\eta_2^2 + \frac{1}{2}Cy^2$ as a Lyapunov function, see Lemma 8.1 in Section 8.4.)

Consider now the error variable

$$z = \hat{\eta} - \eta + \beta(y)$$

with

---

[12] The saturation function is not required in the definition of u^*, since from (8.34) the argument of the saturation is always in the interval $[\epsilon, 1]$. However, it is required in the construction of the output feedback controller.

$$\beta(y) = \begin{bmatrix} \gamma_1 y \\ \gamma_2 y \end{bmatrix},$$

where γ_1 and γ_2 are positive constants, and define the dynamic output feedback controller

$$\begin{aligned}
\dot{\hat{\eta}}_1 &= -\frac{\gamma_1}{C}\left(u\left(\hat{\eta}_2 + \gamma_2 y\right) - Gy\right), \\
\dot{\hat{\eta}}_2 &= -\frac{\gamma_2}{C}\left(u\left(\hat{\eta}_2 + \gamma_2 y\right) - Gy\right) + \frac{1}{L}\left(\hat{\eta}_1 + \gamma_1 y - uy\right), \\
u &= \mathrm{sat}_{[\epsilon,1]}\!\left(\frac{\hat{\eta}_1 + \gamma_1 y}{y^*}\right).
\end{aligned} \qquad (8.36)$$

As a result, the dynamics of z are given by the equation

$$\dot{z} = \begin{bmatrix} 0 & -\dfrac{\gamma_1}{C}u \\ \dfrac{1}{L} & -\dfrac{\gamma_2}{C}u \end{bmatrix} z. \qquad (8.37)$$

To complete the proof, it is necessary to show that, for any $0 < \epsilon < 1$, it is possible to select $\gamma_1 > 0$ and $\gamma_2 > 0$ such that the system (8.37) is uniformly asymptotically stable. To this end, consider the partial change of co-ordinates

$$\bar{z}_1 = z_1 - \frac{\gamma_1}{\gamma_2} z_2$$

and note that the system (8.37) can be written as

$$\begin{aligned}
\dot{\bar{z}}_1 &= -\frac{\gamma_1}{\gamma_2 L}\bar{z}_1 - \frac{\gamma_1^2}{\gamma_2^2 L} z_2 \\
\dot{z}_2 &= \frac{1}{L}\bar{z}_1 - \left(\frac{\gamma_2 u}{C} - \frac{\gamma_1}{\gamma_2 L}\right) z_2,
\end{aligned}$$

which has a globally asymptotically stable equilibrium at the origin, provided

$$\frac{\gamma_1 C}{\gamma_2^2 L} < \epsilon. \qquad (8.38)$$

This concludes the proof of Proposition 8.7. □

Condition (8.38) can be read in two directions. Given $\gamma_1 > 0$ and $\gamma_2 > 0$, it is possible to compute a set of values for ϵ for which the system (8.37) is uniformly asymptotically stable. Given $\epsilon > 0$, it is possible to compute $\gamma_1 > 0$ and $\gamma_2 > 0$ such that the system (8.37) is uniformly asymptotically stable. Note that for any fixed ϵ there are infinite pairs (γ_1, γ_2) such that (8.38) holds and these can be easily parameterised.

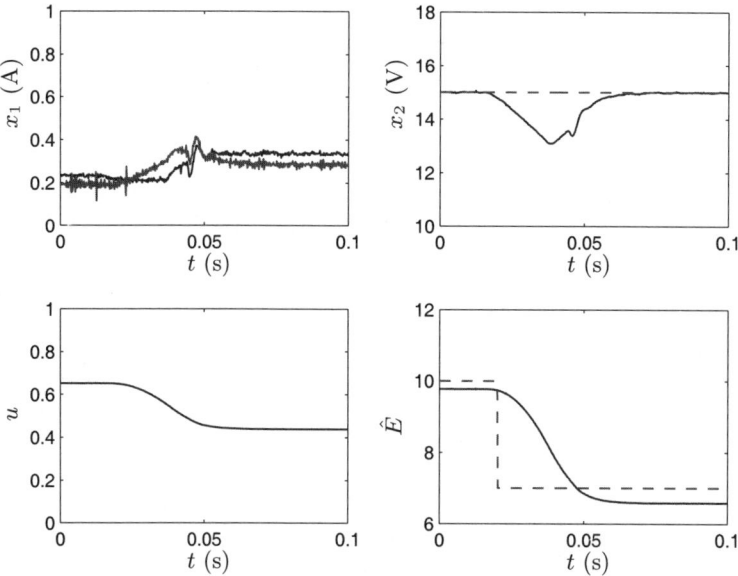

Fig. 8.11. State and control histories for a change in the input voltage from $E = 10$ V to $E = 7$ V.

8.3.4 Experimental Results

Laboratory experiments have been carried out on a DC–DC boost converter with the physical parameters[13] $C = 94$ μF, $L = 1.36$ mH and $G = 1/120$ Ω^{-1}. The switching frequency of the PWM which controls the switch is 20 kHz. The desired output voltage is $y^* = 15$ V. The control lower bound ϵ has been fixed at 0.1 and $\gamma_1 = 0.2$ and $\gamma_2 = 1$ have been set to satisfy condition (8.38). In all experiments we have taken $\lambda = 0$, i.e., the controller does not introduce any additional damping into the system.

The first experiment comprises a change in the input voltage E from 10 V to 7 V, at $t = 0.02$ s. Figure 8.11 shows the inductor current x_1, together with its estimate $\hat{x}_1$, the output voltage x_2, the control signal u and the input voltage E, together with its estimate $\hat{E}$. Observe that the output voltage tracks the desired value, despite partial state measurement and the change in the input voltage. In addition, the inductor current and the input voltage estimates approach the true values with a static error, which is due to unmodelled parasitic elements. Note that, even in the presence of this error, the

---

[13] The particular converter used in the experiments is described in detail in [27].

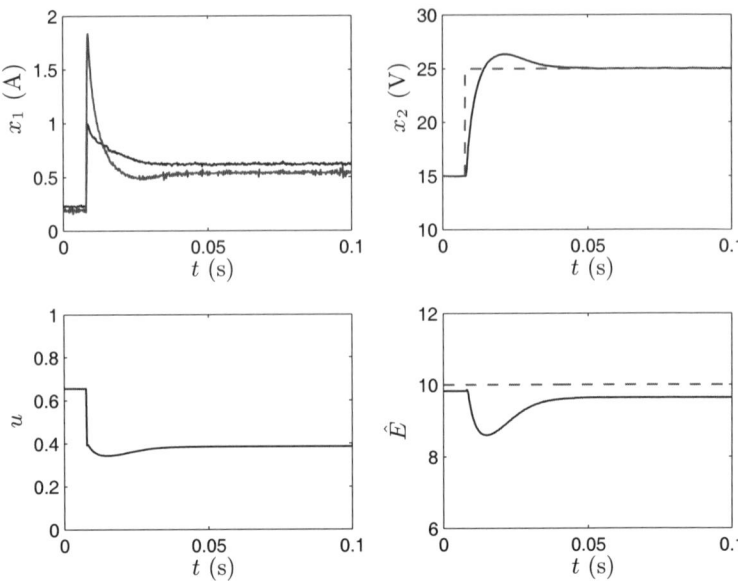

Fig. 8.12. State and control histories for a change in the output voltage reference from $y^* = 15$ V to $y^* = 25$ V.

(static) error in the output voltage is zero. This very useful property is due to the inherent integral action in the controller (8.36)[14].

Figure 8.12 shows the response of the system to a voltage reference change from $y^* = 15$ V to $y^* = 25$ V at $t = 0.01$ s, when $E = 10$ V. Notice that the output voltage tracks the desired reference value with a small overshoot.

8.3.5 A Remark on Robustness

In contrast with the full-information control law (8.35) with $\lambda = 0$, the controller (8.36) requires knowledge of the parameter G. This is a potential drawback, since in most applications the load is not known precisely. Surprisingly, it can be shown that it is not necessary to know the precise value of G, but a constant estimate (within a certain range) can be used in the implementation of the control law (8.36), *i.e.*, output voltage tracking is (locally) achieved despite the uncertainty in the parameter G.

To prove this fact, consider the system (8.30) in closed loop with the adaptive output feedback control law described by the equations

---

[14] Note that from (8.36), ignoring the saturation, we obtain $\dot{\hat{\eta}}_1 - \dfrac{\gamma_1}{\gamma_2}\dot{\hat{\eta}}_2 = \dfrac{\gamma_1(\hat{\eta}_1 + \gamma_1 y)}{\gamma_2 L y^*}(y - y^*)$.

8.3 Output Feedback Control of the DC–DC Boost Converter

$$\dot{\hat{\eta}}_1 = -\frac{\gamma_1}{C}\left(u\left(\hat{\eta}_2 + \gamma_2 y\right) - \hat{G}y\right),$$

$$\dot{\hat{\eta}}_2 = -\frac{\gamma_2}{C}\left(u\left(\hat{\eta}_2 + \gamma_2 y\right) - \hat{G}y\right) + \frac{1}{L}\left(\hat{\eta}_1 + \gamma_1 y - uy\right), \quad (8.39)$$

$$u = \mathrm{sat}_{[\epsilon,1]}\left(\frac{\hat{\eta}_1 + \gamma_1 y}{y^*}\right),$$

where $\hat{G}$ is a positive constant (estimate of G). The z-system is now written as

$$\dot{z} = \begin{bmatrix} 0 & -\dfrac{\gamma_1}{C}u \\ \dfrac{1}{L} & -\dfrac{\gamma_2}{C}u \end{bmatrix} z + \begin{bmatrix} \dfrac{\gamma_1}{C}\tilde{G} \\ \dfrac{\gamma_2}{C}\tilde{G} \end{bmatrix} y, \quad (8.40)$$

where $\tilde{G} = \hat{G} - G$. The Jacobian matrix of the closed-loop system (8.30), (8.40) at the equilibrium point

$$(x_1, x_2, z_1, z_2) = \left(Gy^{*2}/E, y^*, 0, \tilde{G}y^{*2}/E\right) \quad (8.41)$$

is given by

$$\tilde{A} = \begin{bmatrix} 0 & -\dfrac{E}{Ly^*} & -\dfrac{1}{L} & 0 \\ \dfrac{E}{Cy^*} & -\dfrac{G}{C} & \dfrac{Gy^*}{CE} & 0 \\ 0 & \dfrac{\gamma_1 \tilde{G}}{C} & -\dfrac{\gamma_1 \tilde{G}y^*}{CE} & -\dfrac{\gamma_1 E}{Cy^*} \\ 0 & \dfrac{\gamma_2 \tilde{G}}{C} & \dfrac{1}{L} - \dfrac{\gamma_2 \tilde{G}y^*}{CE} & -\dfrac{\gamma_2 E}{Cy^*} \end{bmatrix}. \quad (8.42)$$

Note that, for $\tilde{G} = 0$, the matrix (8.42) reduces to a Hurwitz matrix, denoted by A, and $\tilde{A} = A + B\tilde{G}K$, where $B = [0, 0, \gamma_1/C, \gamma_2/C]^\top$ and $K = [0, 1, -y^*/E, 0]$, i.e., the linearisation of the system (8.30), (8.40) at the equilibrium point (8.41) can be written as the feedback interconnection of an asymptotically stable LTI system, defined by (A, B, K), and a constant gain $\tilde{G}$, hence stability is preserved provided $\tilde{G}$ is sufficiently small.

To verify the above result by means of an example, assume that $\hat{G} = 1/120$ Ω^{-1} and let $L = 1.36$ mH, $C = 94$ μF, $E = 10$ V, $y^* = 15$ V, $\gamma_1 = 0.2$ and $\gamma_2 = 1$. Then the eigenvalues of the matrix $\tilde{A}$ in (8.42) remain strictly in the left half-plane for any G that satisfies $1.152 \times 10^{-4} \leq G \leq 2$. Hence, for any load G in this range, the closed-loop system has a locally asymptotically stable equilibrium at the desired set-point.

Figure 8.13 shows the response of the system to a step change in the load from 120 Ω to 60 Ω at $t = 0.01$ s, for $E = 10$ V and $y^* = 15$ V. Observe that the estimate of E is not affected significantly; as a result, the output voltage is almost insensitive to load changes.

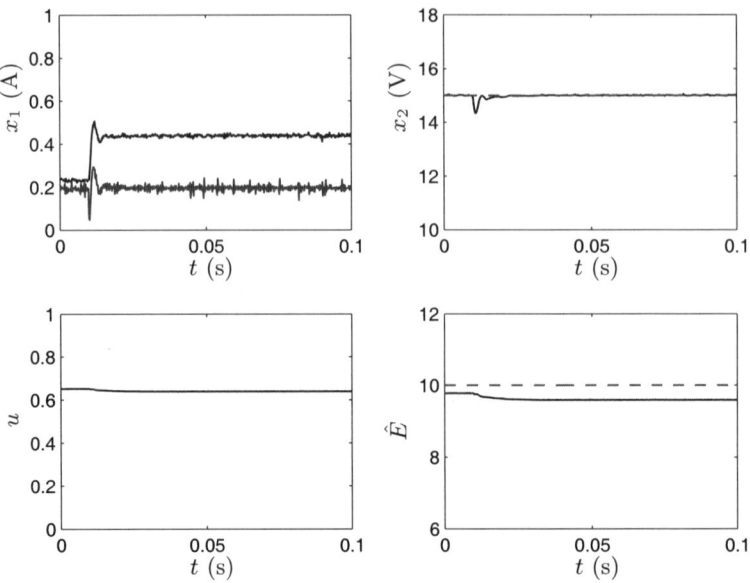

Fig. 8.13. State and control histories for a change in the load from 120 Ω to 60 Ω.

8.4 Adaptive Control of the Power Factor Precompensator

Power factor precompensators (PFP) are an important class of switched AC–DC converters. As their name suggests, their main function is to achieve a nearly unit power factor by drawing a sinusoidal current that is in phase with the source voltage, thus eliminating the reactive power and the harmonic interference with other equipment operating off the same source.

More precisely, the control objective is twofold. First, the input current should track a sinusoidal reference signal that is in phase with the input voltage. Second, the output voltage should be driven to a desired constant level. An additional requirement is robustness against variation of the system parameters and in particular of the load, which is usually unknown.

Since the amplitude of the input current determines explicitly the DC output voltage, one may satisfy both objectives in a single *current control* loop, which typically comprises a hysteresis or sliding-mode controller[15].

The main drawback of these controllers is that they require very high switching frequency (typically few hundred kHz) leading to high converter losses. For this reason, this section focuses on pulse-width modulation (PWM)

---

[15] See [145, Chapter 18] for a general description and [147] for a design example.

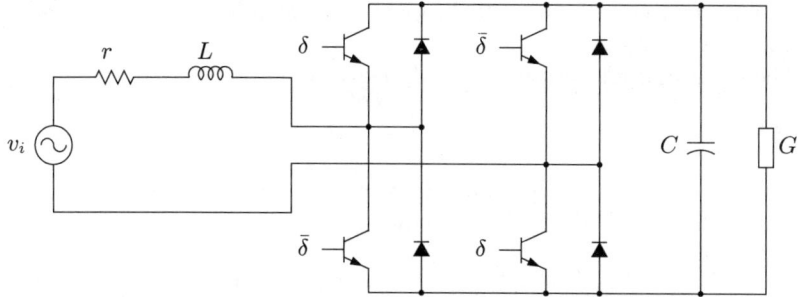

Fig. 8.14. Power factor precompensator (PFP) circuit.

control techniques, which are based on *averaged models*[16] and can be implemented using lower switching frequency (*e.g.*, 10 kHz). Furthermore, the issue of robustness against variations of the load is treated by appropriate adaptive schemes. This allows to avoid the use of an output current sensor, thus making the practical implementation more attractive.

In this section the I&I methodology is used to design an adaptive state feedback regulator for a power factor compensation circuit driving an unknown resistive load. The design follows the results in Chapter 3 and is based on the reconstruction of a full-information control law which has been obtained using feedback linearisation. However, the proposed adaptive scheme can be used in conjunction with any stabilising full-information control law[17].

8.4.1 Problem Formulation

Consider the *full-bridge boost* PFP circuit shown in Figure 8.14, which consists of two pairs of transistor-diode switches working in a complementary way. The switching signal δ is generated by a PWM circuit and takes values in the finite set $\{-1, 1\}$.

The averaged model of the PFP can be obtained using Kirchhoff's laws and is given by the equations

$$L\dot{x}_1 = -rx_1 - ux_2 + v_i(t), \\ C\dot{x}_2 = -Gx_2 + ux_1, \tag{8.43}$$

where $v_i(t) = E\sin(\omega t)$ is the source voltage, x_1 is the input (inductor) current, $x_2 > 0$ is the output (capacitor) voltage, L, C, r and G are positive constants, representing the inductance, capacitance, parasitic resistance and load conductance respectively, and $u \in [-1, 1]$ is the modulating signal of

---

[16] See [188, 107].
[17] An experimental comparison of several such schemes can be found in [105].

the PWM and acts as the control input[18]. For simplicity, in the sequel it is assumed that u is unconstrained, i.e., $u \in \mathbb{R}$.

The desired input current in steady-state is

$$x_1^*(t) = I_d \sin(\omega t), \qquad (8.44)$$

for some $I_d > 0$ yet to be specified. Substituting equation (8.44) into the first equation in (8.43) yields the steady-state value for ux_2, namely

$$u^*(t)x_2^*(t) = v_i(t) - rx_1^*(t) - L\dot{x}_1^*(t). \qquad (8.45)$$

Substituting equations (8.44) and (8.45) into the second equation in (8.43) written for $x_2 = x_2^*$ yields the (linear in x_2^{*2}) dynamics

$$C\dot{x}_2^* x_2^* = -Gx_2^{*2} + \frac{(E - rI_d)I_d}{2} - \frac{(E - rI_d)I_d}{2}\cos(2\omega t) - \frac{LI_d^2 \omega}{2}\sin(2\omega t).$$

The steady-state solution of the above equation can be directly computed and is given by

$$x_2^{*2}(t) = \frac{(E - rI_d)I_d}{2G} + A\sin(2\omega t + \varphi), \qquad (8.46)$$

where

$$A = \frac{I_d}{2}\sqrt{\frac{(LI_d\omega)^2 + (E - rI_d)^2}{G^2 + (C\omega)^2}}, \qquad \varphi = \arctan\left(\frac{G(E - rI_d) - C\omega LI_d\omega}{GLI_d\omega + C\omega(E - rI_d)}\right).$$

Hence, in steady-state, $x_2^{*2}(t)$ consists only of a DC term and a second-order harmonic. Note that in practice the second-order harmonic in equation (8.46) can be neglected, since its amplitude is much smaller than the DC term, and so the average of $x_2^*(t)$ can be approximated by the RMS value, which is given by[19]

$$V_d = \sqrt{\frac{(E - rI_d)I_d}{2G}}. \qquad (8.47)$$

From (8.47), solving for I_d yields the solutions

$$I_d = \frac{E}{2r} \pm \sqrt{\frac{E^2}{4r^2} - \frac{2GV_d^2}{r}},$$

which are real if and only if $E/V_d \geq \sqrt{8rG}$. Selecting the smallest solution which corresponds to minimum power (i.e., minimum converter losses), the

---

[18] Note that the exact model of the PFP is described by the same equations with u replaced by δ.

[19] For instance, for the set of parameters used in the experiments (see Section 8.4.4) and for $V_d = 200$, we have that $A = 1330 \ll V_d^2 = 40000$ and the average of $x_2^*(t)$ is equal to $199.986 \approx V_d = 200$.

8.4 Adaptive Control of the Power Factor Precompensator

amplitude of the input current that drives the output voltage to the desired level V_d is given by[20]

$$I_d = \frac{E}{2r} - \sqrt{\frac{E^2}{4r^2} - \frac{2GV_d^2}{r}}. \tag{8.48}$$

It follows that, by controlling the input current so that it tracks the signal (8.44), where I_d is given by (8.48), the two control objectives, namely unit power factor and output voltage regulation, can be achieved simultaneously.

The main drawback of this approach, as the identities (8.47) and (8.48) reveal, is the sensitivity of the output voltage to the parameters r and G. The dependence on G poses a significant problem, since in many applications the load is unknown or time-varying. This obstacle can be overcome by adding an adaptation scheme for these parameters.

8.4.2 Full-information Controllers

In this section we present two full-information control laws that achieve the desired tracking objective. These are then augmented with an adaptation scheme to deal with the uncertainty in the parameters r and G. We first establish the following result, which allows to decouple the design of the controller from the design of the estimator.

Lemma 8.1. *Consider the system (8.43) with $r > 0$ and assume that $|v_i(t)| \leq E$, for some constant $E > 0$ and for all t. Then, for any $u \in \mathbb{R}$, all trajectories of (8.43) are bounded.*

Proof. Consider the positive-definite function

$$V(x_1, x_2) = \frac{1}{2}Lx_1^2 + \frac{1}{2}Cx_2^2, \tag{8.49}$$

which corresponds to the electrical energy stored in the system. The time-derivative of (8.49) along the trajectories of (8.43) is

$$\dot{V} = -rx_1^2 + v_i x_1 - Gx_2^2 \leq -\frac{r}{2}x_1^2 - Gx_2^2 + \frac{E^2}{2r},$$

where we have used Young's inequality and the bound on $v_i(t)$. Note that, since $r > 0$, the term $\frac{E^2}{2r}$ is bounded. Hence, all trajectories are bounded and converge to the set $\Omega = \{(x_1, x_2) \in \mathbb{R} \times \mathbb{R} \mid \frac{r}{2}x_1^2 + Gx_2^2 < \frac{E^2}{2r}\}$. $\square$

A full-information *feedback linearising* control law can be obtained as follows.

---

[20] Note that the same solution has been obtained by a similar analysis in [53].

Proposition 8.8. *Consider the system (8.43) in closed loop with the control law*

$$u = \frac{1}{x_2}\left[v_i - rx_1^* - L\dot{x}_1^* + K(x_1 - x_1^*)\right],\qquad(8.50)$$

with $K > 0$, where x_1^* is given by (8.44) and (8.48). Then all trajectories satisfying $x_2(t) > 0$, for all t, are bounded and $\lim_{t\to\infty}(x_1(t) - x_1^*(t)) = 0$.

Proof. Substituting (8.50) into the first equation in (8.43) and defining the tracking error $\tilde{x}_1 = x_1 - x_1^*$ yields the linearised system

$$L\dot{\tilde{x}}_1 = -(r+K)\tilde{x}_1,$$

which is exponentially stable for any $K > -r$. Hence, the time response of x_1 is given by

$$x_1(t) = x_1(0)\exp(-\frac{r+K}{L}t) + I_d\sin(\omega t).$$

The proof is completed by noting that, from Lemma 8.1 and the fact that u is well defined by assumption, all trajectories remain bounded. □

Note that, from the analysis in Section 8.4.1 it follows that, under the foregoing control law, x_2 converges (on average) to the desired voltage V_d, which is given by (8.47).

If instead of x_1 we take as output the signal $\zeta = ux_2$, which corresponds to the input voltage of the transistor–diode bridge, then the system can be linearised by means of a *dynamic* controller, as the following statement shows.

Proposition 8.9. *Consider the system (8.43) in closed loop with the dynamic control law*

$$\dot{u} = \frac{1}{x_2}\left(-\frac{u^2 x_1}{C} + w\right),\qquad(8.51)$$

where w is such that[21]

$$w(s) = k\frac{s^2 + as + b}{s^2 + \omega^2}e(s),\qquad(8.52)$$

with $k > 0$, $a > 0$ and $b > 0$ design parameters, and

$$e = v_i - rx_1^* - L\dot{x}_1^* + K(x_1 - x_1^*) - ux_2,$$

with $K \geq 0$, where x_1^* is given by (8.44) and (8.48). Then there exists k (sufficiently large) such that all trajectories satisfying $x_2(t) > 0$, for all t, are bounded and $\lim_{t\to\infty}(x_1(t) - x_1^*(t)) = 0$.

---

[21] With a slight abuse of notation, $w(s)$ denotes the Laplace transform of $w(t)$.

8.4 Adaptive Control of the Power Factor Precompensator

Proof. Consider the variable $\zeta = ux_2$, whose dynamics are given by

$$\dot{\zeta} = \dot{u}x_2 - \frac{G}{C}\zeta + \frac{u^2 x_1}{C},$$

and note that applying the control law (8.51) yields the system

$$\begin{bmatrix} \dot{x}_1 \\ \dot{\zeta} \end{bmatrix} = \begin{bmatrix} -\frac{r}{L} & -\frac{1}{L} \\ 0 & -\frac{G}{C} \end{bmatrix} \begin{bmatrix} x_1 \\ \zeta \end{bmatrix} + \begin{bmatrix} \frac{v_i}{L} \\ w \end{bmatrix}.$$

It is clear from the analysis in Section 8.4.1 that the control objective can be achieved by forcing ζ to track the signal

$$\zeta^* = v_i - rx_1^* - L\dot{x}_1^* + K(x_1 - x_1^*),$$

where the last term has been added to speed-up convergence. To this end, consider the output

$$y = \begin{bmatrix} -K & 1 \end{bmatrix} \begin{bmatrix} x_1 \\ z \end{bmatrix}$$

and the reference signal

$$y^* = v_i - rx_1^* - L\dot{x}_1^* - Kx_1^*,$$

and note that the transfer function from w to y is given by

$$H(s) = \frac{s + (r+K)/L}{(s+r/L)(s+G/C)}.$$

It can be straightforwardly verified (*e.g.*, using the root locus) that, for sufficiently large $k > 0$, the closed-loop system is asymptotically stable. Moreover, by the presence of the poles at $s = \pm j\omega$, we conclude that y converges to y^* and hence x_1 converges to x_1^*. □

The choice of the controller (8.52) is motivated by the internal model principle and the fact that the reference signal y^* is a sinusoid of frequency ω. Furthermore, the zeros have been inserted to ensure stability.

8.4.3 I&I Adaptation

So far we have assumed that the parameters r and G were known and so we could control the output voltage indirectly via formula (8.48). To overcome this robustness problem, in this section we design an asymptotic estimator for both r and G using the I&I approach.

Proposition 8.10. *Consider the system (8.43) and assume that $x_1(t) \notin \mathcal{L}_2$[22]. Then the estimator*

$$\dot{\hat{\theta}}_1 = -\frac{2}{L}\gamma_1 x_1(ux_2 + \hat{r}x_1 - v_i), \qquad \hat{r} = \hat{\theta}_1 - \gamma_1 x_1^2,$$

$$\dot{\hat{\theta}}_2 = -\frac{1}{C}\gamma_2(-ux_1 + \hat{G}x_2), \qquad \hat{G} = \hat{\theta}_2 - \gamma_2 x_2, \qquad (8.53)$$

with $\gamma_1 > 0$, $\gamma_2 > 0$, is such that

$$\lim_{t\to\infty}(\hat{r} - r) = 0, \qquad \lim_{t\to\infty}(\hat{G} - G) = 0.$$

Proof. Define the error variables $z_1 = \hat{\theta}_1 - r + \beta_1(x_1)$, $z_2 = \hat{\theta}_2 - G + \beta_2(x_2)$, where $\beta_1(\cdot)$ and $\beta_2(\cdot)$ are continuous functions to be defined. Selecting the update laws

$$\dot{\hat{\theta}}_1 = \frac{\partial \beta_1}{\partial x_1}\frac{1}{L}\left[ux_2 + (\hat{\theta}_1 + \beta_1(x_1))x_1 - v_i\right],$$

$$\dot{\hat{\theta}}_2 = \frac{\partial \beta_2}{\partial x_2}\frac{1}{C}\left[-ux_1 + (\hat{\theta}_2 + \beta_2(x_2))x_2\right],$$

yields the error dynamics

$$\dot{z}_1 = \frac{\partial \beta_1}{\partial x_1}\frac{1}{L}x_1 z_1,$$

$$\dot{z}_2 = \frac{\partial \beta_2}{\partial x_2}\frac{1}{C}x_2 z_2.$$

Hence, choosing $\beta_1(x_1) = -\gamma_1 x_1^2$ and $\beta_2(x_2) = -\gamma_2 x_2$, with $\gamma_1 > 0$, $\gamma_2 > 0$ constants, and recalling that $x_2(t) > 0$, for all t, and $x_1(t) \notin \mathcal{L}_2$, yields the dynamics

$$\dot{z}_1 = -\frac{2\gamma_1 x_1^2}{L}z_1,$$

$$\dot{z}_2 = -\frac{\gamma_2 x_2}{C}z_2,$$

which have an asymptotically stable equilibrium at zero. Hence, $\hat{r} = \hat{\theta}_1 + \beta_1(x_1)$ converges to r and $\hat{G} = \hat{\theta}_2 + \beta_2(x_2)$ converges to G. □

In view of Lemma 8.1, an adaptive state feedback control law can be constructed by replacing the parameters r and G in the full-information control laws (8.50) or (8.51) with their respective estimates $\hat{r}$ and $\hat{G}$ obtained from (8.53). Note that the current reference is given by

$$x_1^* = \frac{E}{2\hat{r}} - \sqrt{\frac{E^2}{4\hat{r}^2} - \frac{2\hat{G}V_d^2}{\hat{r}}}.$$

---

[22] Recall that x_1 is the AC input current, hence this assumption is always satisfied in practice.

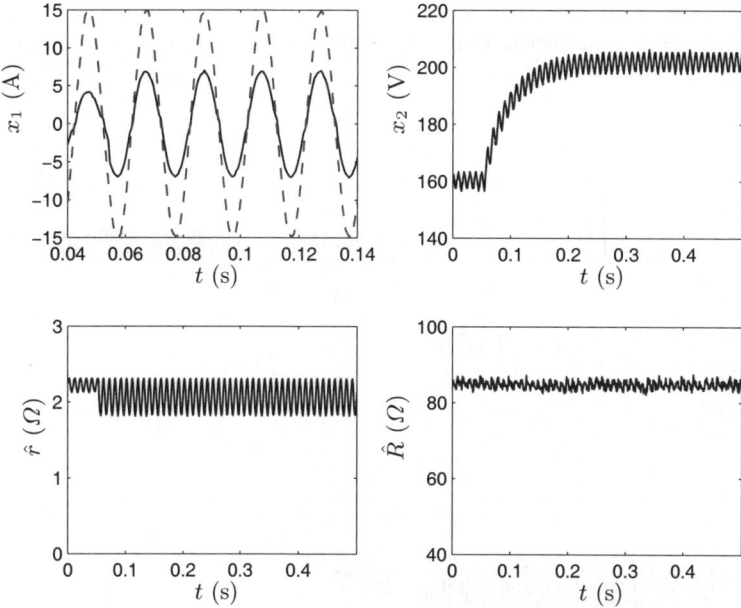

Fig. 8.15. Response of the PFP to a voltage reference change from $V_d = 160$ V to $V_d = 200$ V. Top left: Input current (solid line) and scaled input voltage (dashed line). Top right: Output voltage. Bottom left: Estimated resistance $\hat{r}$. Bottom right: Estimated load $\hat{R} = 1/\hat{G}$.

8.4.4 Experimental Results

The adaptive state feedback control laws described above have been implemented on an experimental system[23] with the parameters $E = 150$ V, $\omega = 100\pi$ rad/s, $r = 2.2$ Ω, $L = 2.13$ mH, $C = 1100$ μF and $G = 1/87$ Ω^{-1}. The design parameters have been selected as $K = 15$, $\gamma_1 = 0.01$ and $\gamma_2 = 0.0002$. The switching frequency of the PWM is 13 kHz. The measurements are filtered using low-pass filters with cut-off frequency of 7 kHz. The experiments aim to exhibit the behaviour of the proposed controllers[24] with respect to step changes in the desired output voltage V_d and in the load G. In particular, the set-points for the output voltage are $V_d = 160$ V and $V_d = 200$ V, while the load resistance can be switched from 87 Ω to 51 Ω.

---

[23] The experiments have been carried out at the Laboratoire de Génie Electrique de Paris, on a 2 kW power converter.

[24] We present experimental results only for the adaptive controller obtained using Proposition 8.9. The results for the feedback linearising controller (8.50) are very similar and are not displayed (see [105]).

Figure 8.15 shows the response of the system to a voltage reference change from $V_d = 160$ V to $V_d = 200$ V at $t = 0.05$ s. Observe that both x_1 and x_2 track the desired references, while the estimates of r and G remain close to the correct values.

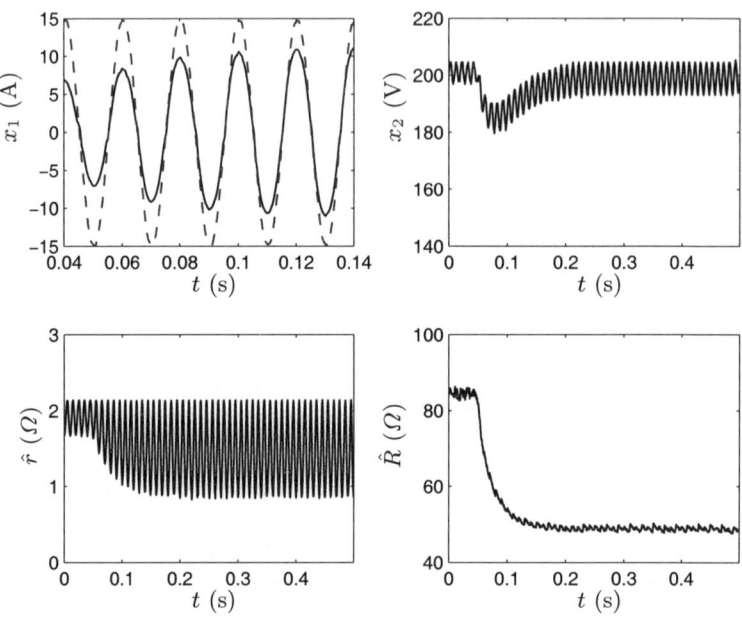

Fig. 8.16. Response of the PFP to a load change from 87 Ω to 51 Ω. Top left: Input current (solid line) and scaled input voltage (dashed line). Top right: Output voltage. Bottom left: Estimated resistance $\hat{r}$. Bottom right: Estimated load $\hat{R} = 1/\hat{G}$.

The action of the adaptation mechanism can be observed in Figure 8.16, which shows the response of the system to a load change from 87 Ω to 51 Ω at $t = 0.05$ s. Notice that the output voltage remains at the desired level after a relatively small undershoot. For comparison, Figure 8.17 shows the response of the system when no adaptation is used (*i.e.*, $\gamma_1 = \gamma_2 = 0$). Note that in this case the steady-state error in the output voltage is unacceptably large.

Finally, Figure 8.18 shows the harmonic content of the input current for load 51 Ω compared to the European electromagnetic compatibility (EMC) standard EN61000-3-2, which specifies the harmonic current limits for equipment drawing up to 16 A per phase. Observe that the proposed controller achieves with a large margin the requirements imposed by this standard.

8.4 Adaptive Control of the Power Factor Precompensator

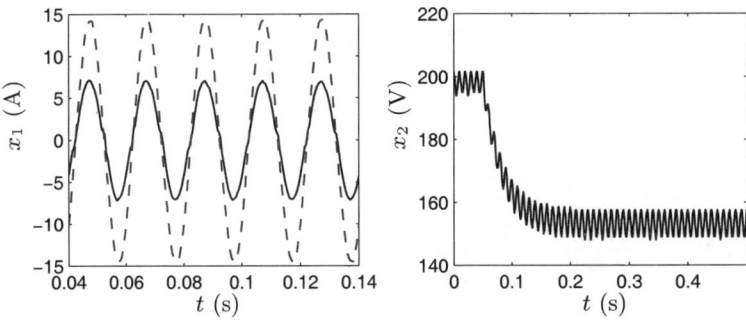

Fig. 8.17. Response to a load change from 87 Ω to 51 Ω when no adaptation is used. Left: Input current (solid line) and scaled input voltage (dashed line). Right: Output voltage.

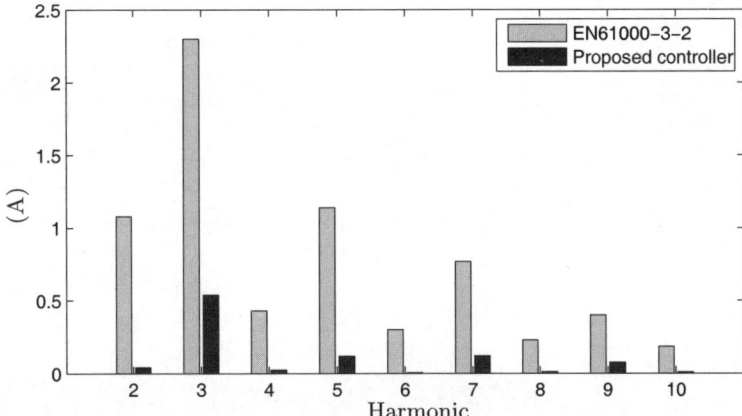

Fig. 8.18. Amplitude of the input current harmonics for load 51 Ω compared to the standard EN61000-3-2 (the amplitude of the first harmonic is 12 A).

9

Mechanical Systems

In this chapter we revisit some of the examples of mechanical systems introduced in Chapters 2 and 5, and design controllers using the I&I approach. First, we extend the result presented in Proposition 2.4 to a more general class of flexible joints robots. In particular, we show that I&I provides a solution to the problem of global tracking for flexible joints robots with *full* inertia matrix.

Then the observer design of Section 5.4 is used to construct an output feedback controller for a two-link robot manipulator and the resulting scheme is tested via simulations.

Finally, we present a solution to the important problem of controlling the attitude of a (fully-actuated) rigid body in the presence of unknown external forces, and we use it to design an autopilot for an autonomous aircraft.

9.1 Control of Flexible Joints Robots

Consider an open-chained robot arm with n elastic rotational joints, and let $q \in \mathbb{R}^n$ denote the link positions and $\theta \in \mathbb{R}^n$ the actuator positions, as reflected through the gear reduction ratios. The difference $q_i - \theta_i$ represents the ith joint deformation. The rotors of the actuators are modelled as uniform bodies having their centre of mass on the axis of rotation, implying that both the inertia matrix and the gravity term in the dynamic model do not depend upon the actuator positions[1].

Under the above assumption the robot dynamics are described by the equations

$$\mathcal{D}(q)\ddot{y} + \mathcal{C}(q,\dot{y})\dot{y} + \mathcal{G}(q) + \mathcal{K}(q-\theta) = \begin{bmatrix} 0 \\ \tau \end{bmatrix}, \qquad (9.1)$$

with $y = \mathrm{col}(q, \theta)$,

---

[1] See [47, 219].

$$\mathcal{D}(q) = \begin{bmatrix} B_1(q) & B_2(q) \\ B_2(q)^\top & J \end{bmatrix}, \quad \mathcal{C}(q,\dot y) = \begin{bmatrix} C_A(q,\dot\theta) + C_{B1}(q,\dot q) & C_{B2}(q,\dot q) \\ C_{B3}(q,\dot q) & 0 \end{bmatrix},$$

$$\mathcal{G}(q) = \begin{bmatrix} G(q) \\ 0 \end{bmatrix}, \quad \mathcal{K} = \begin{bmatrix} K & -K \\ -K & K \end{bmatrix}, \tag{9.2}$$

where $B_1(q)$ denotes the inertia matrix of the rigid arm,

$$J = \operatorname{diag}(J_{m1}r_1^2, \ldots, J_{mn}r_n^2)$$

is the matrix of effective actuator inertias (J_{mi} is the axial inertia and r_i is the gear ratio of the ith actuator), $B_2(q)$ denotes the inertia couplings between actuators and links, $\mathcal{C}(q,\dot y)$ is the matrix related to the Coriolis and centrifugal terms C_A and C_{Bi}, $G(q)$ is the gravity vector of the rigid arm, $K = \operatorname{diag}(k_1, \ldots, k_n)$ is the constant joint stiffness matrix, and τ is the n-dimensional vector of torques. Furthermore, we assume that the matrix $B_2(q)$ has the upper triangular structure

$$B_2(q) = \begin{bmatrix} 0 & b_{12}(q_1) & b_{13}(q_1,q_2) & \cdots & b_{1n}(q_1,\ldots,q_{n-1}) \\ 0 & 0 & b_{23}(q_2) & \cdots & b_{2n}(q_2,\ldots,q_{n-1}) \\ \vdots & \vdots & \vdots & \ddots & \vdots \\ 0 & 0 & 0 & \cdots & b_{n-1,n}(q_{n-1}) \\ 0 & 0 & 0 & \cdots & 0 \end{bmatrix}. \tag{9.3}$$

The vector $\mathcal{C}(q,\dot y)\dot y$ can be given the form[2]

$$\mathcal{C}(q,\dot y)\dot y = \begin{bmatrix} C_1(q,\dot q) & C_2(q,\dot q) \\ C_3(q,\dot q) & 0 \end{bmatrix} \begin{bmatrix} \dot q \\ \dot\theta \end{bmatrix},$$

where

$$C_2(q,\dot q) = \begin{bmatrix} 0 & 0 & c_{13}(q,\dot q) & \cdots & c_{1n}(q,\dot q) \\ 0 & 0 & c_{23}(q,\dot q) & \cdots & c_{2n}(q,\dot q) \\ \vdots & \vdots & \vdots & \ddots & \vdots \\ 0 & 0 & 0 & \cdots & c_{n-1,n}(q,\dot q) \\ 0 & 0 & 0 & \cdots & 0 \end{bmatrix}. \tag{9.4}$$

In summary, the flexible joints robot dynamics (9.1) can be described by equations of the form

$$\begin{aligned} B_1(q)\ddot q + B_2(q)\ddot\theta + C_1(q,\dot q)\dot q + C_2(q,\dot q)\dot\theta + G(q) + K(q-\theta) &= 0, \\ B_2(q)^\top \ddot q + J\ddot\theta + C_3(q,\dot q)\dot q + K(\theta - q) &= \tau. \end{aligned} \tag{9.5}$$

---

[2] See [185].

Note that, for robots with two links, the Coriolis and centrifugal terms do not depend on $\dot{\theta}$. This fact, which has general validity, is not exposed in the model (9.1), but it is apparent in the model (9.5). Moreover, if $B_2(q)$ is a constant matrix then $C_2(q,\dot{q}) = 0$. Finally, if $B_2(q) = 0$ then (9.5) reduces to the simplified model (1.9) considered in Examples 1.2 and 2.5.

9.1.1 Control Design

To design a global tracking controller for system (9.5) we exploit the ideas of Example 2.5. To this end, we first redefine the input signal, performing a partial feedback linearisation, as

$$\tau = Jw + B_2(q)^\top \ddot{q} + C_3(q,\dot{q})\dot{q} + K(\theta - q), \qquad (9.6)$$

with $w \in \mathbb{R}^n$ a new control input. As a result, the system (9.5), (9.6) can be written as

$$\begin{aligned} B_1(q)\ddot{q} + C_1(q,\dot{q})\dot{q} + C_2(q,\dot{q})\dot{\theta} + G(q) + K(q - \theta) &= -B_2(q)w, \\ \ddot{\theta} &= w. \end{aligned} \qquad (9.7)$$

Note that the feedback (9.6), which *linearises* the motor subsystem, is implementable by state feedback.

Similarly to what is discussed in Example 2.5, the control signal w has to be designed to immerse the system (9.7) into a controlled rigid robot model. To this end, consider the variable

$$z = K(q - \theta) + C_2(q,\dot{q})\dot{\theta} + u(q,\dot{q},t) + B_2(q)w, \qquad (9.8)$$

where $u(q,\dot{q},t)$ has been selected in such a way that the system

$$B_1(\xi)\ddot{\xi} + C_1(\xi,\dot{\xi})\dot{\xi} + G(\xi) = u(\xi,\dot{\xi},t) + \delta \qquad (9.9)$$

satisfies the following assumption.

Assumption 9.1. For any bounded reference signal $\xi^*(t)$, with bounded derivatives, and for any initial condition $(\xi(0),\dot{\xi}(0))$, the trajectories of the system (9.9) with $\delta = 0$ are bounded for all t and are such that

$$\lim_{t \to \infty} (\xi(t) - \xi^*(t)) = 0.$$

Moreover, the system (9.9) is ISS with respect to the input δ.

Replacing equation (9.8) in the model (9.7) yields

$$\begin{aligned} B_1(q)\ddot{q} + C_1(q,\dot{q})\dot{q} + G(q) + z &= u(q,\dot{q},t), \\ \ddot{\theta} &= w. \end{aligned} \qquad (9.10)$$

As a result, the link position tracking problem is solvable if it is possible to find a control law w such that

$$\lim_{t \to \infty} z(t) = 0 \tag{9.11}$$

and such that $\theta(t)$ and $\dot{\theta}(t)$ remain bounded.

In general, the output map (9.8) depends explicitly on the control signal w. Hence, the design of a control law such that (9.11) holds requires special attention. However, as in Example 2.5, if $B_2(q) \equiv 0$ then the signal w does not appear in the variable (9.8).

Note that if the mapping $u = u(q, \dot{q}, t)$ in equation (9.9) is selected as

$$u = B_1(q)\ddot{q}_r + C_1(q, \dot{q})\dot{q}_r + G(q) - K_d s - K_p \tilde{q}, \tag{9.12}$$

with $q^* = q^*(t)$ a desired link trajectory, $\tilde{q} = q - q^*$, $\dot{q}_r = \dot{q}^* - \Lambda \tilde{q}$, $s = \dot{\tilde{q}} + \Lambda \tilde{q}$, and K_p, K_d and Λ properly selected gain matrices, then system (9.9) is such that Assumption 9.1 holds[3].

We are now ready to state and prove the main results of this section.

Lemma 9.1. *Consider the flexible joints robot dynamics described by the equations (9.10) and the variable (9.8). Suppose the rank of the matrix $B_2(q)$ is equal to $n-1$ for all q and the mapping $u(\cdot)$ is such that Assumption 9.1 holds. Then there exists a state feedback $w = w(q, \dot{q}, \theta, \dot{\theta}, t)$ such that condition (9.11) holds and all internal signals are bounded.*

Proof. Note that, by assumption, the states q and $\dot{q}$ of system (9.10) are bounded if z is bounded. Therefore, to prove the claim, it is enough to show that there exists a static state feedback control law w such that condition (9.11) holds and θ and $\dot{\theta}$ are bounded for all t. Such a control law can be explicitly constructed as follows.

By assumption, and the structure of the matrix $B_2(q)$, one has

$$B_2(q) = \begin{bmatrix} 0 & B_2'(q_a) \\ 0 & 0 \end{bmatrix}, \tag{9.13}$$

with $B_2'(q_a) \in \mathbb{R}^{(n-1) \times (n-1)}$ a full rank upper triangular matrix and $q_a = [q_1, \ldots, q_{n-1}]^\top$. This partition induces the partition

$$z = \begin{bmatrix} z_a \\ z_n \end{bmatrix} = \begin{bmatrix} K_a(q_a - \theta_a) + C_2'(q, \dot{q})\dot{\theta}_b + u_a(q, \dot{q}, t) + B_2'(q_a)w_b \\ k_n(q_n - \theta_n) + u_n(q, \dot{q}, t) \end{bmatrix}, \tag{9.14}$$

where $w_b = [w_2, \cdots, w_n]^\top$, $u_a = [u_1, \cdots, u_{n-1}]^\top$, $K_a = \text{diag}(k_1, \ldots, k_{n-1})$, $\theta_a = [\theta_1, \cdots, \theta_{n-1}]^\top$ $\theta_b = [\theta_2, \cdots, \theta_n]^\top$, and $C_2'(q, \dot{q})$ is a sub-matrix of (9.4) such that

---
[3] See [163].

9.1 Control of Flexible Joints Robots

$$C_2(q,\dot{q}) = \begin{bmatrix} 0 & C_2'(q,\dot{q}) \\ 0 & 0 \end{bmatrix}.$$

By assumption there exists w_b solving the equation $z_a = 0$, namely

$$w_b = -\left(B_2'(q_a)\right)^{-1}\left(K_a(q_a - \theta_a) + C_2'(q,\dot{q})\dot{\theta}_b + u_a(q,\dot{q},t)\right), \quad (9.15)$$

i.e., it is possible to select w_b such that z_a is identically equal to zero for all t. Consider now the system (9.10), with w_b as in equation (9.15), namely

$$\begin{aligned} B_1(q)\ddot{q} + C_1(q,\dot{q})\dot{q} + G(q) + \begin{bmatrix} 0 & \cdots & 0 & z_n \end{bmatrix}^T &= u(q,\dot{q},t), \\ \ddot{\theta}_1 &= w_1, \\ \ddot{\theta}_b &= -\left(B_2'(q_a)\right)^{-1}\left(K_a(q_a - \theta_a) + C_2'(q,\dot{q})\dot{\theta}_b + u_a(q,\dot{q},t)\right), \end{aligned} \quad (9.16)$$

and note that the system (9.16) with output z_n defined in (9.14) has a (globally) defined relative degree. To see this, rewrite z_n as

$$z_n = -k_n \theta_n + \phi(q,\dot{q},t),$$

with $\phi(\cdot)$ function of its arguments, and $\ddot{\theta}_b$ as

$$\begin{bmatrix} \ddot{\theta}_2 \\ \vdots \\ \ddot{\theta}_{n-1} \\ \ddot{\theta}_n \end{bmatrix} = \begin{bmatrix} b_{12}^I & b_{13}^I & \cdots & b_{1n}^I \\ 0 & b_{22}^I & \cdots & b_{2n}^I \\ 0 & 0 & \ddots & \vdots \\ 0 & \cdots & 0 & b_{n-1,n}^I \end{bmatrix} \begin{bmatrix} k_1\theta_1 + \varphi_1(q,\dot{q},\dot{\theta}_3,\cdots,\dot{\theta}_n,z_n,t) \\ \vdots \\ k_{n-2}\theta_{n-2} + \varphi_{n-2}(q,\dot{q},\dot{\theta}_n,z_n,t) \\ k_{n-1}\theta_{n-1} + \varphi_{n-1}(q,\dot{q},z_n,t) \end{bmatrix},$$

where b_{ij}^I are the elements of the inverse of $B_2'(q_a)$ and $\varphi_i(\cdot)$ are functions of their arguments. Hence,

$$z_n^{(2i)} = -\prod_{j=0}^{i} k_{n-j} \prod_{j=1}^{i} b_{n-j,n-j+1}^I \theta_{n-i} + \sigma_{2i}(q,\dot{q},\dot{\Theta}_i,z_n,t),$$

for $i = 1,\ldots,n-1$, with Θ_1 empty and $\Theta_i = [\theta_{n-i+2},\ldots,\theta_n]$, and for some $\sigma_{2i}(\cdot)$, and

$$z_n^{(2n)} = -\prod_{j=0}^{n-1} k_{n-j} \prod_{j=1}^{n-1} b_{n-j,n-j+1}^I w_1 + \sigma_{2n}(q,\dot{q},\theta,\dot{\theta},z_n,t), \quad (9.17)$$

for some $\sigma_{2n}(\cdot)$. The coefficient of w_1 in equation (9.17), namely

$$M = -\prod_{j=0}^{n-1} k_{n-j} \prod_{j=1}^{n-1} b_{n-j,n-j+1}^I,$$

is nonzero and, in addition, the variables $(q, \dot q, z_n, \dot z_n \ldots, z_n^{(2n-1)})$ describe a set of globally defined co-ordinates, in which the system (9.16) is described by equations of the form

$$B_1(q)\ddot q + C_1(q,\dot q)\dot q + G(q)\begin{bmatrix} 0 & \cdots & 0 & z_n \end{bmatrix}^\top = u(q,\dot q,t)$$

$$\begin{bmatrix} \dot z_n \\ \vdots \\ \dot z_n^{(2n-1)} \end{bmatrix} = \begin{bmatrix} 0 & 1 & \cdots & 0 \\ \vdots & \vdots & \ddots & \vdots \\ 0 & 0 & \cdots & 1 \\ 0 & 0 & \cdots & 0 \end{bmatrix} \begin{bmatrix} z_n \\ \vdots \\ z_n^{(2n-1)} \end{bmatrix} + e_{2n-1}\bigl(Mw_1 + \sigma_{2n}(q,\dot q, \theta, \dot\theta, z_n, t)\bigr),$$

(9.18)

hence there is a feedback $w_1 = w_1(q,\dot q, \theta, \dot\theta, t)$ of the form

$$w_1 = M^{-1}\left(\sigma_{2n}(q,\dot q, \theta, \dot\theta, z_n, t) + \gamma(z_n, \ldots, z_n^{(2n-1)})\right), \qquad (9.19)$$

with $\gamma(\cdot)$ a properly selected function of its arguments, such that the states $z_n, \ldots, z_n^{(2n-1)}$ are bounded and converge to zero, which implies that θ and $\dot\theta$ are bounded. This completes the proof. $\square$

Lemma 9.1 leads to the following (global) tracking result for a class of flexible joints robots.

Proposition 9.1. *Consider the flexible joints robot model (9.7). Let $q^*(t)$ be a bounded link reference trajectory with bounded derivatives. Under the Assumptions of Lemma 9.1, and for any initial condition $(q(0), \dot q(0), \theta(0), \dot\theta(0))$, there exists a time-varying, static, state feedback control law*

$$w = w(q, \dot q, \theta, \dot\theta, t) \qquad (9.20)$$

such that all states of the closed-loop system (9.7), (9.20) are bounded and

$$\lim_{t \to \infty}(q(t) - q^*(t)) = 0.$$

9.1.2 A 2-DOF Flexible Joints Robot

Consider the flexible joints robot dynamics described by

$$B_1(q) = \begin{bmatrix} a_1 + 2a_3\cos(q_2) & a_2 + a_3\cos(q_2) \\ a_2 + a_3\cos(q_2) & a_2 \end{bmatrix}, \quad B_2(q) = \begin{bmatrix} 0 & b_{12} \\ 0 & 0 \end{bmatrix},$$

$$C_1(q,\dot q) = \begin{bmatrix} -2a_3\sin(q_2)\dot q_2 & -a_3\sin(q_2)\dot q_2 \\ a_3\sin(q_2)\dot q_1 & 0 \end{bmatrix},$$

for which the rank condition in Lemma 9.1 holds. For such a system the variable (9.8) is given by

$$z_1 = k_1(q_1 - \theta_1) + u_1(q, \dot{q}, t) + b_{12}w_2,$$
$$z_2 = k_2(q_2 - \theta_2) + u_2(q, \dot{q}, t),$$
(9.21)

where the $u_i(\cdot)$ have been selected so that Assumption 9.1 holds. According to the construction in Lemma 9.1, set

$$w_2 = -\frac{1}{b_{12}}(k_1(q_1 - \theta_1) + u_1(q, \dot{q}, t))$$

and note that

$$z_2^{(4)} = k_2\left(q_2^{(4)} - \frac{k_2 k_1}{b_{12}}w_1\right) + u_2^{(4)}(q, \dot{q}, \theta, z_2, t),$$

hence w_1 can be selected as

$$w_1 = \frac{b_{12}}{k_2 k_1}\left(k_2 q_2^{(4)} + u_2^{(4)}(q, \dot{q}, \theta, z_2, t) + \gamma_1(z_2^{(3)}, \ddot{z}_2, \dot{z}_2, z_2)\right),$$

where $\gamma_1(\cdot)$ is such that all trajectories of the system

$$z_2^{(4)} = \gamma_1(z_2^{(3)}, \ddot{z}_2, \dot{z}_2, z_2)$$

are bounded and converge to zero. This completes the design of a global tracking controller.

9.1.3 A 3-DOF Flexible Joints Robot

The proposed control design procedure is applicable even if the rank condition in Lemma 9.1 does not hold, provided the rank of the matrix $B_2(q)$ is constant, as illustrated by the following example.

Consider a 3-DOF flexible joints robot[4] with

$$B_1(q) = \begin{bmatrix} b_{11}(q_2, q_3) & 0 & 0 \\ 0 & b_{22}(q_3) & b_{23}(q_3) \\ 0 & b_{23}(q_3) & b_{33} \end{bmatrix}, \quad B_2(q) = \begin{bmatrix} 0 & 0 & 0 \\ 0 & 0 & J_{m_3}r_3 \\ 0 & 0 & 0 \end{bmatrix},$$

$$C_1(q, \dot{q}) = \begin{bmatrix} -\frac{1}{2}r(q_2, q_3)\dot{q}_2 - \frac{1}{2}s(q_2, q_3)\dot{q}_3 & -\frac{1}{2}r(q_2, q_3)\dot{q}_1 & -\frac{1}{2}s(q_2, q_3)\dot{q}_1 \\ -\frac{1}{2}r(q_2, q_3)\dot{q}_1 & -\frac{1}{2}t(q_3)\dot{q}_3 & -\frac{1}{2}t(q_3)(\dot{q}_2 + \dot{q}_3) \\ \frac{1}{2}s(q_2, q_3)\dot{q}_1 & \frac{1}{2}t(q_3)\dot{q}_2 & 0 \end{bmatrix},$$

$C_2(q, \dot{q}) = 0$ and $J = \text{diag}(J_{m_1}r_1^2, J_{m_2}r_2^2, J_{m_3}r_3^2)$, where $r(\cdot)$, $s(\cdot)$ and $t(\cdot)$ are suitably defined functions. Note that $B_2(q)$ has rank equal to one, for all q, and that $C_2(q, \dot{q}) = 0$. As a result, the variable (9.8) is described by the equations

---
[4]See [219].

$$\begin{bmatrix} z_1 \\ z_2 \\ z_3 \end{bmatrix} = \begin{bmatrix} k_1(q_1 - \theta_1) + u_1(q, \dot{q}, t) \\ k_2(q_2 - \theta_2) + u_2(q, \dot{q}, t) + J_{m3} r_3 w_3 \\ k_3(q_3 - \theta_3) + u_3(q, \dot{q}, t), \end{bmatrix}, \qquad (9.22)$$

where the $u_i(\cdot)$ have been selected such that Assumption 9.1 holds. From equation (9.22) it is possible to select w_3 such that $z_2(t) = 0$ for all t, i.e.,

$$w_3 = -\frac{1}{J_{m3} r_3} \left(k_2(q_2 - \theta_2) + u_2(q, \dot{q}, t) \right). \qquad (9.23)$$

Consider now the system with w_3 as in equation (9.23), control inputs w_1 and w_2 and outputs z_1 and z_3, as given in equation (9.22). A simple computation yields

$$\ddot{z}_1 = \phi_1(q, \dot{q}, t) - k_1 w_1,$$

where $\phi_1(\cdot)$ is a computable function of its arguments. Set

$$w_1 = \frac{1}{k_1} \left(\phi_1(q, \dot{q}, t) + \gamma_1(z_1, \dot{z}_1) \right), \qquad (9.24)$$

with $\gamma_1(\cdot)$ such that all trajectories of the system

$$\ddot{z}_1 = \gamma_1(z_1, \dot{z}_1)$$

are bounded and converge to zero, and consider the system with w_1 and w_3 as in equations (9.24) and (9.23), input w_2 and output z_3. Straightforward computations yield

$$z_3^{(4)} = \phi_3(q, \dot{q}, \theta_2, \theta_3, \dot{\theta}_2, \dot{\theta}_3, z_2, z_1) - \frac{k_2 k_3}{J_{m3} r_3} w_2,$$

from which w_2 can be defined as

$$w_2 = \frac{J_{m3} r_3}{k_2 k_3} \left(\phi_3(q, \dot{q}, \theta_2, \theta_3, \dot{\theta}_2, \dot{\theta}_3, z_2, z_1) + \gamma_3(z_3, \dot{z}_3, \ddot{z}_3, z_3^{(3)}) \right), \qquad (9.25)$$

with $\gamma_3(\cdot)$ such that all trajectories of the system

$$z_3^{(4)} = \gamma_1(z_3, \dot{z}_3, \ddot{z}_3, z_3^{(3)})$$

are bounded and converge to zero. This concludes the design of a global tracking controller for the considered flexible joints robot.

9.2 Position-feedback Control of a Two-link Manipulator

In this section we construct globally asymptotically stabilising output feedback controllers for Euler–Lagrange systems with two degrees of freedom, where only the positions are measurable. As a special case we consider a two-link manipulator and, using the computed torque control with nonlinear PD-like terms, we show that the closed-loop system with output feedback is ISS with respect to the estimation error, hence Theorem 6.2 applies.

9.2 Position-feedback Control of a Two-link Manipulator

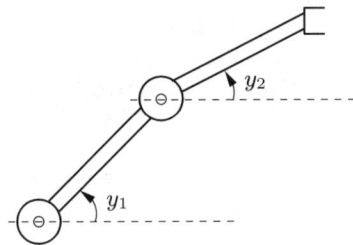

Fig. 9.1. Two-link manipulator with revolute joints.

Note 9.1. For an overview of robot manipulators and the associated control problems the reader is referred to the monographs [196, 206]. In [163, Section 5.4.1] it has been shown that position-only feedback laws can be designed using a PD-type controller with "dirty derivative", yielding semiglobal results. However, global properties of systems controlled with computed torque plus PD-type controllers have not been established for the class of Euler–Lagrange systems considered here. ◁

9.2.1 Observer Design

Consider a revolute-joints arm with two links (depicted in Figure 9.1) and with a mass attached to the extremum of each link[5], described by the equations

$$\begin{aligned}\dot{y} &= \eta, \\ \dot{\eta} &= -M(y)^{-1}\left(C(y,\eta)\eta + G(y) - \tau\right).\end{aligned} \quad (9.26)$$

Note that, comparing with the more general form (5.34), we have neglected the dissipative force vector $F(y, \eta)$. An observer for the angular velocities of the links can be constructed by applying the result in Section 5.4.3. The inertia matrix $M(y)$ and the vector $G(y)$ of the two-link manipulator are given by

$$M(y) = \begin{bmatrix} a_1 + a_2 \cos(y_2) & \frac{1}{2}a_2 \cos(y_2) + a_3 \\ \frac{1}{2}a_2 \cos(y_2) + a_3 & a_3 \end{bmatrix},$$

$$G(y) = \begin{bmatrix} g_1 \cos(y_1) + g_2 \cos(y_1 + y_2) \\ g_2 \cos(y_1 + y_2) \end{bmatrix},$$

where y_i is the angle measured at the ith revolute joint with respect to the horizontal. The model parameters are given by

$$\begin{aligned} a_1 &= I_1 + I_2 + m_1 L_{c1}^2 + m_2(L_1^2 + L_{c2}^2), \\ a_2 &= 2m_2 L_{c2} L_1, \quad a_3 = m_2 L_{c2}^2 + I_2, \\ g_1 &= (L_{c1}(m_1 + M_1) + L_1(m_2 + M_2))g, \\ g_2 &= L_{c2}(m_2 + M_2)g, \end{aligned} \quad (9.27)$$

---

[5] See [206] for details on this system.

where g is the gravitational acceleration, M_i is the mass attached to link i and m_i, L_i, L_{ci}, and I_i denote the mass, length, centre of mass, and moment of inertia for each link i, respectively. The matrix for the Coriolis and centrifugal forces evaluated through the Christoffel symbols is

$$C(y,\eta) = \begin{bmatrix} -\frac{1}{2}a_2 \sin(y_2)\eta_2 & -\frac{1}{2}a_2 \sin(y_2)(\eta_1+\eta_2) \\ \frac{1}{2}a_2 \sin(y_2)\eta_1 & 0 \end{bmatrix}.$$

The transformation matrix $T(y)$ is given by

$$T(y) = \begin{bmatrix} \sqrt{a_1+a_2\cos(y_2)} & \frac{a_2\cos(y_2)+2a_3}{2\sqrt{a_1+a_2\cos(y_2)}} \\ 0 & \frac{1}{2}\sqrt{\frac{4a_3a_1-a_2^2(\cos(y_2))^2-4a_3^2}{a_1+a_2\cos(y_2)}} \end{bmatrix}.$$

The functions $\beta_1(\xi,y)$ and $\beta_2(\xi,y)$, evaluated directly via (5.40) and (5.41), are

$$\beta_1(\xi,y) = \sqrt{a_1+a_2\cos(y_2)}$$
$$\times \left[k_1\left(k_2\pi+\xi_2^2\right)\left(y_1+\int_{y_{20}}^{y_2} \frac{\frac{1}{2}a_2\cos(\zeta)+a_3}{a_1+a_2\cos(\zeta)}d\zeta\right) + \xi_1\right],$$
$$\beta_2(\xi,y) = k_2 \arctan(y_2) + \xi_2.$$

Assuming that there is no *a priori* knowledge about initial velocities of the links, the initial conditions of the observer states (ξ_{10},ξ_{20}) are chosen so that the initial estimate $\hat{x}_0$ is zero, *i.e.*,

$$\xi_{20} = -k_2 \arctan(y_{20}),$$
$$\xi_{10} = -k_1\left(k_2\pi+\xi_{20}^2\right)y_{10}.$$

Only the case of zero friction is considered, *i.e.*, $\nu=0$. The model parameters are set to $I_1 = 0.06$ kg m^2, $L_1 = 0.6$ m, $m_1 = 1$ kg, $M_1 = 1$ kg, $L_{c1} = 0.45$ m, $I_2 = 0.019$ kg m^2, $L_2 = 0.4$ m, $m_2 = 0.6$ kg, $M_2 = 0.8$ kg, $L_{c2} = 0.314$ m, and $g = 9.81$ m/s^2, yielding $a_1 = 0.558$, $a_2 = 0.226$, $a_3 = 0.078$, $g_1 = 17.07$ and $g_2 = 4.32$.

Two simulation results are shown for different observer gains. The initial states of the two-link manipulator are $(y_{10},y_{20},\eta_{10},\eta_{20}) = (0,0,1.5,1.5)$, with motor torque $\tau = 0$. The state trajectories of the two-link arm with the given initial conditions are depicted in Figure 9.2. Figure 9.3 shows the estimation errors for different values of k_1 and k_2, with $c_1 = 1.08$. It is evident, as suggested by the Lyapunov function derivative (5.38), that the observer gain k_1 mainly affects the convergence to zero of the error z_1, whereas k_2 affects the convergence of both z_1 and z_2.

9.2.2 State Feedback Controller

A state feedback control law that stabilises the origin of the system (9.26) is the standard computed torque control with nonlinear PD-like terms, namely

9.2 Position-feedback Control of a Two-link Manipulator

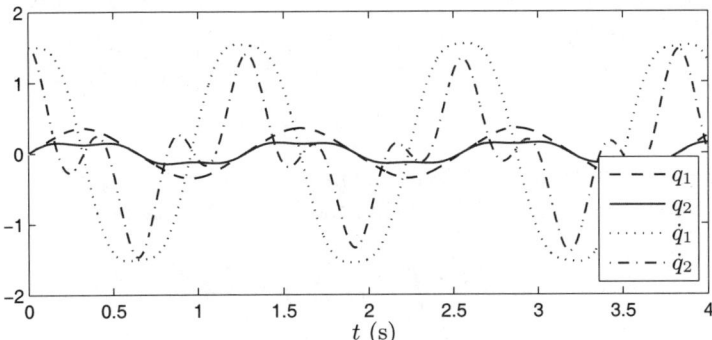

Fig. 9.2. Unforced trajectories of the two-link manipulator.

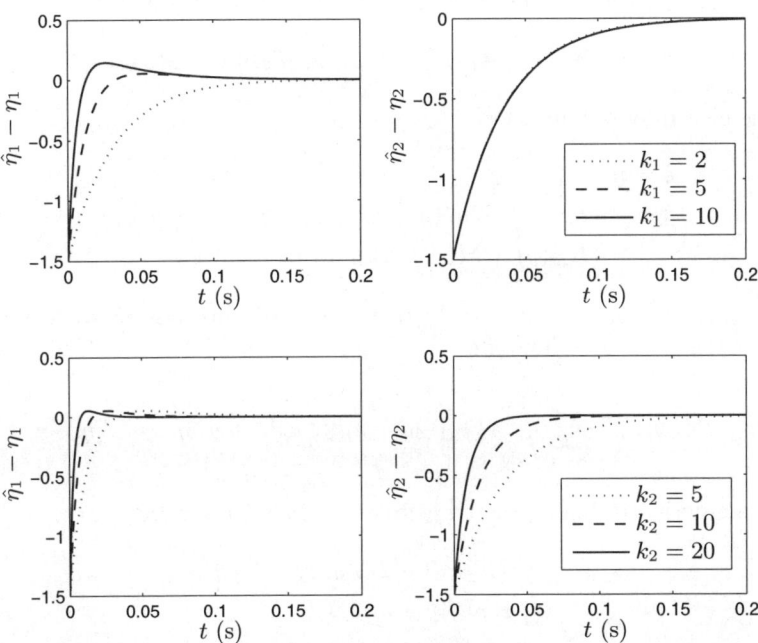

Fig. 9.3. Time histories of the observer errors $\hat{\eta}_1 - \eta_1$ and $\hat{\eta}_2 - \eta_2$ for fixed $k_2 = 5$ (top graphs) and for fixed $k_1 = 5$ (bottom graphs).

$$\tau_s = C(y,\eta)\eta + G(y) - M(y)\left(K_D(y)\eta + K_P(y)y\right), \qquad (9.28)$$

where

$$K_D(y) = \begin{bmatrix} k_{D1}(y_1) & 0 \\ 0 & k_{D2}(y_2) \end{bmatrix}, \quad K_P(y) = \begin{bmatrix} k_{P1}(y_1) & 0 \\ 0 & k_{P2}(y_2) \end{bmatrix} \qquad (9.29)$$

are uniformly positive-definite matrices. The dynamics of the closed-loop system with the state feedback τ_s are given by

$$\dot{y} = \eta,$$
$$\dot{\eta} = -K_D(y)\eta - K_P(y)y. \tag{9.30}$$

The system (9.30) has a globally asymptotically stable equilibrium at the origin, which can be established using the Lyapunov function

$$V_s = \sum_{i=1}^{2} \int_{y_{i0}}^{y_i} k_{Pi}(\zeta)\zeta d\zeta + \frac{1}{2}\eta^\top \eta.$$

9.2.3 Output Feedback Design

Substituting η with its estimate $\hat{\eta}$ in (9.28) yields the output feedback control

$$\tau_o = C(y,\hat{\eta})\hat{\eta} + G(y) - M(y)\left(K_D(y)\hat{\eta} + K_P(y)y\right) \tag{9.31}$$

and the closed-loop dynamics

$$\dot{y} = \eta,$$
$$\dot{\eta} = -\left(K_D(y) - 2M(y)^{-1}C(y,L(y)z)\right)\eta - K_P(y)y \tag{9.32}$$
$$- \left(K_D(y) - M(y)^{-1}C(y,L(y)z)\right)L(y)z,$$

where $L(y) = T(y)^{-1}$ and $z = T(y)(\hat{\eta} - \eta)$. For this system with input z consider the Lyapunov function

$$V_o(y,\eta) = \sum_{i=1}^{2} \int_{y_{i0}}^{y_i} (k_{Di}(\zeta)\zeta + k_{Pi}(\zeta)\zeta) d\zeta + y^\top \eta + \frac{1}{2}\eta^\top \eta,$$

with time-derivative along the trajectories of (9.32) given by

$$\dot{V}_o = -\eta^\top \left(K_D(y) - 2\Delta(y,z) - I_2\right)\eta$$
$$- y^\top K_P(y)y + y^\top 2\Delta(y,z)\eta$$
$$- (y+\eta)^\top \left(K_D(y) - \Delta(y,z)\right)L(y)z,$$

where $\Delta(y,z) = M(y)^{-1}C(y,L(y)z)$ is bounded with respect to y and linear in z, by the properties of the matrix $C(\cdot)$. Note now that $\dot{V}_o$ is bounded by

$$\dot{V}_o \leq -\sum_{i=1}^{2} \left(\eta_i^2(k_{Di}(y_i) - 2\delta_0 - 1) + y_i^2 k_{Pi}(y_i)\right) + 2\delta_0|y^\top \eta|$$
$$+ \sum_{i=1}^{2} |y_i + \eta_i|(k_{Di}(y_i) + \delta_0) c_0,$$

where $\delta_0 = \sigma_{\max}(\Delta(y,z))$ is the maximum singular value of $\Delta(y,z)$ and $c_0 = \sigma_{\max}(L(y))|z|$. Selecting the diagonal matrices $K_D(y)$ and $K_P(y)$ so that

$$\lim_{|y_i| \to \infty} k_{Di}(y_i) = +\infty, \qquad \frac{k_{Pi}(y_i)}{k_{Di}(y_i)} \geq 1,$$

for $i = 1, 2$, and recalling that, by Proposition 5.1, the system (5.35) has a uniformly globally stable equilibrium at the origin, yields the next result.

Proposition 9.2. *Consider the system (5.34) in closed loop with the reduced-order observer (5.2) and the output feedback control (9.31). Then there exist positive-definite diagonal matrices $K_D(y)$ and $K_P(y)$ as in (9.29) such that the origin of the system (5.35), (9.32) is globally asymptotically stable.*

A tracking version for bounded references q_d (with bounded derivatives) can be easily derived using the control law

$$\tau_o = C(y, \hat{\eta})\hat{\eta} + G(y) + M(y)\left(\ddot{q}_d + K_D(y)(\dot{q}_d - \eta) + K_P(y)(q_d - y)\right).$$

9.2.4 Simulation Results

The simulations are performed with a two-link manipulator having the parameters $a_1 = 0.541$, $a_2 = 0.45$, $a_3 = 0.21$, $g_1 = 5.87$ and $g_2 = 1.47$, which have been evaluated from (9.27) with $I_1 = 0.16$ kg m², $L_1 = 0.3$ m, $m_1 = 2$ kg, $M_1 = 0$ kg, $L_{c1} = 0.15$ m, $I_2 = 0.008$ kg m², $L_2 = 0.3$ m, $m_2 = 1$ kg, $M_2 = 0$ kg, $L_{c2} = 0.15$ m, and $g = 9.81$ m/s².

The computed torque plus nonlinear PD-like feedback control law in (9.31), with the estimated joint velocities $\hat{\eta}$ evaluated by the observer (5.2), is considered.

The simulation results of the proposed output feedback control law are shown for both constant and sinusoidal references in Figures 9.4–9.7. The initial conditions are set to $(y_{10}, y_{20}, \dot{q}_{10}, \dot{q}_{20}) = (0.5, -0.5, 0.5, -1)$. The reference signals are $q_{1d} = \pi/2$, $q_{2d} = 0$ and $q_{1d} = -\sin(t)$, $q_{2d} = \sin(2t)$, respectively. Assuming that the gains k_{Di} and k_{Pi} are constant until large values of $|y_i|$ are reached, they have been fixed equal to $k_{Di} = 40$ and $k_{Pi} = 30$. The observer gains are chosen as $k_1 = k_2 = 5$. The initial conditions of the observer state are selected so that the initial estimates $\hat{\eta}_1$ and $\hat{\eta}_2$ are zero, namely

$$\xi_{10} = -k_{20}\arctan(y_{20}), \qquad \xi_{20} = -(k_1 y_{10}(k_2\pi + \xi_{20}^2)).$$

It is apparent from the figures that the output feedback controller yields closed-loop properties similar to that resulting from the use of state feedback.

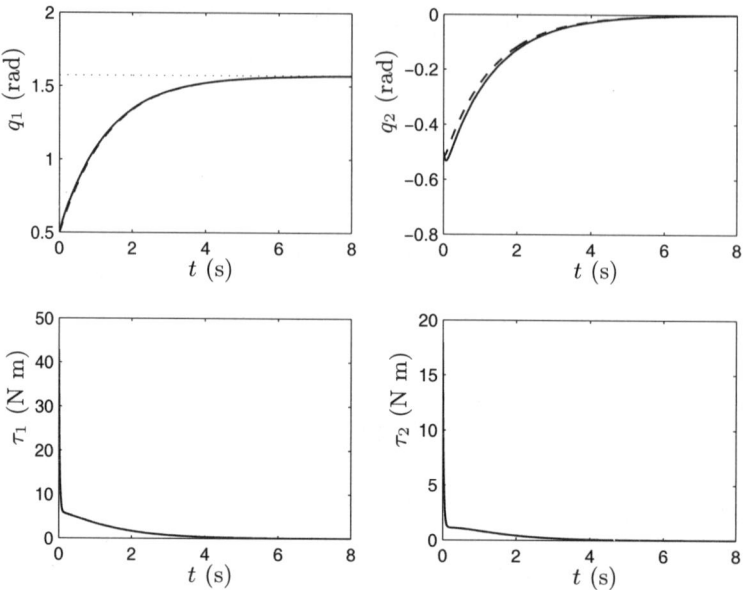

Fig. 9.4. Time histories of the link positions y (top) and the control torques τ (bottom) for the regulation case. Dotted line: Reference signal. Dashed line: State feedback controller. Solid line: Output feedback controller.

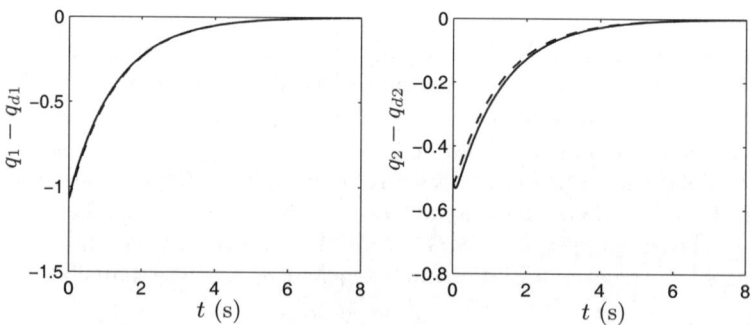

Fig. 9.5. Position errors for constant references $(q_{1d}, q_{2d}) = (\pi/2, 0)$. Dashed line: State feedback controller. Solid line: Output feedback controller.

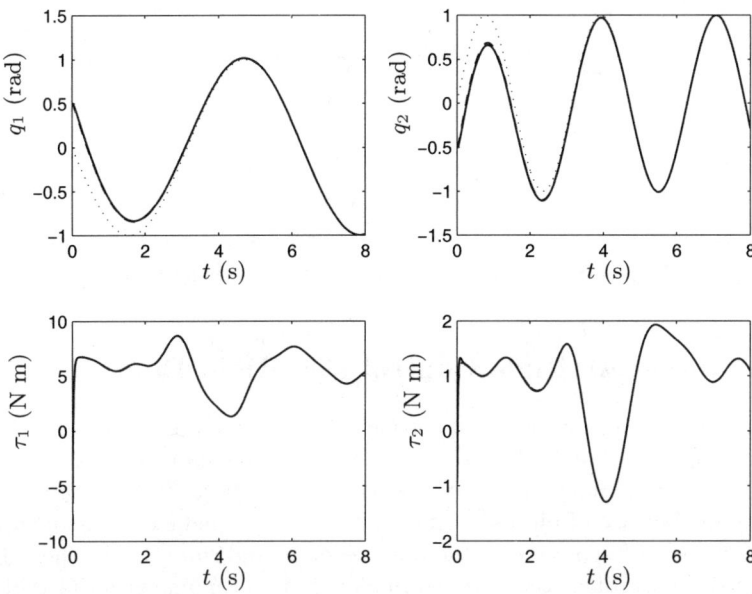

Fig. 9.6. Time histories of the link positions y (top) and the control torques τ (bottom) for the tracking case. Dotted line: Reference signal. Dashed line: State feedback controller. Solid line: Output feedback controller.

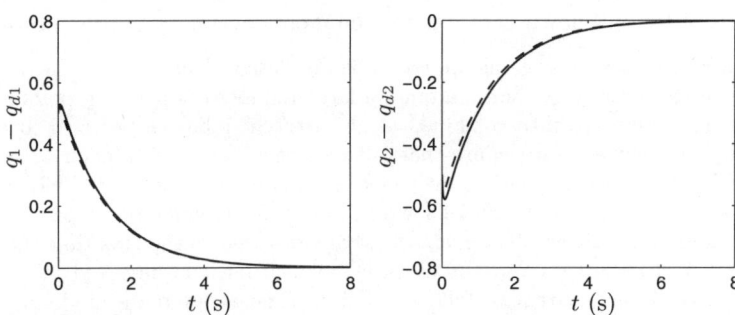

Fig. 9.7. Position errors for sinusoidal references $(q_{1d}, q_{2d}) = (-\sin(t), \sin(2t))$. Dashed line: State feedback controller. Solid line: Output feedback controller.

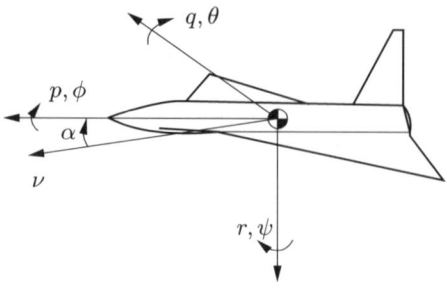

Fig. 9.8. Diagram of the aircraft body axes system.

9.3 Adaptive Attitude Control of a Rigid Body

In this section we apply the I&I adaptive control tools to the problem of attitude tracking for a rigid body with three degrees of freedom and with unknown forces acting on all axes. It is assumed that the body is fully actuated. (For example, in the case of aircraft, this implies one independent control surface for each degree of freedom, *i.e.*, ailerons, elevators and rudder.) We then apply the proposed controller to a detailed model of an autonomous aerial vehicle.

We consider only the state feedback case, *i.e.*, the variables that are assumed to be measurable are Euler angles and angular rates, based on a conventional body axes system, depicted in Figure 9.8. The proposed adaptive attitude controller achieves asymptotic tracking of desired roll, pitch and yaw angles. In addition, the adaptive scheme generates asymptotically converging estimates of the unknown aerodynamic moments acting upon the aircraft.

Note 9.2. Flight control systems are traditionally designed based on *linear* approximations of the aircraft dynamics around a large number of operating points [208]. These local designs are then combined using gain scheduling in order to cover the entire flight envelope, a procedure that is time-consuming and difficult to iterate. This fact has led in recent years to the wider use of nonlinear control techniques such as feedback linearisation [78], also known as nonlinear dynamic inversion [52, 151], and backstepping [123, 66]. The main drawback of these methods is that they rely on *exact* knowledge of the aircraft dynamics, which must be inverted in order to force the closed-loop system to follow a desired (linear) reference model, and are therefore particularly sensitive to modelling errors. This problem can be alleviated by representing the uncertainties in the model as nonlinear functions with *unknown* coefficients and applying nonlinear adaptive control techniques [197, 37, 96]. ◁

9.3.1 Model Description

Consider a rigid body of constant mass whose attitude is expressed in the standard Euler angles, see Figure 9.8. The equations of rotational motion are given by[6]

---
[6] See [56] for a derivation of these equations.

9.3 Adaptive Attitude Control of a Rigid Body

$$\dot{\varphi} = R(\varphi)\omega, \tag{9.33}$$
$$I\dot{\omega} = S(\omega)I\omega + \mathcal{M}, \tag{9.34}$$

where $\varphi = [\phi, \theta, \psi]^\top$ are the Euler angles, $\omega = [p, q, r]^\top$ is the angular velocity vector, $I = [I_1^\top, I_2^\top, I_3^\top]^\top$ with $I_i^\top \in \mathbb{R}^3$ is the inertia matrix,

$$R(\varphi) = \begin{bmatrix} 1 & \sin\phi\tan\theta & \cos\phi\tan\theta \\ 0 & \cos\phi & -\sin\phi \\ 0 & \sin\phi/\cos\theta & \cos\phi/\cos\theta \end{bmatrix}, \quad S(\omega) = \begin{bmatrix} 0 & r & -q \\ -r & 0 & p \\ q & -p & 0 \end{bmatrix},$$

and $\mathcal{M} = [L, M, N]^\top$ are the (unknown) aerodynamic moments acting on the body. For simplicity (and to ensure that the adaptive control problem is solvable) it is assumed that the aerodynamic moments can be described by equations of the form

$$\mathcal{M} = \begin{bmatrix} \vartheta_1^\top \rho_1(I_1\omega) \\ \vartheta_2^\top \rho_2(I_2\omega) \\ \vartheta_3^\top \rho_3(I_3\omega) \end{bmatrix} + B\delta, \tag{9.35}$$

where $\delta = [\delta_a, \delta_e, \delta_r] \in \mathbb{R}^3$ are the control inputs[7], $\rho_i(\cdot)$ are continuous functions, $B \in \mathbb{R}^{3\times 3}$ is a known matrix-valued function and ϑ_i are *unknown* parameter vectors.

9.3.2 Controller Design

We consider the problem of forcing the aircraft attitude φ to follow a smooth reference signal $\varphi_d = \varphi_d(t)$ by controlling δ. Define the "tracking errors"

$$\tilde{\varphi} = \varphi - \varphi_d, \qquad \tilde{\omega} = \omega - R(\varphi)^{-1}\dot{\varphi}_d,$$

where

$$R(\varphi)^{-1} = \begin{bmatrix} 1 & 0 & -\sin\theta \\ 0 & \cos\phi & \sin\phi\cos\theta \\ 0 & -\sin\phi & \cos\phi\cos\theta \end{bmatrix},$$

and the energy function

$$H(\tilde{\varphi}, \tilde{\omega}) = \frac{1}{2}\tilde{\varphi}^\top K_1 \tilde{\varphi} + \frac{1}{2}\tilde{\omega}^\top I^2 \tilde{\omega}, \tag{9.36}$$

where $K_1 = K_1^\top > 0$ is a constant matrix, and note that in the new coordinates the system (9.33), (9.34) can be written in the *perturbed Hamiltonian form*[8]

---

[7]The notation used here is common in flight control systems, where the control inputs correspond to the ailerons, elevators and rudder deflection angles, respectively.

[8]See, *e.g.*, [167].

$$\begin{bmatrix} \dot{\tilde{\varphi}} \\ I\dot{\tilde{\omega}} \end{bmatrix} = \begin{bmatrix} 0 & R(\varphi)I^{-1} \\ -I^{-1}R(\varphi)^\top & S(\omega) \end{bmatrix} \begin{bmatrix} \frac{\partial H^\top}{\partial \tilde{\varphi}} \\ \frac{\partial H^\top}{\partial (I\tilde{\omega})} \end{bmatrix} + \begin{bmatrix} 0 \\ \mathcal{M} + \kappa(\varphi,\omega,t) \end{bmatrix}, \qquad (9.37)$$

where

$$\kappa(\varphi,\omega,t) = I^{-1}R(\varphi)^\top K_1 \tilde{\varphi} + S(\omega)IR(\varphi)^{-1}\dot{\varphi}_d - I\dot{R}(\varphi)^{-1}\dot{\varphi}_d - IR(\varphi)^{-1}\ddot{\varphi}_d.$$

Consider now the parameter estimation errors

$$z_i = \hat{\vartheta}_i - \vartheta_i + \beta_i(I_i\omega), \qquad (9.38)$$

for $i = 1, 2, 3$, where $\hat{\vartheta}_i$ are new states and $\beta_i(\cdot)$ are continuous functions to be defined. A control law u, which drives to zero the Hamiltonian function (9.36) along the trajectories of (9.37) when $z_i = 0$, is given by

$$\delta = -B^{-1} \begin{bmatrix} \left(\hat{\vartheta}_1 + \beta_1(I_1\omega)\right)^\top \rho_1(I_1\omega) \\ \left(\hat{\vartheta}_2 + \beta_2(I_2\omega)\right)^\top \rho_2(I_2\omega) \\ \left(\hat{\vartheta}_3 + \beta_3(I_3\omega)\right)^\top \rho_3(I_3\omega) \end{bmatrix} - B^{-1}\left(\kappa(\varphi,\omega,t) + K_2 I\tilde{\omega}\right),$$

where $K_2 = K_2(\varphi,\omega)$ is a positive-definite matrix-valued function.

The resulting closed-loop system can be written in the "perturbed" Hamiltonian form

$$\begin{bmatrix} \dot{\tilde{\varphi}} \\ I\dot{\tilde{\omega}} \end{bmatrix} = \begin{bmatrix} 0 & R(\varphi)I^{-1} \\ -I^{-1}R(\varphi)^\top & S(\omega) - K_2 \end{bmatrix} \begin{bmatrix} \frac{\partial H^\top}{\partial \tilde{\varphi}} \\ \frac{\partial H^\top}{\partial (I\tilde{\omega})} \end{bmatrix} - \begin{bmatrix} 0 \\ \Delta \end{bmatrix}, \qquad (9.39)$$

where each element of the perturbation vector Δ is given by

$$\Delta_i = z_i^\top \rho_i(I_i\omega). \qquad (9.40)$$

Note that the system (9.39) is *converging-input converging-state* stable with respect to Δ, for any $K_2 > 0$, and has an asymptotically stable equilibrium at the origin when $\Delta = 0$.

9.3.3 Estimator Design

We now design the update laws $\dot{\hat{\vartheta}}_i$ and the functions $\beta_i(\cdot)$ so that the perturbation (9.40) is driven asymptotically to zero. To this end, consider the estimation errors (9.38) and the dynamic update laws

$$\dot{\hat{\vartheta}}_i = -\frac{\partial \beta_i}{\partial I_i\omega}\left[I_i\dot{\omega} + \Delta_i\right]. \qquad (9.41)$$

Note that the term in brackets is a function of φ, ω and the first and second derivative of φ_d, therefore it is measurable. Using (9.41), the dynamics of (9.38) along the trajectories of (9.39) are given by

$$\dot{z}_i = -\frac{\partial \beta_i}{\partial I_i \omega} \rho_i(I_i\omega)^\top z_i.$$

Following the results in Chapter 4, we select the functions $\beta_i(\cdot)$ according to

$$\beta_i(I_i\omega) = \gamma_i \int_0^{I_i\omega} \rho_i(y) dy$$

with $\gamma_i > 0$, which implies that $\frac{\partial \beta_i}{\partial (I_i\omega)} = \gamma_i \rho_i(I_i\omega)$, yielding the system

$$\dot{z}_i = -\gamma_i \rho_i(I_i\omega) \rho_i(I_i\omega)^\top z_i, \tag{9.42}$$

which has a uniformly globally stable equilibrium at zero. In particular, the Lyapunov function $W(z) = \sum_{i=1}^{3} \frac{1}{\gamma_i} |z_i|^2$ is such that $\dot{W} = -2|\Delta|^2$, which implies that the perturbation signal Δ is square-integrable. Global stability of the equilibrium $(\tilde{\varphi}, \tilde{\omega}, z) = (0, 0, 0)$ and convergence of $\tilde{\varphi}$ and $\tilde{\omega}$ to zero follows by considering the Lyapunov function $H(\tilde{\varphi}, \tilde{\omega}) + W(z)$ and invoking similar arguments to those in Chapter 4.

9.3.4 Simulation Results

Simulations of the proposed controller combined with a detailed model of an autonomous aircraft are presented in Figure 9.9, which displays the response of the aircraft to *filtered step changes* in the commanded attitude φ_d. (In the following section we present results where the aircraft is required to track *time-varying* attitude references, which are generated by the guidance algorithm.)

Note that we only apply filtered step inputs to the roll and pitch references (which are typically commanded by the pilot's stick), while the yaw reference is calculated based on "turn co-ordination" considerations (see Section 9.4).

9.4 Trajectory Tracking for Autonomous Aerial Vehicles

In this section we employ the immersion and invariance ideas of Chapter 2 to control an autonomous aerial vehicle so that it tracks a predefined *geometric* path. Our "control inputs" in this case are the attitude (Euler) angles φ. In other words, the problem considered here is to design a reference φ_d such that, when $\varphi = \varphi_d$, the aircraft converges to a predefined geometric path. This can then be combined with the attitude controller of Section 9.3 to obtain a complete flight control system. We also consider the problem of speed regulation and propose a solution using the adaptive control tools developed in this book.

For simplicity, we assume that the path to be tracked consists of *straight lines* or *arcs of circles* in the x-y plane, while the desired altitude is constant. Note, however, that the proposed approach can also be used to track more complex paths.

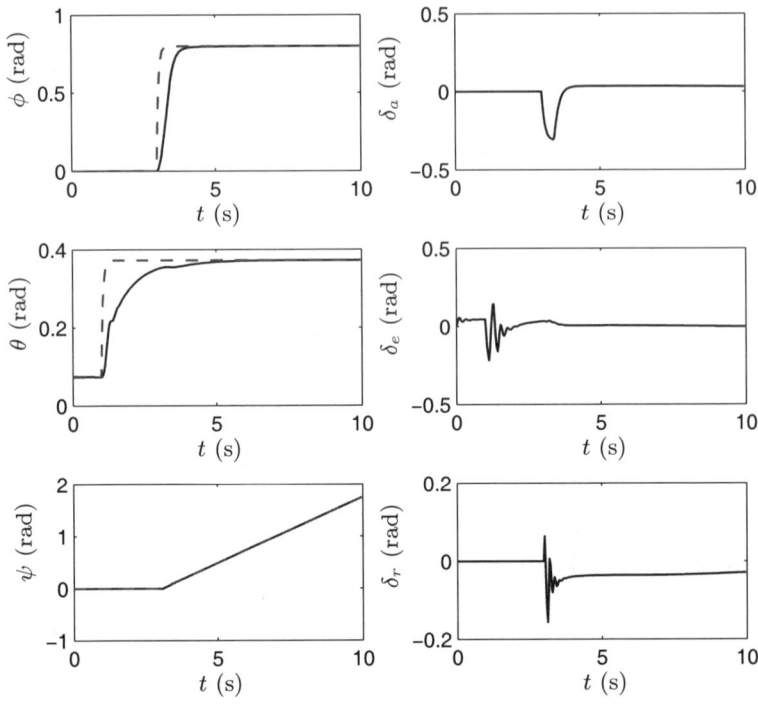

Fig. 9.9. Aircraft response to roll and pitch commands. Left: Attitude angles (solid lines) and references (dashed lines). Right: Control surface deflections.

9.4.1 Model Description

The translational motion of an aircraft whose attitude is expressed in the standard Euler angles $\varphi = [\phi, \theta, \psi]^\top$ is described by the equations[9]

$$\dot{x} = (u\cos\theta + v\sin\phi\sin\theta + w\cos\phi\sin\theta)\cos\psi - (v\cos\phi - w\sin\phi)\sin\psi,$$
$$\dot{y} = (u\cos\theta + v\sin\phi\sin\theta + w\cos\phi\sin\theta)\sin\psi + (v\cos\phi - w\sin\phi)\cos\psi,$$
$$\dot{h} = u\sin\theta - (v\sin\phi + w\cos\phi)\cos\theta,$$

where x, y and h are the co-ordinates of the aircraft's centre of gravity relative to an inertial reference frame (with h corresponding to the altitude) and u, v, w are the components of the groundspeed. Using the identity

$$a\sin\psi + b\cos\psi = \sqrt{a^2 + b^2}\sin(\psi + \arctan(\frac{b}{a})) = \sqrt{a^2 + b^2}\cos(\psi - \arctan(\frac{a}{b})) \quad (9.43)$$

---

[9] See [56] or [209] for a derivation of these equations.

the equations of motion can be rewritten as

$$\dot{x} = V_1 \cos\gamma \cos\psi - (v\cos\phi - w\sin\phi)\sin\psi,$$
$$\dot{y} = V_1 \cos\gamma \sin\psi + (v\cos\phi - w\sin\phi)\cos\psi,$$
$$\dot{h} = V_1 \sin\gamma,$$

where
$$V_1 = \sqrt{u^2 + (v\sin\phi + w\cos\phi)^2}$$

and
$$\gamma = \theta - \arctan(\frac{v\sin\phi + w\cos\phi}{u}). \qquad (9.44)$$

Note that for wings-level, non-sideslipping flight (*i.e.*, $\phi = 0$, $v = 0$), the variable γ represents the *path angle*, while V_1 is equal to the total airspeed V. Applying the identity (9.43) once more yields the system

$$\dot{x} = V_2 \cos\lambda, \qquad (9.45)$$
$$\dot{y} = V_2 \sin\lambda, \qquad (9.46)$$
$$\dot{h} = V_1 \sin\gamma, \qquad (9.47)$$

where
$$V_2 = \sqrt{(V_1 \cos\gamma)^2 + (v\cos\phi - w\sin\phi)^2}$$

and
$$\lambda = \psi + \arctan(\frac{v\cos\phi - w\sin\phi}{V_1 \cos\gamma}). \qquad (9.48)$$

Note that for wings-level, non-sideslipping flight, the variable λ represents the *heading*, while V_2 is equal to $V\cos\gamma$.

9.4.2 Controller Design

Consider the planar system (9.45), (9.46) and the problem of finding a control law[10] λ such that all trajectories $(x(t), y(t))$ converge towards a manifold defined by equations of the form

$$e_s \triangleq ax + by + c = 0 \qquad (9.49)$$

or
$$e_c \triangleq (x-a)^2 + (y-b)^2 - c^2 = 0, \qquad (9.50)$$

for some constant a, b and c. Note that in the case of equation (9.49) we assume that $a^2 + b^2 > 0$, while in the case of equation (9.50) we assume $c > 0$.

---

[10] It must be noted that the trajectory tracking problem for the planar system (9.45), (9.46), *i.e.*, the problem of tracking a given time-varying reference x_d, y_d by controlling λ, has been studied extensively, see *e.g.*, [85, 183].

In addition, we require that the altitude h is regulated around a constant reference h_d, i.e., $\lim_{t\to\infty} h(t) = h_d$.

We first design λ so that the lines (9.49) and (9.50) are invariant and globally attractive. To this end, consider the dynamics of e_s and e_c which are given, respectively, by

$$\dot{e}_s = V_2\sqrt{a^2+b^2}\sin(\lambda + \arctan(\frac{a}{b})), \tag{9.51}$$

and

$$\dot{e}_c = 2V_2\sqrt{e_c+c^2}\sin(\lambda + \arctan(\frac{x-a}{y-b})), \tag{9.52}$$

where we have used the identity (9.43). Selecting λ equal to

$$\lambda_s = -\arctan(\frac{a}{b}) - \arctan(k_p e_s),$$

or

$$\lambda_s = -\arctan(\frac{a}{b}) + \pi + \arctan(k_p e_s),$$

with $k_p > 0$, ensures that e_s converges asymptotically to zero, while selecting λ equal to

$$\lambda_c = -\arctan(\frac{x-a}{y-b}) - \arctan(k_p e_c),$$

or

$$\lambda_c = -\arctan(\frac{x-a}{y-b}) + \pi + \arctan(k_p e_c),$$

with $k_p > 0$, ensures convergence of e_c to zero. Note that for each line there are two different control laws that drive the system onto the desired path, in opposite directions.

Consider now the system (9.47) and the "control law"

$$\gamma = -\arctan(k_h(h - h_d)),$$

with $k_h > 0$, which yields the closed-loop dynamics

$$\dot{h} = -V_1 \frac{k_h(h-h_d)}{\sqrt{1+(k_h(h-h_d))^2}}.$$

It is straightforward to show that the above system has a globally asymptotically stable equilibrium at $h = h_d$, hence the altitude converges to its reference value.

Note finally that, given λ and γ, it is possible to select θ and ψ so that (9.44) and (9.48) hold, with ϕ a free variable. For straight flight we typically choose $\phi = 0$ (or, more generally, $\phi = n\pi$, where n is an integer), while in the case of a steady *co-ordinated* turn along the path (9.50) we select ϕ to satisfy the equations

$$\mathcal{L}\cos\phi = mg,$$
$$\mathcal{L}\cos\alpha\sin\phi = \frac{mV^2}{c},$$

where $\mathcal{L}$ is the lift applied to the aircraft and the terms on the right-hand side correspond to the gravitational and centripetal force, respectively. Dividing the two foregoing equations yields

$$\phi = \arctan\left(\frac{G}{\cos\alpha}\right), \tag{9.53}$$

where $G = V^2/(gc)$ is the centripetal acceleration (in gs)[11].

9.4.3 Airspeed Regulation

In this section we design an adaptive controller that asymptotically regulates the aircraft speed to a desired set-point in the presence of unknown aerodynamic forces acting on the body.

The translational dynamics of the aircraft are described by the equations

$$m\dot{\nu} = S(\omega)m\nu + mg\varepsilon(\phi,\theta) + \mathcal{F} + \mathcal{T}, \tag{9.54}$$

where $\nu = [u,v,w]^\top$ is the velocity vector, m is the aircraft mass, g is the gravitational acceleration, $\varepsilon(\phi,\theta) = [-\sin\theta, \sin\phi\cos\theta, \cos\phi\cos\theta]^\top$, $\mathcal{F} = [X,Y,Z]^\top$ are the aerodynamic forces, and $\mathcal{T} = [T_x,0,0]^\top$ is the thrust (along the x body axis). We consider the problem of finding a control law for the thrust T_x that regulates the total airspeed $V = |\nu|$ to a desired constant V_d. We assume that the aerodynamic forces acting on the aircraft are constant and unknown.

To begin with, define the *kinetic energy* error

$$E = \frac{1}{2}m\left(V^2 - V_d^2\right) \tag{9.55}$$

and note that, from (9.54), the dynamics of E are given by

$$\dot{E} = \nu^\top mg\varepsilon(\phi,\theta) + \nu^\top \mathcal{F} + uT_x. \tag{9.56}$$

Define now the estimation error $z = \hat{F} - \mathcal{F} + \beta(\nu)$, where $\hat{F}$ is a new state and $\beta(\cdot)$ is a continuous function to be defined, and the control law

$$T_x = -\frac{1}{u}\left[\nu^\top\left(\hat{F} + \beta(\nu)\right) + \nu^\top mg\varepsilon(\phi,\theta) + \kappa(V)E\right],$$

where $\kappa(V)$ is a positive function. Selecting the update law

---
[11] Note that (9.53) is the condition for turn co-ordination when the sideslip and path angle are zero. For a more general condition, see [209, p. 190].

$$\dot{\hat{F}} = -\frac{\partial f}{\partial \nu}\left[S(\omega)\nu + g\varepsilon(\phi,\theta) + \frac{1}{m}\left(\hat{F} + \beta(\nu) + \mathcal{T}\right)\right]$$

yields the closed-loop system

$$\begin{aligned}\dot{E} &= -\kappa(V)E - \nu^\top z,\\ \dot{z} &= -\frac{1}{m}\frac{\partial \beta}{\partial \nu}z.\end{aligned} \qquad (9.57)$$

Finally, an appropriate selection of $\kappa(V)$ and $\beta(\nu)$ ensures that the *cascaded* system (9.57) has a globally stable equilibrium at the origin with E converging to zero. Two such selections are

$$\kappa(V) = k, \qquad \beta_i(\nu) = \gamma_i \nu_i^3$$

and

$$\kappa(V) = kV^2, \qquad \beta_i(\nu) = \gamma_i \nu_i,$$

with $k > 0$, $\gamma_i > 0$. Note that the first selection, similarly to the attitude control case, ensures that the *perturbation* signal $\nu^\top z$ in (9.57) is square-integrable.

9.4.4 Simulation Results

The proposed controller has been implemented in a MATLAB®/SIMULINK® environment and combined with a detailed model of the Eclipse flight demonstrator developed at Cranfield University. Note that the model includes also actuator dynamics, which have been neglected in the control design. Moreover, we have assumed that the aerodynamic moments are linear functions, i.e., $\rho_i(I_i\omega) = [1, I_i\omega]^\top$ in (9.35), and all parameter estimates are initialised at zero.

The flight control system is composed of the adaptive attitude controller developed in Section 9.3 and the adaptive speed controller and the trajectory tracking controller developed in this section. A diagram of the overall control scheme is shown in Figure 9.10.

The purpose of the simulations is to verify the asymptotic tracking properties of the controller, both during standard demonstration manoeuvres (climb, steady turn, dive), and non-standard ones (such as a 180-degree roll). The flight plan includes the following sectors.

1. Initial speed 26 m/s. Climb from 30 m to 60 m at a rate of 12 m/s.
2. Accelerate to 42 m/s.
3. Perform a 180-degree turn of radius 200 m (at roll angle 43 deg).
4. Perform a 180-degree roll.
5. Perform a 180-degree turn of radius 200 m (at roll angle 137 deg).
6. Resume level flight.
7. Perform two successive 270-degree turns of radius 110 m (at roll angle 60 deg).

9.4 Trajectory Tracking for Autonomous Aerial Vehicles   235

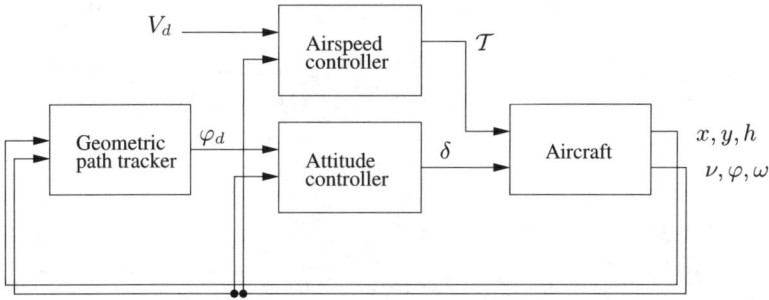

Fig. 9.10. Diagram of the flight control system.

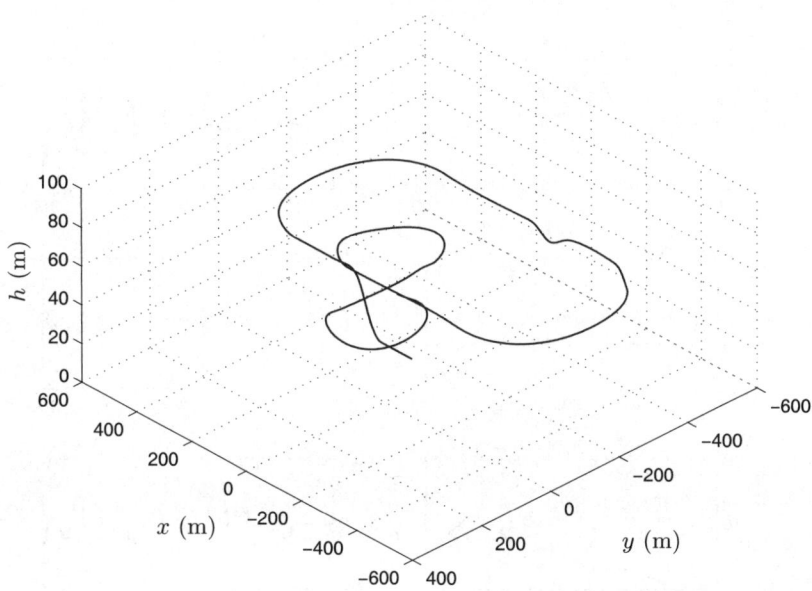

Fig. 9.11. Trajectory of the aircraft.

The trajectory of the aircraft is shown in Figure 9.11. Figure 9.12 shows the time histories of the airspeed and altitude, and of their references, while Figure 9.13 shows the time histories of the Euler angles and their corresponding references, together with the flap deflection angles. Note that the maximum available thrust is 180 N, while the maximum deflection for the flaps is 0.313 rad. Observe that the aircraft follows the desired trajectory with zero tracking error for the airspeed and attitude and zero steady-state error for the altitude, despite the lack of information on the aerodynamics and despite the fact that we have used a simplified aerodynamic model for the design.

236   9 Mechanical Systems

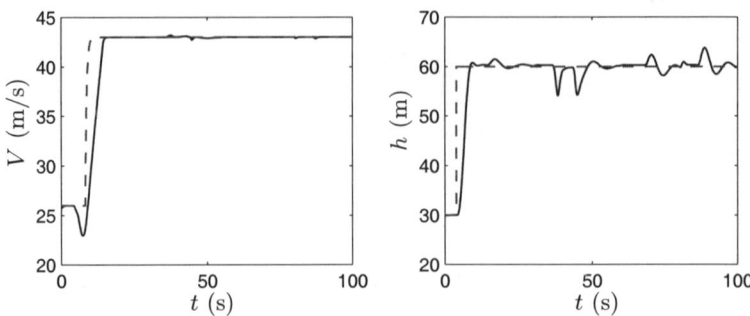

Fig. 9.12. Time histories of the airspeed V and altitude h (solid lines) and their references V_* and h_* (dashed lines).

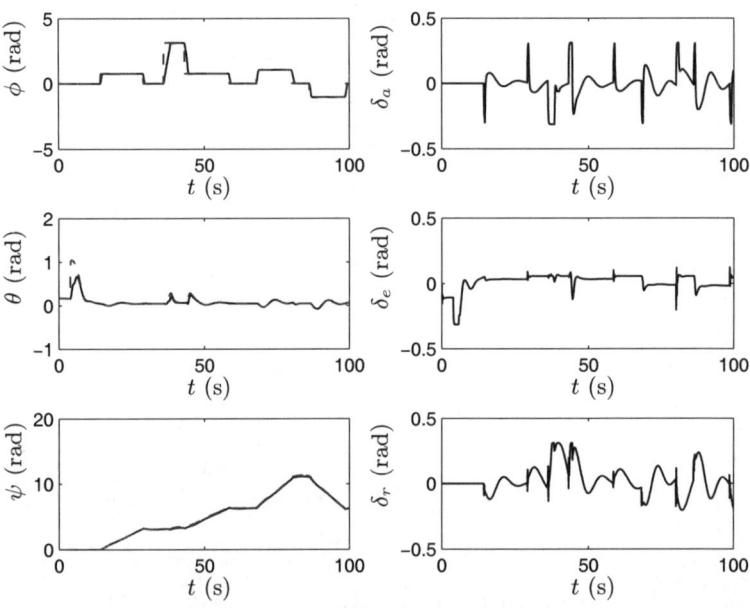

Fig. 9.13. Time histories of the attitude angles (left) and flap deflection angles (right).

10
Electromechanical Systems

In this chapter we apply the I&I methodology to three examples of electromechanical systems. The first is a simple model of a generator connected to an infinite bus, where the objective is to estimate the rotor angle and the total inductance of the bus. This model has been used extensively in the study of the transient stability of power systems[1] and, despite its simplicity, it poses a challenging observer design problem due to the fact that it is nonlinear in the unmeasured states.

Then, in the last two sections, we develop adaptive output feedback controllers for the induction motor, which has been considered as the workhorse of industry, but it is also increasingly used in high dynamical performance applications due to its low cost and reliability in comparison with permanent magnet and DC motors. The induction motor is accurately described by a sixth-order nonlinear model but it contains highly uncertain parameters and only part of the state is measurable. For these reasons it has become one of the benchmark examples for nonlinear controller design. In both examples we construct a controller by combining a state feedback passivity-based control law with an I&I adaptive scheme[2].

10.1 Observer Design for Single-machine Infinite-bus Systems

Transient stability of power systems is a key problem in all distribution networks and pertains to the ability of the system to reach an acceptable steady state following a fault, *e.g.*, a short circuit or a generator outage, that is cleared after some time by the protective system operation.

---

[1] The term "transient stability" in this context refers to the ability of the system to return to an equilibrium following a disturbance.

[2] For a comprehensive account of the passivity-based control methodology, see [163] and [22].

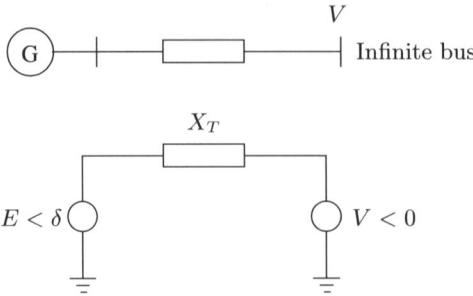

Fig. 10.1. Single-machine infinite-bus system and equivalent circuit.

In this section we address the problem of estimating the states of a single-machine power system with non-negligible transfer conductances. We assume that the system is described by the so-called "swing equation" model[3] and is controlled by varying the total inductance of the bus[4].

First, to motivate the observer design, we propose a feedback controller that depends on the unmeasured states. Then, using the approach of Chapter 5, we design a reduced-order nonlinear observer that asymptotically reconstructs the unmeasured states, namely the total inductance of the bus and a function of the rotor slip angle of the generator, from measurements of the rotor velocity.

10.1.1 Model Description

We consider a single-machine infinite-bus system with the equivalent circuit shown in Figure 10.1, where E is the generator voltage, X_T is the effective reactance of the bus, V is the bus voltage and δ is the rotor angle, *i.e.*, the angle by which E leads V. Assuming that the effective reactance X_T of the bus is controlled via a first-order model, the dynamics of the system are described by the equations[5]

$$\dot{\delta} = \omega,$$
$$\dot{\omega} = \frac{\omega_0}{2H}\left(P_m - D\omega - EV\lambda\sin(\delta)\right), \quad (10.1)$$
$$\dot{\lambda} = \frac{1}{T}(-\lambda + \lambda_* + u),$$

---

[3] See [125] for details on this problem and [160] for a state feedback control design using field excitation and the more general "flux decay" model.

[4] This can be achieved, for instance, using the thyristor-controlled series capacitor (TCSC) of Section 8.1, which is a special class of Flexible AC Transmission Systems (FACTS) and is extensively used for enhancing transient stability in power systems, see, *e.g.*, [225, 210].

[5] See [125, p. 830].

where $\delta \in [0, 2\pi)$ is the rotor angle, ω is the deviation of the rotor speed from the synchronous speed ω_0, $\lambda = 1/X_T$ is the effective susceptance of the bus, H, P_m and D are positive constants representing the inertia (in MW·s/MVA), mechanical power (in p.u.), and damping coefficient, respectively, T is the time constant of the control circuit, and u is the control signal. Note that the quantity $P_{\max} = EV\lambda$ represents the maximum electrical power (in p.u.).

It is assumed that only ω is available for measurement. The operating point for the system (10.1) is defined as $(\delta_*, 0, \lambda_*)$. Note that, for this point to be an equilibrium, λ_* must satisfy the condition $\lambda_* = P_m/(EV \sin(\delta_*))$.

10.1.2 Controller Design

To motivate the observer design, we first develop a state feedback control law using arguments pertaining to Hamiltonian systems and passivity-based control. To this end, define the state vector $x = [\delta, \omega, \lambda]^\top$ and note that the system (10.1) can be put in the perturbed Hamiltonian form

$$\dot{x} = (J(\delta) - R)\frac{\partial H(x)}{\partial x} + e_3\left(\frac{1}{T}u - EV\sin(\delta)\omega\right), \tag{10.2}$$

where

$$J(\delta) = \begin{bmatrix} 0 & \frac{\omega_0}{2H} & 0 \\ -\frac{\omega_0}{2H} & 0 & -\frac{EV\omega_0}{2H}\sin(\delta) \\ 0 & \frac{EV\omega_0}{2H}\sin(\delta) & 0 \end{bmatrix}, \quad R = \begin{bmatrix} 0 & 0 & 0 \\ 0 & \frac{D\omega_0^2}{4H^2} & 0 \\ 0 & 0 & \frac{1}{T} \end{bmatrix},$$

and

$$H(x) = -P_m(\delta - \delta_*) - EV\lambda_*(\cos(\delta) - \cos(\delta_*)) + \frac{H}{\omega_0}\omega^2 + \frac{1}{2}(\lambda - \lambda_*)^2. \tag{10.3}$$

Note that the Hamiltonian function (10.3) has a minimum at the desired equilibrium $(\delta_*, 0, \lambda_*)$. Hence, the control law

$$u = TEV\sin(\delta)\omega - K(\lambda - \lambda_*), \tag{10.4}$$

with $K > -1$, is such that the closed-loop system (10.1), (10.4) has a locally asymptotically stable equilibrium at $(\delta_*, 0, \lambda_*)$. Note that the controller (10.4) requires measurement of $\sin(\delta)$ and λ, which are assumed unknown.

10.1.3 Observer Design

We now proceed to the design of an observer for the unmeasured states δ and λ using the result in Theorem 5.1. Let

$$\phi(\delta, \lambda) = \begin{bmatrix} \sin(\delta) & \cos(\delta) & \lambda \end{bmatrix}^\top \tag{10.5}$$

and define the error variable

$$z = \beta(\xi, \omega) - \phi(\delta, \lambda), \tag{10.6}$$

where $\xi \in \mathbb{R}^3$ is the observer state and $\beta(\xi, \omega) = [\beta_1(\xi, \omega), \beta_2(\xi, \omega), \beta_3(\xi, \omega)]^\top$ is a mapping to be defined. The dynamics of z are given by

$$\dot{z} = \frac{\partial \beta}{\partial \xi}\dot{\xi} + \frac{\partial \beta}{\partial \omega}\frac{\omega_0}{2H}\left(P_m - D\omega - EV\left(\beta_3(\xi, \omega) - z_3\right)(\beta_1(\xi, \omega) - z_1)\right)$$

$$- \begin{bmatrix} \omega\left(\beta_2(\xi, \omega) - z_2\right) \\ -\omega\left(\beta_1(\xi, \omega) - z_1\right) \\ \frac{1}{T}\left(z_3 - \beta_3(\xi, \omega) + \lambda_* + u\right) \end{bmatrix}.$$

Provided that the Jacobian matrix $\frac{\partial \beta}{\partial \xi}$ is invertible, the observer dynamics can be selected as

$$\dot{\xi} = -\left(\frac{\partial \beta}{\partial \xi}\right)^{-1}\frac{\partial \beta}{\partial \omega}\frac{\omega_0}{2H}\left(P_m - D\omega - EV\beta_3(\xi, \omega)\beta_1(\xi, \omega)\right)$$

$$+ \left(\frac{\partial \beta}{\partial \xi}\right)^{-1}\begin{bmatrix} \omega\beta_2(\xi, \omega) \\ -\omega\beta_1(\xi, \omega) \\ \frac{1}{T}\left(-\beta_3(\xi, \omega) + \lambda_* + u\right) \end{bmatrix}, \tag{10.7}$$

yielding the error dynamics

$$\dot{z} = \begin{bmatrix} 0 & \omega & 0 \\ -\omega & 0 & 0 \\ 0 & 0 & -\frac{1}{T} \end{bmatrix} z + \frac{EV\omega_0}{2H}\frac{\partial \beta}{\partial \omega}\begin{bmatrix} \beta_3(\xi, \omega) & 0 & \sin(\delta) \end{bmatrix} z.$$

The above system is simplified by selecting

$$\beta_2(\xi, \omega) = \xi_2, \qquad \beta_3(\xi, \omega) = \xi_3,$$

which yields

$$\dot{z} = \begin{bmatrix} \frac{\partial \beta_1}{\partial \omega}\frac{EV\omega_0}{2H}\xi_3 & \omega & \frac{\partial \beta_1}{\partial \omega}\frac{EV\omega_0}{2H}\sin(\delta) \\ -\omega & 0 & 0 \\ 0 & 0 & -\frac{1}{T} \end{bmatrix} z. \tag{10.8}$$

Selecting the function $\beta_1(\cdot)$ as

$$\beta_1(\xi, \omega) = \xi_1 - \gamma\frac{2H}{EV\omega_0}\xi_3\omega, \tag{10.9}$$

10.1 Observer Design for Single-machine Infinite-bus Systems   241

ensures that the invertibility condition on the Jacobian of $\beta(\xi,\omega)$ is satisfied and that $\beta(\xi,\omega)$ is invertible in its first argument. Moreover, if $\gamma < 4/T$, using the candidate Lyapunov function $V(z) = \frac{1}{2}|z|^2$ it can easily be shown that the equilibrium $z = 0$ is uniformly globally stable and

$$\lim_{t\to\infty} z_1 = 0, \qquad \lim_{t\to\infty} z_3 = 0.$$

As a result, from (10.5) and (10.6) an asymptotically converging estimate of $\sin(\delta)$ and λ is given by $\beta_1(\xi,\omega)$, defined in (10.9), and $\beta_3(\xi,\omega) = \xi_3$, respectively.

10.1.4 Simulation Results

To test the performance of the proposed observer, the system (10.1) with $u = 0$ (which implies $\lambda(t) > 0$ for all t) and the observer in (10.7), namely

$$\dot{\xi}_1 = \gamma \xi_3 \frac{1}{EV}\left(P_m - D\omega - EV\xi_3\left(\xi_1 - \gamma\frac{2H}{EV\omega_0}\xi_3\omega\right)\right)$$
$$+\omega\xi_2 + \gamma\frac{2H}{EV\omega_0}\omega\frac{1}{T}(-\xi_3 + \lambda_* + u),$$
$$\dot{\xi}_2 = -\omega\left(\xi_1 - \gamma\frac{2H}{EV\omega_0}\xi_3\omega\right),$$
$$\dot{\xi}_3 = \frac{1}{T}(-\xi_3 + \lambda_* + u),$$

has been simulated using the parameters[6] $P_m = 0.8$ p.u., $D = 1$ p.u., $E = 1.06679$ p.u., $V = 1$ p.u., $2H/\omega_0 = 0.05$ s, and $T = 0.05$ s. The rotor angle at the operating point is $\delta_* = \pi/13.671$ rad. The initial conditions have been set to $\xi(0) = [0, 1, 1]^\top$. The system is considered to be at an equilibrium at $t = 0$ s. At $t = 0.4$ s a three-phase fault occurs at the generator bus and it is cleared at $t = 0.6$ s.

Figure 10.2 shows the time histories of the rotor angle and velocity and of the observation errors for two different values of the observer gain γ, namely $\gamma = 5$ and $\gamma = 50$. Observe that z_1 and z_3 converge asymptotically to zero at a rate that depends on γ. Note that during the fault the stability properties of z_1 and z_2 are lost, since the dynamics of the machine become

$$\dot{\delta} = \omega, \qquad \dot{\omega} = \frac{\omega_0}{2H}(P_m - D\omega),$$

hence δ becomes undetectable. However, after the fault has been cleared, z_1 and z_3 converge asymptotically to zero.

---
[6]These parameters are taken from [210].

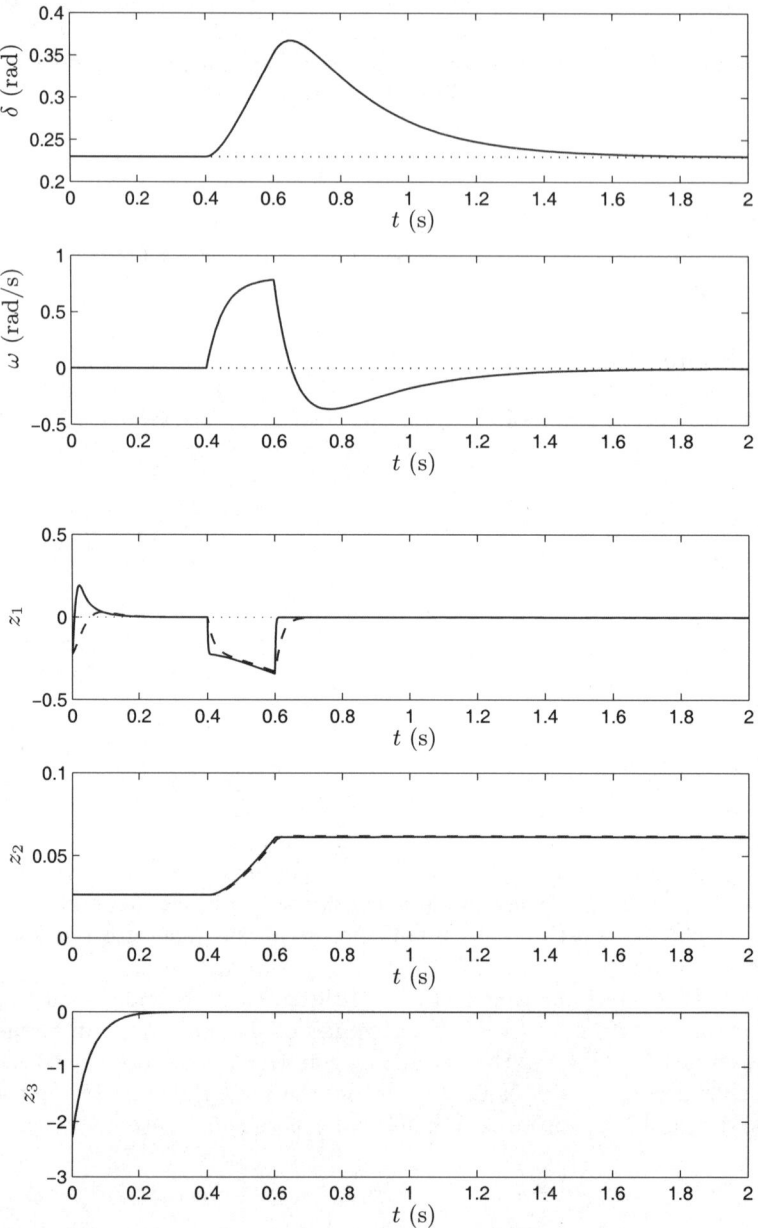

Fig. 10.2. Time histories of the rotor angle δ and rotor velocity ω, and of the observation errors z_1, z_2 and z_3, for $\gamma = 5$ (dashed lines) and $\gamma = 50$ (solid lines).

10.2 Adaptive Control of Current-fed Induction Motors

The industry standard in induction motor control is the so-called *indirect field-oriented control* (IFOC), which provides asymptotic regulation of the rotor speed and flux modulus around constant references and does not require rotor flux sensors (or estimators). The popularity of IFOC stems from its extreme simplicity and intuitive operation that permits independent tuning of the flux and torque control loops. IFOC relies on the assumption that the stator *currents* are available as control inputs—an assumption that is often justified in practice by the use of high-gain current control loops.

A drawback of the standard IFOC scheme is that it requires accurate knowledge of the rotor resistance, which may vary significantly with temperature, frequency and current amplitude. Even though it has been shown[7] that stability is preserved for very large errors in rotor resistance estimation, this mismatch seriously affects the performance: it degrades the flux regulation, which may lead to saturation or under-excitation, slows down the torque response and induces a steady-state error. It is fair to say that, from the practitioners' viewpoint, the development of a plug-in scheme to incorporate adaptation to IFOC is one of the most relevant open problems in induction motor control[8].

In this section the I&I approach is applied to the problem of estimating the rotor resistance and load torque of current-fed induction motors using measurements of the rotor speed, the electromagnetic torque and the rotor flux norm. The proposed estimator does not require persistence of excitation and achieves asymptotic convergence even in the case of zero rotor speed and/or low torque. Furthermore, the problem of torque/speed regulation by means of output feedback is addressed by combining the proposed estimator with the standard IFOC scheme.

Note 10.1. For further detail on IFOC the reader is referred to [48, 163], see also the books [128, 46, 44] and the survey paper [214]. The first globally stable adaptive IFOC, with the only assumption of the rotor resistance belonging to a discrete known (but arbitrarily large) set, was reported in [42]. A globally stable output feedback adaptive design for the current-fed machine was reported in [133]—however, the proposed controller is much more complicated than the basic IFOC and is difficult to implement and tune, see [211] for some experimental evidence. The problem of estimating the rotor resistance of induction motors is also important in fault detection applications and has been studied extensively, see, *e.g.*, [132, 171] and references therein. ◁

---

[7] See [48].

[8] Another important problem is the so-called sensorless control problem, where measurement of mechanical co-ordinates is avoided.

10.2.1 Problem Formulation

The dynamical model of an induction motor in the stator reference frame (also known as a-b or *two-phase equivalent* model) is given by the equations[9]

$$\dot{i}_{ab} = -\left(\frac{R_s}{\sigma L_s} + \frac{M^2 R_r}{\sigma L_s L_r^2}\right) i_{ab} + \frac{M}{\sigma L_s L_r}\left(\frac{R_r}{L_r}I - n_p \omega J\right)\lambda_{ab} + \frac{1}{\sigma L_s}v_{ab},$$

$$\dot{\lambda}_{ab} = \frac{MR_r}{L_r} i_{ab} - \left(\frac{R_r}{L_r}I - n_p \omega J\right)\lambda_{ab}, \qquad (10.10)$$

$$\dot{\omega} = \frac{n_p M}{J_m L_r} i_{ab}^\top J \lambda_{ab} - \frac{\tau}{J_m},$$

where $i_{ab} = [i_a, i_b]^\top \in \mathbb{R}^2$ is the stator current vector, $v_{ab} = [v_a, v_b]^\top \in \mathbb{R}^2$ is the stator voltage vector, $\lambda_{ab} = [\lambda_a, \lambda_b]^\top \in \mathbb{R}^2$ is the rotor flux vector, ω is the rotor speed, R_r, L_r, M, n_p, J_m, R_s and L_s are positive constants representing the rotor resistance, rotor inductance, mutual inductance, number of pole pairs, moment of inertia, stator resistance and stator inductance, respectively, $\sigma = 1 - M^2/(L_s L_r)$ is the leakage parameter, τ is the load torque, and

$$I = \begin{bmatrix} 1 & 0 \\ 0 & 1 \end{bmatrix}, \qquad J = \begin{bmatrix} 0 & -1 \\ 1 & 0 \end{bmatrix}.$$

The dynamical model of the current-fed induction motor is obtained from (10.10) by taking the stator current vector i_{ab} as the control input, *i.e.*, by neglecting the electromagnetic dynamics in the stator circuit.

Defining the rotation matrix

$$e^{-Jn_p q} = \begin{bmatrix} \cos(n_p q) & \sin(n_p q) \\ -\sin(n_p q) & \cos(n_p q) \end{bmatrix},$$

where q is the rotor shaft angle, and the transformations

$$\lambda_r = \begin{bmatrix} \lambda_{r1} \\ \lambda_{r2} \end{bmatrix} = e^{-Jn_p q}\lambda_{ab}, \qquad i_s = \begin{bmatrix} i_{s1} \\ i_{s2} \end{bmatrix} = e^{-Jn_p q} i_{ab},$$

where $\lambda_r \in \mathbb{R}^2$ is the transformed rotor flux vector and $i_s \in \mathbb{R}^2$ is the transformed stator current vector, yields the system

$$\dot{\lambda}_r = -\frac{R_r}{L_r}\lambda_r + \frac{MR_r}{L_r} i_s,$$

$$\dot{\omega} = \frac{n_p M}{J_m L_r} i_s^\top J \lambda_r - \frac{\tau}{J_m}.$$

Note that the above system describes the dynamic behaviour of the current-fed induction motor in a frame rotating with angular speed $n_p \omega$.

---
[9] For a derivation of this model, see [163] or [46].

10.2 Adaptive Control of Current-fed Induction Motors

In the sequel, to simplify the presentation and without loss of generality, it is assumed that all constants are equal to one, except for the rotor resistance and the load torque which are considered unknown. Defining the control input as $u = i_s$ yields the simplified model

$$\dot{\lambda}_r = -R_r \lambda_r + R_r u,$$
$$\dot{\omega} = u^T J \lambda_r - \tau. \tag{10.11}$$

Consider now the variables

$$y_1 = u^T J \lambda_r, \qquad y_2 = u^T \lambda_r, \qquad \eta = |\lambda_r|,$$

where y_1 represents the generated electromagnetic torque and η is the flux norm, and suppose that y_1 and y_2 are available for measurement[10].

The indirect field-oriented controller (IFOC) is described by the equations

$$u = e^{J\rho_d} \begin{bmatrix} \eta_d \\ \tau_d/\eta_d \end{bmatrix}, \tag{10.12}$$

$$\dot{\rho}_d = \frac{\bar{R}_r \tau_d}{\eta_d^2}, \tag{10.13}$$

where η_d and τ_d are the reference values of the flux norm and torque, respectively, $\bar{R}_r$ is an estimate of the rotor resistance and

$$e^{J\rho_d} = \begin{bmatrix} \cos(\rho_d) & -\sin(\rho_d) \\ \sin(\rho_d) & \cos(\rho_d) \end{bmatrix}.$$

In what follows the focus is (mainly) on the case of torque regulation, where τ_d is a constant reference, as opposed to speed regulation, where τ_d is the output of a PI controller, i.e.,

$$\tau_d = -\left(K_P + \frac{K_I}{s}\right)(\omega - \omega_d), \tag{10.14}$$

where ω_d is the speed reference, s denotes the Laplace operator and K_P, K_I are constant gains. Note, however, that the two cases are comparable if the PI is sufficiently slow.

The closed-loop system (10.11), (10.12) can be rewritten in the y_1, y_2 and ω co-ordinates as

$$\dot{y}_1 = -R_r y_1 + \dot{\rho}_d y_2 + \frac{\dot{\tau}_d}{c}\left(\frac{\tau_d}{\eta_d^2} y_1 + y_2\right),$$
$$\dot{y}_2 = -\dot{\rho}_d y_1 - R_r y_2 + R_r c + \frac{\dot{\tau}_d}{c}\left(-y_1 + \frac{\tau_d}{\eta_d^2} y_2\right), \tag{10.15}$$
$$\dot{\omega} = y_1 - \tau,$$

---
[10] In practice these two quantities are not measurable, but they can be estimated, see Section 10.2.3.

where $c = \eta_d^2 + (\tau_d/\eta_d)^2$, while the dynamics of the flux norm are given by

$$\dot{\eta} = -R_r\eta + \frac{1}{\eta}R_r y_2. \tag{10.16}$$

Note that, for the case of the torque regulation problem, $\dot{\tau}_d = 0$, hence the first two equations in (10.15) reduce to

$$\dot{y}_1 = -R_r y_1 + \dot{\rho}_d y_2,$$
$$\dot{y}_2 = -\dot{\rho}_d y_1 - R_r y_2 + R_r c.$$

The design objective is to obtain asymptotic estimates of the rotor resistance R_r and the load torque τ using measurements of y_1, y_2 and ω.

10.2.2 Estimator Design

Following the general theory of Chapter 5, we define the error variables

$$z_1 = \hat{\tau} - \tau + \beta_1(\omega), \tag{10.17}$$
$$z_2 = \hat{R}_r - R_r + \beta_2(y_1), \tag{10.18}$$

where $\beta_1(\cdot)$ and $\beta_2(\cdot)$ are continuous functions yet to be specified, and the update laws

$$\dot{\hat{\tau}}_l = -\frac{\partial \beta_1}{\partial \omega}(y_1 - \hat{\tau} - \beta_1(\omega)), \tag{10.19}$$

$$\dot{\hat{R}}_r = -\frac{\partial \beta_2}{\partial y_1}\left[-\left(\hat{R}_r + \beta_2(y_1)\right)y_1 + \dot{\rho}_d y_2 + \frac{\dot{\tau}_d}{c}\left(\frac{\tau_d}{\eta_d^2}y_1 + y_2\right)\right]. \tag{10.20}$$

The resulting error dynamics are described by the equations

$$\dot{z}_1 = \frac{\partial \beta_1}{\partial \omega} z_1, \tag{10.21}$$
$$\dot{z}_2 = \frac{\partial \beta_2}{\partial y_1} y_1 z_2. \tag{10.22}$$

Selecting the function $\beta_1(\cdot)$ as

$$\beta_1(\omega) = -k_1\omega, \tag{10.23}$$

with $k_1 > 0$, yields the error system

$$\dot{z}_1 = -k_1 z_1, \tag{10.24}$$

which has an asymptotically stable equilibrium at the origin, hence z_1 converges to zero. As a result, from (10.17), an asymptotically converging estimate of the load τ is given by

$$\bar{\tau} = \hat{\tau} + \beta_1(\omega). \tag{10.25}$$

10.2 Adaptive Control of Current-fed Induction Motors

Consider now the problem of finding a function $\beta_2(\cdot)$ such that the system (10.22) has a stable equilibrium[11] at $z_2 = 0$. A possible selection is

$$\beta_2(y_1) = \frac{k_2}{2} \frac{1}{1 + k_3 y_1^2}, \qquad (10.26)$$

with $k_2 > 0$ and $k_3 > 0$ constants, yielding the error system

$$\dot{z}_2 = -\frac{k_2 k_3 y_1^2}{(1 + k_3 y_1^2)^2} z_2, \qquad (10.27)$$

which has a globally stable (uniformly in y_1) equilibrium at zero. From (10.18), and assuming that

$$\frac{y_1(t)}{1 + k_3 y_1(t)^2} \notin \mathcal{L}_2, \qquad (10.28)$$

an asymptotically converging estimate of the rotor resistance R_r is given by

$$\bar{R}_r = \hat{R}_r + \beta_2(y_1). \qquad (10.29)$$

The advantage of (10.26) over the more obvious selection $\beta_2(y_1) = \frac{k_2}{2} y_1^2$ is that it ensures boundedness of the estimate $\bar{R}_r$ for any y_1. This turns out to be particularly useful in the following section to prove stability of a suitable equilibrium of the adaptive closed-loop system.

Summarising, the proposed (second-order) estimator is given by the equations (10.19), (10.20), (10.23), (10.26) and (10.29).

10.2.3 Rotor Flux Estimation

A practical limitation of the proposed scheme is that it relies on measurements of y_1 and y_2, which, in turn, require knowledge of the rotor flux. To overcome this problem it is necessary to devise a method for estimating the flux vector λ_{ab}. To this end, note that from the first two equations in (10.10) we obtain

$$\dot{\lambda}_{ab} = -\frac{\sigma L_s L_r}{M} \dot{i}_{ab} - \frac{R_s L_r}{M} i_{ab} + \frac{L_r}{M} v_{ab},$$

where $\dot{i}_{ab}$ can be computed using (10.12) and the transformation $u = e^{-Jn_p q} i_{ab}$, i.e.,

$$\dot{i}_{ab} = n_p \omega J i_{ab} + e^{J n_p q} \left(\dot{\rho}_d J u + e^{J \rho_d} [0, \dot{\tau}_d / \eta_d] \right).$$

Hence, the (open-loop) observer

---

[11] It is interesting to note that, for any function $\beta_2(\cdot)$, when $y_1 = 0$ the system (10.22) reduces to $\dot{z}_2 = 0$. This implies that it is not possible to estimate the rotor resistance when the torque is identically equal to zero.

$$\dot{\hat{\lambda}}_{ab} = -\frac{\sigma L_s L_r}{M} \dot{i}_{ab} - \frac{R_s L_r}{M} i_{ab} + \frac{L_r}{M} v_{ab}$$

is such that $\hat{\lambda}_{ab} - \lambda_{ab} = \text{const}$, i.e., $\hat{\lambda}_{ab}$ is an exact estimate of λ_{ab} up to a constant error term. It is clear that open-loop observation is a fragile operation whose use is hard to justify in practice. However, in this particular application it may be successful because under normal operation the flux has zero mean, hence a constant error can be practically removed by filtering out the DC component of $\hat{\lambda}_{ab}$.

10.2.4 Controller Design

Consider again the closed-loop system (10.11)–(10.13), where $\bar{R}_r$ is given by (10.29). It has been shown in the previous section that the estimate $\bar{R}_r$ remains bounded and asymptotically converges to the true value R_r, provided (10.28) holds. It is now shown that, for the torque regulation case, the rotor flux and the generated torque remain bounded and asymptotically converge to the reference values.

Proposition 10.1. *Consider the IFOC-driven current-fed induction motor described by the equations (10.15), (10.16), where $\dot{\rho}_d$ is given by (10.13) and η_d and τ_d are constant references, in closed loop with the estimator given by the equations (10.19), (10.20), (10.23), (10.26) and*

$$\bar{R}_r = \max(\hat{R}_r + \beta_2(y_1), R_{min}), \qquad (10.30)$$

with $R_{min} > 0$ an arbitrarily small lower bound on R_r. Then, for all initial conditions, $y_1(t) \in \mathcal{L}_\infty$, $y_2(t) \in \mathcal{L}_\infty$ and

$$\lim_{t \to \infty} y_1(t) = \tau_d, \qquad \lim_{t \to \infty} y_2(t) = \eta_d^2, \qquad \lim_{t \to \infty} \eta(t) = \eta_d.$$

Proof. Define the error variables

$$x_1 = y_1 - \tau_d, \qquad x_2 = y_2 - \eta_d^2$$

and note that the system (10.15) can be rewritten in the x_1, x_2 and ω coordinates as

$$\begin{aligned}
\dot{x}_1 &= -R_r x_1 + \frac{\bar{R}_r \tau_d}{\eta_d^2} x_2 + \left(\bar{R}_r - R_r\right) \tau_d, \\
\dot{x}_2 &= -\frac{\bar{R}_r \tau_d}{\eta_d^2} x_1 - R_r x_2 - \left(\bar{R}_r - R_r\right) \frac{\tau_d^2}{\eta_d^2}, \\
\dot{\omega} &= x_1 + \tau_d - \tau.
\end{aligned} \qquad (10.31)$$

Consider now the Lyapunov function $V(x_1, x_2) = \frac{1}{2} x_1^2 + \frac{1}{2} x_2^2$, whose time-derivative along the trajectories of (10.31) is

$$\dot{V} = -R_r \left(x_1^2 + x_2^2\right) + \tau_d \left(\bar{R}_r - R_r\right) \left(x_1 - \frac{\tau_d}{\eta_d^2} x_2\right).$$

10.2 Adaptive Control of Current-fed Induction Motors

A simple application of Corollary A.3 shows that there exist constants $\epsilon > 0$ and $\delta > 0$ such that

$$\dot{V} \leq -\epsilon V + \delta \left(\bar{R}_r - R_r\right)^2,$$

hence the (x_1, x_2)-subsystem of (10.31) is ISS with respect to $\bar{R}_r - R_r$. It remains to prove that the error $\bar{R}_r - R_r$ is bounded and asymptotically converges to zero. To this end, recall first that the dynamics of the error variable z_2 defined in (10.18) are described by (10.27), hence $z_2(t) \in \mathcal{L}_\infty$ and

$$\frac{y_1 z_2}{1 + k_3 y_1^2} \in \mathcal{L}_2.$$

Since z_2 is bounded, $\bar{R}_r$ is also bounded, hence y_1 and y_2 are bounded and, from (10.15) and (10.27), $\dot{y}_1$, $\dot{y}_2$ and $\dot{z}_2$ are also bounded. From Lemma A.2, this implies that

$$\lim_{t \to \infty} \frac{y_1(t) z_2(t)}{1 + k_3 y_1(t)^2} = 0,$$

hence either z_2 converges to zero or y_1 converges to zero and z_2 converges to a nonzero constant. Due to the dynamics (10.15) the latter is only possible if $\dot{\rho}_d = 0$. But from (10.13) and (10.30) it follows that

$$|\dot{\rho}_d| \geq \frac{R_{min}|\tau_d|}{\eta_d^2} > 0.$$

Hence, z_2 and therefore $\bar{R}_r - R_r$ converge to zero. As a result, x_1 and x_2 are bounded and asymptotically converge to zero, which proves the claims. □

10.2.5 Simulation Results

The model of the induction motor (10.10) has been simulated using the parameters $R_r = 2$, $\tau = 2$ and assuming that all other constants are equal to one. A high-gain current control loop has been implemented as

$$v_{ab} = K_c \left(e^{Jn_p q} u - i_{ab}\right),$$

with $K_c = 500$, where u is given by (10.12), (10.13), with $\bar{R}_r$ given by (10.30). The initial conditions are defined as $\lambda_{ab}(0) = [1, 0]^\top$, $\omega(0) = 0$ and $i_{ab}(0) = [0, 0]^\top$. The reference of the flux norm has been set to $\eta_d = 1$. The adaptive gains have been set to $k_1 = k_2 = 10$ and $k_3 = 1$. In order to compare the response with the non-adaptive case, during the first 5 seconds we replace $\bar{R}_r$ in (10.13) with a fixed estimate. In this case we have selected $\bar{R}_r = R_r/2$.

Torque regulation

We first consider the torque regulation problem, where the torque reference is fixed at $\tau_d = 2$. Figure 10.3 (a) shows the time histories of the flux vector λ_r

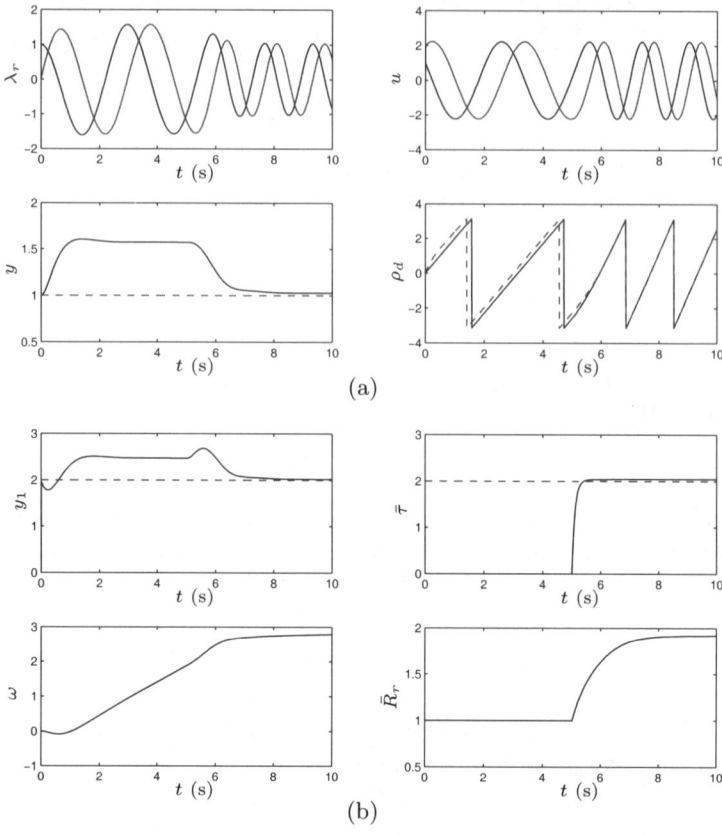

Fig. 10.3. Time histories for the torque regulation case. (a) Top left: Rotor flux vector λ_r. Bottom left: Flux norm y. Top right: Stator current vector u. Bottom right: Controller state ρ_d (solid line) and flux angle (dashed line). (b) Top left: Generated torque y_1. Bottom left: Rotor speed ω. Top right: Load torque estimate $\bar{\tau}$. Bottom right: Rotor resistance estimate $\bar{R}_r$.

and flux norm η, together with control input u, the controller state ρ_d and the flux angle. A plot of the generated torque y_1 is shown in Figure 10.3 (b). Notice that, as expected from the theory[12], the mismatch in the estimate of the rotor resistance during the first 5 seconds results in a significant steady-state error both in the flux level and in the generated torque. The convergence of the estimates $\bar{\tau}$ and $\bar{R}_r$ to the true values τ and R_r is shown in Figure 10.3 (b). Notice that there is a small residual error which is due to the filtering of the flux observations (see Section 10.2.3).

---

[12] See the analysis in [48].

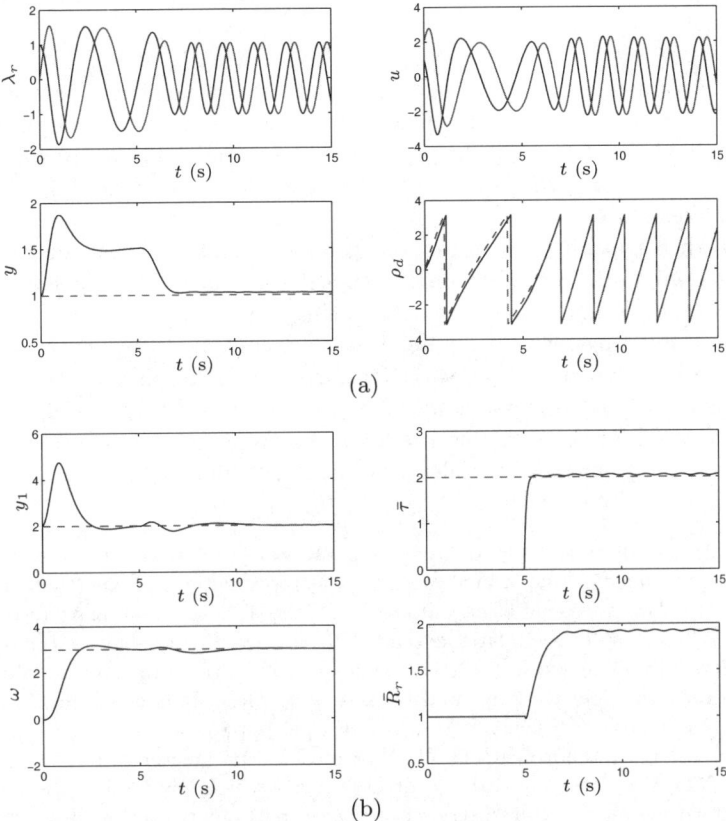

Fig. 10.4. Time histories for the speed regulation case. (a) Top left: Rotor flux vector λ_r. Bottom left: Flux norm y. Top right: Stator current vector u. Bottom right: Controller state ρ_d (solid line) and flux angle (dashed line). (b) Top left: Generated torque y_1. Bottom left: Rotor speed ω. Top right: Load torque estimate $\bar{\tau}$. Bottom right: Rotor resistance estimate $\bar{R}_r$.

Speed regulation

For the speed regulation case, recall that τ_d is given by the PI control law (10.14), where we have selected $K_P = K_I = 1$. A plot of the flux vector λ_r and the flux norm y is shown in Figure 10.4 (a), together with the time histories of the control input u, the state ρ_d and the flux angle. The time histories of the generated torque y_1 and the rotor speed ω are shown in Figure 10.4 (b). Again we see that all signals converge to their respective reference values. Figure 10.4 (b) also shows the convergence of the estimates $\bar{\tau}$ and $\bar{R}_r$ to the true values τ and R_r with a small steady-state error due to the inaccuracy in the flux observations.

10.3 Speed/Flux Tracking for Voltage-fed Induction Motors

In this section an adaptive output feedback control scheme for the voltage-fed model of the induction motor is presented. The proposed controller uses measurements of the stator currents and rotor speed and yields global asymptotic convergence of the rotor speed and flux magnitude to their desired (time-varying) references.

The distinguishing feature of the proposed scheme, which makes it appealing from a practical viewpoint, is that it consists of a simple *static output feedback* control law which is derived using standard arguments pertaining to Hamiltonian systems and passivity-based control. In order to make the scheme robust to load torque variations, an adaptation mechanism is added, which yields asymptotic estimates of the unknown load torque and a rigorous stability proof for the resulting *cascaded* closed-loop system is provided. The efficacy of the proposed controller is verified via simulations and experimental tests.

Note 10.2. The first globally convergent IFOC-based controller for the full-order (voltage-fed) model of the induction motor was reported in [164] and then extended in [154] to a larger class of electrical machines, see [163] for a summary of the main results. These results were later extended to asymptotic tracking of *time-varying* speed and flux references in [173, 172] using a combination of proportional-integral (PI) current and velocity loops and backstepping ideas. It is worth noting that in all the aforementioned schemes the load torque applied to the rotor is assumed constant though unknown, while all other machine parameters are assumed to be known. Adaptive output feedback control schemes which rely on flux observers and can deal with additional parametric uncertainty (particularly in the rotor resistance) have been reported in [71, 46, 133, 134]. More recent efforts have focused on speed-sensorless schemes, where the load torque is assumed known but the rotor speed is not available for feedback, see for instance [57, 139]. ◁

10.3.1 Problem Formulation

The two-phase equivalent model of an induction motor in the stator $(a\text{-}b)$ reference frame is described by the equations (10.10), which we rewrite for convenience as

$$\begin{aligned}
\dot{i}_{ab} &= -a_0 i_{ab} + a_1 \left(I - T_r n_p \omega J\right) \lambda_{ab} + a_2 u_{ab}, \\
\dot{\lambda}_{ab} &= \frac{M}{T_r} i_{ab} - \left(\frac{1}{T_r} I - n_p \omega J\right) \lambda_{ab}, \\
\dot{\omega} &= a_3 i_{ab}^\top J \lambda_{ab} - \frac{\tau}{J_m},
\end{aligned} \qquad (10.32)$$

where $i_{ab} \in \mathbb{R}^2$ is the stator current vector, $u_{ab} \in \mathbb{R}^2$ is the stator voltage vector, $\lambda_{ab} \in \mathbb{R}^2$ is the rotor flux vector, $\omega \in \mathbb{R}$ is the rotor speed, τ represents the unknown (constant) load torque, and

10.3 Speed/Flux Tracking for Voltage-fed Induction Motors

$$I = \begin{bmatrix} 1 & 0 \\ 0 & 1 \end{bmatrix}, \quad J = \begin{bmatrix} 0 & -1 \\ 1 & 0 \end{bmatrix}.$$

The parameters a_0, a_1, a_2, a_3 are known positive constants defined as

$$a_0 = \frac{R_s}{\sigma L_s} + \frac{M^2}{\sigma L_s L_r T_r}, \quad \sigma = 1 - \frac{M^2}{L_s L_r}, \quad T_r = \frac{L_r}{R_r},$$

$$a_1 = \frac{M}{\sigma L_s L_r T_r}, \quad a_2 = \frac{1}{\sigma L_s}, \quad a_3 = \frac{n_p M}{J_m L_r},$$

where R_r, L_r, M, n_p, J_m, R_s and L_s represent the rotor resistance, rotor inductance, mutual inductance, number of pole pairs, moment of inertia, stator resistance and stator inductance, respectively. It is assumed that only the stator currents i_{ab} and the rotor speed ω are available for measurement, while the control input is represented by the stator voltages u_{ab}.

The system (10.32) can be alternatively expressed in the, so-called, d-q reference frame via the change of co-ordinates

$$i_{ab} = e^{J\theta_s} i_{dq}, \quad \lambda_{ab} = e^{J\theta_s} \lambda_{dq}, \quad u_{ab} = e^{J\theta_s} u_{dq}, \qquad (10.33)$$

where θ_s is the angle between the d-axis and the a-axis and

$$e^{J\theta_s} = \begin{bmatrix} \cos(\theta_s) & -\sin(\theta_s) \\ \sin(\theta_s) & \cos(\theta_s) \end{bmatrix}.$$

In the new reference frame the dynamical model of the induction motor is described by the equations

$$\dot{i}_{dq} = -(a_0 I + \omega_s J) i_{dq} + a_1 (I - T_r n_p \omega J) \lambda_{dq} + a_2 u_{dq},$$

$$\dot{\lambda}_{dq} = \frac{M}{T_r} i_{dq} - \left(\frac{1}{T_r} I + (\omega_s - n_p \omega) J\right) \lambda_{dq}, \qquad (10.34)$$

$$\dot{\omega} = a_3 i_{dq}^\top J \lambda_{dq} - \frac{\tau}{J_m},$$

and

$$\dot{\theta}_s = \omega_s, \qquad (10.35)$$

where $i_{dq} \in \mathbb{R}^2$ are the stator currents, $\lambda_{dq} \in \mathbb{R}^2$ are the rotor fluxes, $u_{dq} \in \mathbb{R}^2$ are the stator voltages, and ω_s is the angular velocity of the d-q reference frame and acts as a new control input[13].

Define the state and input vectors

$$x = \begin{bmatrix} x_{12} \\ x_{34} \\ x_5 \end{bmatrix} \triangleq \begin{bmatrix} i_{dq} \\ \lambda_{dq} \\ \omega \end{bmatrix} \in \mathbb{R}^5, \quad u = \begin{bmatrix} u_{12} \\ u_3 \end{bmatrix} \triangleq \begin{bmatrix} u_{dq} \\ \omega_s - n_p \omega \end{bmatrix} \in \mathbb{R}^3,$$

---

[13] This extra degree of freedom, which is used to shape the dynamics of the rotor flux, was first reported in the control literature in [158] and has been widely exploited since.

where the new input u_3 represents the *slip* velocity, and note that the system (10.34) can be rewritten as

$$\begin{aligned}
\dot{x}_{12} &= -(a_0 I + (n_p x_5 + u_3)J)x_{12} + a_1(I - T_r n_p x_5 J)x_{34} + a_2 u_{12}, \\
\dot{x}_{34} &= \frac{M}{T_r}x_{12} - \left(\frac{1}{T_r}I + u_3 J\right)x_{34}, \\
\dot{x}_5 &= a_3 x_{12}^\top J x_{34} - \frac{\tau}{J_m}.
\end{aligned} \qquad (10.36)$$

The vector of variables to be controlled consists of the rotor speed and flux modulus and is denoted by

$$y = \begin{bmatrix} x_5 \\ |x_{34}| \end{bmatrix} = \begin{bmatrix} \omega \\ |\lambda_{ab}| \end{bmatrix}. \qquad (10.37)$$

Along with (10.37) we also consider a vector of (time-varying) reference signals denoted by

$$r = r(t) = \begin{bmatrix} x_5^* \\ y_2^* \end{bmatrix}, \qquad (10.38)$$

where x_5^* is the desired rotor speed and $y_2^* > 0$ is the desired flux modulus.

The speed and flux tracking problem is to find a (dynamic) control law of the form

$$\begin{aligned}
u &= v(x_{12}, x_5, r, \dot{r}, \ddot{r}, \xi), \\
\dot{\xi} &= \alpha(x_{12}, x_5, r, \dot{r}, \xi),
\end{aligned} \qquad (10.39)$$

such that all trajectories of the closed-loop system (10.36), (10.39) remain bounded and

$$\lim_{t \to \infty}(y(t) - r(t)) = 0, \qquad (10.40)$$

where $r(t)$ is bounded with bounded first and second derivatives.

10.3.2 Nonlinear Control Design

In this section we design a *static* output feedback control law that solves the tracking problem *assuming* that the load torque τ is known. This assumption is removed in the following section by considering a dynamic control law of the form (10.39). The basic idea[14] is to transform the system (10.36) into a Hamiltonian system of the form

$$\dot{\tilde{x}} = A_d(\tilde{x}, t)\frac{\partial H_d^\top}{\partial \tilde{x}}, \qquad (10.41)$$

where $A_d + A_d^\top < 0$ and $H_d(\tilde{x})$ is the desired closed-loop *total energy* function which has a minimum at zero. The main requirement is that the mapping from the target system (10.41) to the original system is stability preserving and that

---
[14] See the methodology in [167].

10.3 Speed/Flux Tracking for Voltage-fed Induction Motors

the equilibrium point of the system (10.41) is mapped onto a trajectory of the original system which satisfies condition (10.40) with r given by (10.38).

To begin with, consider the global change of variables

$$\begin{aligned}
\tilde{x}_{12} &= x_{12} - x_{12}^*(x_5, r, \dot{r}), \\
\tilde{x}_{34} &= x_{34} - x_{34}^* = x_{34} - \begin{bmatrix} y_2^* \\ 0 \end{bmatrix}, \\
\tilde{x}_5 &= x_5 - x_5^*,
\end{aligned} \tag{10.42}$$

where $x_{12}^*(\cdot)$ is a mapping from $\mathbb{R}^3$ to $\mathbb{R}^2$ yet to be determined. Note that, following the idea of *field orientation*[15], we have chosen the desired rotor flux vector x_{34}^* to be aligned with the d-axis of the d-q reference frame. The system (10.36) is described in the $\tilde{x}$ co-ordinates by the equations

$$\begin{aligned}
\dot{\tilde{x}}_{12} &= -(a_0 I + \omega_s J)(\tilde{x}_{12} + x_{12}^*) + a_1(I - T_r n_p x_5^* J)(\tilde{x}_{34} + x_{34}^*) \\
&\quad - a_1 T_r n_p \tilde{x}_5 J(\tilde{x}_{34} + x_{34}^*) + a_2 u_{12} - \dot{x}_{12}^*, \\
\dot{\tilde{x}}_{34} &= \frac{M}{T_r}(\tilde{x}_{12} + x_{12}^*) - \left(\frac{1}{T_r} I + u_3 J\right)(\tilde{x}_{34} + x_{34}^*) - \dot{x}_{34}^*, \\
\dot{\tilde{x}}_5 &= a_3 \tilde{x}_{12}^\top J(\tilde{x}_{34} + x_{34}^*) + a_3 x_{12}^{*\top} J \tilde{x}_{34} + \frac{\tau^*}{J_m} - \frac{\tau}{J_m},
\end{aligned} \tag{10.43}$$

where

$$\tau^* = J_m \left(a_3 x_{12}^{*\top} J x_{34}^* - \dot{x}_5^*\right) \tag{10.44}$$

represents the required electromagnetic torque at steady state.

Consider now the total energy function

$$H_d(\tilde{x}) = \frac{m_1}{2}|\tilde{x}_{12}|^2 + \frac{m_2}{2}|\tilde{x}_{34}|^2 + \frac{m_3}{2}\tilde{x}_5^2, \tag{10.45}$$

where m_1, m_2 and m_3 are positive constants, and note that the system (10.43) can be rewritten in the form

$$\dot{\tilde{x}} = (J_d(\tilde{x}, r, \dot{r}) - R_d(\tilde{x}_{12}, \tilde{x}_5, r, \dot{r}, u_3)) \frac{\partial H_d^\top}{\partial \tilde{x}}$$

$$+ F(\tilde{x}_{12}, \tilde{x}_5, r, \dot{r}, u_{12}, u_3) + G(\tilde{x}_5, r, \dot{r}) \left(\frac{\tau^* - \tau}{J_m} + k\tilde{x}_5\right), \tag{10.46}$$

where

$$J_d(\cdot) = \begin{bmatrix} 0 & 0 & -\frac{a_1 T_r n_p}{m_3} J x_{34} \\ 0 & 0 & -\frac{c}{m_3} J^\top x_{12}^* \\ \frac{a_3}{m_1} x_{34}^\top J^\top & \frac{a_3}{m_2} x_{12}^{*\top} J & 0 \end{bmatrix},$$

$$R_d(\cdot) = \begin{bmatrix} \frac{1}{m_1}(K + a_0 I + \omega_s J) & -\frac{a_1}{m_2}(I - T_r n_p x_5^* J) + \frac{\partial x_{12}^*}{\partial \tilde{x}_5} \frac{a_3}{m_2} x_{12}^\top J & 0 \\ -\frac{1}{m_1}\frac{M}{T_r} I & \frac{1}{m_2}\left(\frac{1}{T_r} I + u_3 J\right) & 0 \\ 0 & 0 & \frac{1}{m_3} k \end{bmatrix},$$

---
[15] Field-oriented control was introduced by Blaschke in [29].

10 Electromechanical Systems

$$F(\cdot) = \begin{bmatrix} \frac{M}{T_r}x_{12}^* - \left(\frac{1}{T_r}I + u_3 J\right)x_{34}^* - \dot{x}_{34}^* + cJ^\top x_{12}^* \tilde{x}_5 \\ 0 \end{bmatrix}, \quad G(\cdot) = \begin{bmatrix} -\frac{\partial x_{12}^*}{\partial \tilde{x}_5} \\ 0 \\ 1 \end{bmatrix},$$

$$\star = K\tilde{x}_{12} - (a_0 I + (n_p x_5 + u_3)J)x_{12}^* + a_1(I - T_r n_p x_5^* J)x_{34}^* + a_2 u_{12}$$
$$- \frac{\partial x_{12}^*}{\partial \tilde{x}_5}(a_3 \tilde{x}_{12}^\top J x_{34}^* - k\tilde{x}_5) - \frac{\partial x_{12}^*}{\partial r}\dot{r} - \frac{\partial x_{12}^*}{\partial \dot{r}}\ddot{r},$$

and we have added and subtracted the damping terms $k\tilde{x}_5$ and $K\tilde{x}_{12}$, where $k > 0$ is a design parameter and $K = K(\tilde{x}_{12}, \tilde{x}_5, r, \dot{r})$ is a positive-definite matrix-valued function that is selected to render the matrix $R_d + R_d^\top$ positive-definite. We have also added and subtracted the term $cJ^\top x_{12}^* \tilde{x}_5$, where $c > 0$ is a constant, to render the matrix J_d skew-symmetric. This is achieved by selecting the coefficients of the total energy function (10.45) as

$$m_1 = \frac{1}{a_1 T_r n_p}, \quad m_2 = \frac{1}{c}, \quad m_3 = \frac{1}{a_3}.$$

The system (10.46) takes the desired Hamiltonian form (10.41), if the *matching equations*

$$F(\tilde{x}_{12}, \tilde{x}_5, r, \dot{r}, u_{12}, u_3) + G(\tilde{x}_5, r, \dot{r})\left(\frac{\tau^* - \tau}{J_m} + k\tilde{x}_5\right) = 0,$$

or equivalently

$$\star = 0,$$
$$\frac{M}{T_r}x_{12}^* - \left(\frac{1}{T_r}I + u_3 J\right)x_{34}^* - \dot{x}_{34}^* + cJ^\top x_{12}^* \tilde{x}_5 = 0, \quad (10.47)$$
$$\frac{\tau^* - \tau}{J_m} + k\tilde{x}_5 = 0,$$

are satisfied. The first equation in (10.47) can be solved directly for the control law u_{12} yielding

$$u_{12} = \frac{1}{a_2}\Bigg(-K\tilde{x}_{12} + (a_0 I + \omega_s J)x_{12}^* - a_1(I - T_r n_p x_5^* J)x_{34}^*$$
$$+ \frac{\partial x_{12}^*}{\partial \tilde{x}_5}(a_3 \tilde{x}_{12}^\top J x_{34}^* - k\tilde{x}_5) + \frac{\partial x_{12}^*}{\partial r}\dot{r} + \frac{\partial x_{12}^*}{\partial \dot{r}}\ddot{r}\Bigg). \quad (10.48)$$

Solving the last two equations in (10.47) for x_{12}^* and u_3 yields a unique solution which is given by

$$x_1^* = \frac{1}{M}y_2^* + \frac{T_r}{M}\dot{y}_2^* - \frac{cT_r}{M}\tilde{x}_5 x_2^*,$$
$$x_2^* = \frac{1}{a_3 y_2^*}\left(\frac{\tau}{J_m} - k\tilde{x}_5 + \dot{x}_5^*\right), \quad (10.49)$$

10.3 Speed/Flux Tracking for Voltage-fed Induction Motors

and
$$u_3 = \frac{1}{y_2^*}\left(\frac{M}{T_r}x_2^* - c\tilde{x}_5 x_1^*\right). \tag{10.50}$$

It remains to prove that there exists a function $K(\cdot)$ such that the matrix $R_d + R_d^\top$ is positive-definite. This would imply that the system (10.46) has a uniformly globally asymptotically stable equilibrium at the origin, hence all trajectories are such that condition (10.40) holds. The result is summarised in the following proposition.

Proposition 10.2. *Consider the induction motor model (10.36), (10.37) with the reference signals defined in (10.38), (10.42) and (10.49), and suppose that $r(t)$ and its first and second derivatives are bounded. Then there exists a matrix $K(\cdot)$ such that, for any constant $c > 0$ and $k > 0$, all trajectories of the closed-loop system (10.36), (10.48), (10.50) are bounded and condition (10.40) holds.*

Proof. Let $K = \text{diag}(K_1, K_2)$ and note that equations (10.47) hold, hence the system (10.46) becomes

$$\dot{\tilde{x}} = (J_d(\tilde{x}, r, \dot{r}) - R_d(\tilde{x}_{12}, \tilde{x}_5, r, \dot{r}, u_3))\frac{\partial H_d^\top}{\partial \tilde{x}}, \tag{10.51}$$

with $J_d = -J_d^\top$. Consider now the matrix $S = \frac{1}{2}(R_d + R_d^\top)$ which satisfies

$$-\tilde{x}^\top S \tilde{x} = -a_1 T_r n_p (K_1 + a_0) \tilde{x}_1^2 - a_1 T_r n_p (K_2 + a_0) \tilde{x}_2^2 - \frac{c}{T_r}\tilde{x}_3^2 - \frac{c}{T_r}\tilde{x}_4^2$$
$$- k a_3 \tilde{x}_5^2 + p_1 \tilde{x}_1 \tilde{x}_3 + p_2 \tilde{x}_2 \tilde{x}_4 + p_3 \tilde{x}_2 \tilde{x}_3 + p_4 \tilde{x}_1 \tilde{x}_4$$
$$\leq -\left(a_1 T_r n_p (K_1 + a_0) - \frac{T_r}{c}\left(p_1^2 + p_4^2\right)\right)\tilde{x}_1^2$$
$$- \left(a_1 T_r n_p (K_2 + a_0) - \frac{T_r}{c}\left(p_2^2 + p_3^2\right)\right)\tilde{x}_2^2$$
$$- \frac{c}{2T_r}\tilde{x}_3^2 - \frac{c}{2T_r}\tilde{x}_4^2 - k a_3 \tilde{x}_5^2,$$

where

$$p_1 = a_1 n_p M + c a_1 - c a_3 \frac{\partial x_1^*}{\partial \tilde{x}_5}x_2, \qquad p_2 = a_1 n_p M + c a_1 + c a_3 \frac{\partial x_2^*}{\partial \tilde{x}_5}x_1$$

$$p_3 = -c a_1 T_r n_p x_5^* - c a_3 \frac{\partial x_2^*}{\partial \tilde{x}_5}x_2, \qquad p_4 = c a_1 T_r n_p x_5^* + c a_3 \frac{\partial x_1^*}{\partial \tilde{x}_5}x_1.$$

Selecting
$$K_1 \geq \frac{1}{c a_1 n_p}\left(p_1^2 + p_4^2\right), \qquad K_2 \geq \frac{1}{c a_1 n_p}\left(p_2^2 + p_3^2\right) \tag{10.52}$$

ensures that $S > 0$, which implies that the system (10.51) has a uniformly globally asymptotically stable equilibrium at zero. As a result, by definition (10.42) and the fact that $r, \dot{r}, \ddot{r}$ are bounded, it follows that all trajectories of the system (10.36) are bounded and condition (10.40) holds, which concludes the proof. □

Note 10.3. Not surprisingly, by fixing the desired rotor flux vector to be aligned with the d-axis, the resulting stator current references given in (10.49) reduce to the ones obtained in the IFOC scheme, see *e.g.*, [173]. Note, however, that in comparison with [173] both the voltage control law and the slip velocity have simpler forms and there are significantly fewer parameters to be tuned. In particular, by *fixing* the functions $K_1(\cdot)$ and $K_2(\cdot)$ to their lower bounds given in (10.52), the flux and speed convergence rates can be assigned by tuning the parameters c and k. ◁

Note 10.4. The same approach as in the present section has been used in [159] to obtain an output feedback controller for a *reduced* model consisting only of the first two equations in (10.36) with x_5 treated as a known disturbance and with controlled output

$$y = \begin{bmatrix} a_3 x_{12}^\top J x_{34} \\ |x_{34}| \end{bmatrix},$$

i.e., the torque and flux regulation problem has been considered. Note that the matching equations solved in [159] are given by the last two equations in (10.47) with $k = c = 0$. ◁

10.3.3 Adaptive Control Design

In this section the controller described above is combined with an adaptive scheme to ensure asymptotic convergence even when the parameter τ is unknown. Note that, in this case, the first matching equation in (10.47) cannot be solved, hence there is an additional *perturbation* term entering the closed-loop system equations (10.51). The idea then is to modify the selection of $K(\cdot)$ so that the system is *robust* with respect to this perturbation which is driven to zero by the adaptation mechanism.

Following the I&I methodology we consider again the system (10.36) and define the error variables

$$z_1 = \xi_1 - \eta_1 - \gamma x_5, \qquad z_2 = \xi_2 - x_{34},$$

where $\eta_1 = \epsilon \tan(\tau/(\epsilon J_m))$, $\gamma > 0$ is a design parameter, and $\epsilon > 0$ is an arbitrarily large constant such that $|\tau/(\epsilon J_m)| < \pi/2$. Note that this nonlinear parameterisation of the load torque (assuming an upper bound on its magnitude is known) ensures that the estimate remains bounded independently of the motor state. In particular, by the definition of η_1, we have that $\tau = J_m \epsilon \arctan(\eta_1/\epsilon)$. Hence, when z_1 converges to zero, an asymptotically converging estimate of τ is given by the bounded function

$$\hat{\tau} = J_m \epsilon \arctan(\frac{\eta_1 + z_1}{\epsilon}) = J_m \epsilon \arctan(\frac{\xi_1 - \gamma x_5}{\epsilon}). \tag{10.53}$$

Consider now the update laws

$$\begin{aligned} \dot{\xi}_1 &= \gamma \left(a_3 x_{12}^\top J \xi_2 - \frac{\hat{\tau}}{J_m} \right), \\ \dot{\xi}_2 &= \frac{M}{T_r} x_{12} - \left(\frac{1}{T_r} I + u_3 J \right) \xi_2, \end{aligned} \tag{10.54}$$

10.3 Speed/Flux Tracking for Voltage-fed Induction Motors

and note that the resulting error dynamics are given by

$$\dot{z}_1 = -\gamma \Delta(z_1) + \gamma a_3 x_{12}^T J z_2,$$
$$\dot{z}_2 = -\left(\frac{1}{T_r} I + u_3 J\right) z_2, \tag{10.55}$$

where

$$\Delta(z_1) = \epsilon \arctan\left(\frac{\eta_1 + z_1}{\epsilon}\right) - \epsilon \arctan\left(\frac{\eta_1}{\epsilon}\right). \tag{10.56}$$

The above system has the property that z_2 converges to zero exponentially, while z_1 is bounded and converges to zero whenever x_{12} is bounded, as the following lemma shows.

Lemma 10.1. *Consider the system (10.55), where $\Delta(\cdot)$ is given by (10.56), and suppose that $x_{12}(t) \in \mathcal{L}_\infty$. Then the system (10.55) has a uniformly globally stable equilibrium at zero and, moreover,*

$$\lim_{t \to \infty} z_2(t) = 0, \qquad \lim_{t \to \infty} \Delta(z_1(t)) = 0.$$

Proof. By skew-symmetry of J it follows that the z_2-subsystem has a uniformly globally exponentially stable equilibrium at zero. Consider now the z_1 subsystem and the positive-definite function

$$V(z_1) = \epsilon^2 \left(\frac{\eta_1 + z_1}{\epsilon} \arctan\left(\frac{\eta_1 + z_1}{\epsilon}\right) - \frac{1}{2} \ln\left(\left(\frac{\eta_1 + z_1}{\epsilon}\right)^2 + 1\right) - \frac{z_1}{\epsilon} \arctan\left(\frac{\eta_1}{\epsilon}\right)\right),$$

which is such that

$$\frac{\partial V}{\partial z_1} = \epsilon \left(\arctan\left(\frac{\eta_1 + z_1}{\epsilon}\right) - \arctan\left(\frac{\eta_1}{\epsilon}\right)\right) = \Delta(z_1).$$

Differentiating along the trajectories of (10.55) yields

$$\dot{V}(z_1) = -\gamma \Delta(z_1)^2 + \gamma a_3 x_{12}^T J z_2 \Delta(z_1)$$
$$\leq -\frac{\gamma}{2} \Delta(z_1)^2 + \frac{\gamma}{2} a_3^2 |x_{12}^T J z_2|^2.$$

Hence, from boundedness of x_{12} and the fact that z_2 converges exponentially to zero, it follows that z_1 is bounded and $\lim_{t \to \infty} \Delta(z_1(t)) = 0$, which proves the claim. □

From Lemma 10.1 and the definitions of $\Delta(z_1)$ and η_1 it follows that an asymptotically converging estimate of the unknown load torque τ is given by (10.53).

We now modify the control design to ensure that x_{12} remains bounded despite the perturbation in z_1. To this end, consider instead of (10.48) the control law

$$u_{12} = \frac{1}{a_2}\bigg(-K\tilde{x}_{12} + (a_0 I + \omega_s J)x_{12}^* - a_1(I - T_r n_p x_5^* J)x_{34}^*$$
$$+ \frac{\partial x_{12}^*}{\partial \tilde{x}_5}(a_3 \tilde{x}_{12}^\top J x_{34}^* - k\tilde{x}_5) + \frac{\partial x_{12}^*}{\partial r}\dot{r} + \frac{\partial x_{12}^*}{\partial \dot{r}}\ddot{r} + \frac{\partial x_{12}^*}{\partial \xi_1}\dot{\xi}_1 \bigg) \quad (10.57)$$

and replace (10.47) with the modified matching equations

$$\frac{M}{T_r}x_{12}^* - \left(\frac{1}{T_r}I + u_3 J\right)x_{34}^* - \dot{x}_{34}^* + cJ^\top x_{12}^* \tilde{x}_5 = 0,$$
$$\frac{\tau^* - \hat{\tau}}{J_m} + k\tilde{x}_5 = 0, \quad (10.58)$$

which we solve as before for x_{12}^* and u_3 yielding

$$x_1^* = \frac{1}{M}y_2^* + \frac{T_r}{M}\dot{y}_2^* - \frac{cT_r}{M}\tilde{x}_5 x_2^*,$$
$$x_2^* = \frac{1}{a_3 y_2^*}\left(\frac{\hat{\tau}}{J_m} - k\tilde{x}_5 + \dot{x}_5^*\right), \quad (10.59)$$

and

$$u_3 = \frac{1}{y_2^*}\left(\frac{M}{T_r}x_2^* - c\tilde{x}_5 x_1^*\right). \quad (10.60)$$

Note that the only difference between equations (10.59), (10.60) and (10.49), (10.50) is that we have replaced the unknown parameter τ with its asymptotic estimate obtained from (10.53), (10.54). The result is summarised in the following proposition.

Proposition 10.3. *Consider the induction motor model (10.36), (10.37) with the reference signals defined in (10.38), (10.42) and (10.59), and suppose that $r(t)$ and its first and second derivatives are bounded. Then there exists a matrix $K(\cdot)$ such that, for any constant $c > 0$, $k > 0$ and $\gamma > 0$, all trajectories of the system (10.36) in closed loop with (10.54), (10.57) and (10.60) are bounded and condition (10.40) holds.*

Proof. Substituting the control law (10.57) and the matching equations (10.58) (or equivalently (10.59) and (10.60)) into the system (10.46) yields the perturbed Hamiltonian closed-loop system

$$\dot{\tilde{x}} = (J_d(\tilde{x}, r, \dot{r}) - R_d(\tilde{x}_{12}, \tilde{x}_5, r, \dot{r}, u_3))\frac{\partial H_d^\top}{\partial \tilde{x}} - G(\tilde{x}_5, r, \dot{r})\Delta(z_1), \quad (10.61)$$

with $J_d = -J_d^\top$ and Δ given by (10.56). Notice that, since x_{12}^* now depends also on ξ_1, there is an additional term in $\dot{\xi}_1$ that appears in the $\tilde{x}_{12}$ dynamics. However, this term is cancelled out by the last term in the control law (10.57).

Consider now the total energy function defined in (10.45) and note that its time-derivative along the trajectories of (10.61) satisfies

10.3 Speed/Flux Tracking for Voltage-fed Induction Motors

$$\dot{H}_d(\tilde{x}) = -\frac{\partial H_d}{\partial \tilde{x}} \left(\frac{R_d + R_d^\top}{2}\right) \frac{\partial H_d^\top}{\partial \tilde{x}} - \frac{\partial H_d}{\partial \tilde{x}} G \Delta$$

$$\leq -\frac{\partial H_d}{\partial \tilde{x}} \left(\frac{R_d + R_d^\top}{2} - \delta G G^\top\right) \frac{\partial H_d^\top}{\partial \tilde{x}} + \frac{1}{4\delta}\Delta^2,$$

where $\delta > 0$ is a constant. Since $\Delta(\cdot)$ is bounded by definition, to prove boundedness of trajectories it suffices to show that there exists $K = \mathrm{diag}(K_1, K_2)$ such that the matrix

$$S_\delta = \frac{1}{2}(R_d + R_d^\top) - \delta G G^\top$$

is positive-definite for some δ. To this end, consider the function $\tilde{x}^\top S_\delta \tilde{x}$ which satisfies

$$-\tilde{x}^\top S_\delta \tilde{x} = -a_1 T_r n_p (K_1 + a_0) \tilde{x}_1^2 - a_1 T_r n_p (K_2 + a_0) \tilde{x}_2^2 - \frac{c}{T_r}\tilde{x}_3^2 - \frac{c}{T_r}\tilde{x}_4^2$$
$$- k a_3 \tilde{x}_5^2 + p_1 \tilde{x}_1 \tilde{x}_3 + p_2 \tilde{x}_2 \tilde{x}_4 + p_3 \tilde{x}_2 \tilde{x}_3 + p_4 \tilde{x}_1 \tilde{x}_4 + \delta\left(\tilde{x}^\top G\right)^2$$

$$\leq -\left(a_1 T_r n_p (K_1 + a_0) - \frac{T_r}{c}(p_1^2 + p_4^2) - \delta\left(\frac{\partial x_1^*}{\partial \tilde{x}_5}\right)^2\right) \tilde{x}_1^2$$
$$-\left(a_1 T_r n_p (K_2 + a_0) - \frac{T_r}{c}(p_2^2 + p_3^2) - \delta\left(\frac{\partial x_2^*}{\partial \tilde{x}_5}\right)^2\right) \tilde{x}_2^2$$
$$-\frac{c}{2T_r}\tilde{x}_3^2 - \frac{c}{2T_r}\tilde{x}_4^2 - (ka_3 - \delta)\tilde{x}_5^2,$$

where

$$p_1 = a_1 n_p M + c a_1 - c a_3 \frac{\partial x_1^*}{\partial \tilde{x}_5} x_2, \quad p_2 = a_1 n_p M + c a_1 + c a_3 \frac{\partial x_2^*}{\partial \tilde{x}_5} x_1$$

$$p_3 = -c a_1 T_r n_p x_5^* - c a_3 \frac{\partial x_2^*}{\partial \tilde{x}_5} x_2, \quad p_4 = c a_1 T_r n_p x_5^* + c a_3 \frac{\partial x_1^*}{\partial \tilde{x}_5} x_1.$$

Selecting

$$K_1 \geq \frac{1}{c a_1 n_p}(p_1^2 + p_4^2) + \frac{\delta}{a_1 T_r n_p}\left(\frac{\partial x_1^*}{\partial \tilde{x}_5}\right)^2,$$
$$K_2 \geq \frac{1}{c a_1 n_p}(p_2^2 + p_3^2) + \frac{\delta}{a_1 T_r n_p}\left(\frac{\partial x_2^*}{\partial \tilde{x}_5}\right)^2, \quad (10.62)$$

and $\delta < k a_3$, ensures that $S_\delta > 0$, which implies that all trajectories of the system (10.61) are bounded and converge to the set

$$\Omega = \{\tilde{x} \in \mathbb{R}^5 : \frac{\partial H_d}{\partial \tilde{x}} S_\delta \frac{\partial H_d^\top}{\partial \tilde{x}} \leq \frac{1}{4\delta}\Delta^2\}.$$

Hence, by definition (10.42) and the fact that $r, \dot{r}, \ddot{r}$ are bounded, it follows that all trajectories of the system (10.36) are bounded and Lemma 10.1 holds. As a result, $\lim_{t\to\infty} \Delta(z_1(t)) = 0$, hence all trajectories of the system (10.61) converge to the origin and condition (10.40) holds, which proves the claim. $\square$

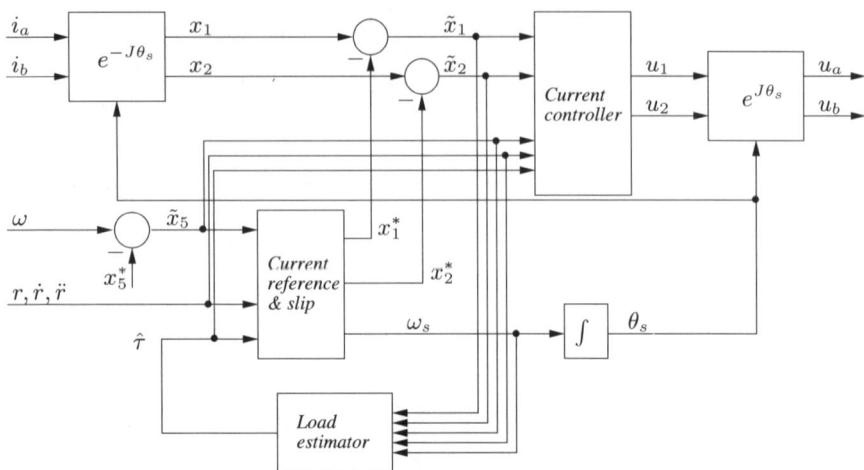

Fig. 10.5. Block diagram of the proposed adaptive controller.

By selecting the function $K = \text{diag}(K_1, K_2)$ as in the proof of Proposition 10.3, the underlying controller described by the equations (10.57), (10.59), (10.60) and (10.53) is such that the system (10.61) is ISS with respect to the perturbation $\Delta(z_1)$. Moreover, from Lemma 10.1, this perturbation remains bounded and is driven to zero by the dynamic update law (10.54) with $\gamma > 0$. Due to this *cascaded* structure of the adaptive controller, we can easily tune the design parameters to achieve the desired performance. In particular, it is clear that the convergence rate of the rotor speed is determined by the parameter k, while γ determines the convergence rate of the estimation error.

A diagram of the proposed adaptive controller is shown in Figure 10.5.

10.3.4 Simulations and Experimental Results

Experimental tests have been carried out using the three-phase induction motor depicted in Figure 10.6 with the parameters given in Table 10.1[16]. The proposed controller is implemented on a dSPACE DS1104 platform using a fourth-order Runge–Kutta integration algorithm and a sampling time of 200 μs. The control voltages applied to the stator are obtained from a three-phase PWM inverter linked to a rectifier with input voltage 230 V. The modulation frequency is 13 kHz. As shown in Figure 10.6, the rotor is connected to a variable load which is initially set to zero.

The stator currents in the *a-b* reference frame are obtained from the three-phase measurements via the transformation

---

[16] See [211] for details on this machine.

10.3 Speed/Flux Tracking for Voltage-fed Induction Motors

Fig. 10.6. The induction motor connected to a variable load.

Table 10.1. Motor parameters

Parameter	Symbol	Value
Rated power	P_n	0.5 kW
Rated torque	τ_n	7 N m
Rated speed	ω_n	1500 rpm
Stator resistance	R_s	8 Ω
Stator inductance	L_s	0.47 H
Rotor resistance	R_r	2.76 Ω
Rotor inductance	L_r	0.4183 H
Mutual inductance	M	0.4183 H
Total leakage	σ	0.11
Number of pole pairs	n_p	2
Moment of inertia	J_m	0.06 kg m^2

$$\begin{bmatrix} i_a \\ i_b \end{bmatrix} = \begin{bmatrix} \frac{\sqrt{2}}{\sqrt{3}} & -\frac{1}{\sqrt{6}} & -\frac{1}{\sqrt{6}} \\ 0 & \frac{1}{\sqrt{2}} & -\frac{1}{\sqrt{2}} \end{bmatrix} \begin{bmatrix} i_1 \\ i_2 \\ i_3 \end{bmatrix}.$$

The transpose transformation is used to compute the three-phase stator reference voltages from the control signal u_{ab}. The speed measurements are obtained from a position transducer mounted on the shaft.

For comparison we have also considered a simplified *torque* and flux controller (see Note 10.4). In order to use it for speed regulation purposes, a proportional control loop has been added around the torque reference. Note that this is *equivalent* to the last equation in (10.47) with $\tau = 0$, where k corresponds to the proportional gain.

Experimental results

The first experiment consists in changing the reference rotor speed from 500 rpm (or 52.36 rad/s) to 1000 rpm, while the reference of the rotor flux magnitude is set to $y_2^* = 1.07$ Wb, which is the rated value for this motor.

Figure 10.7 (a) shows the time histories of the stator voltages and currents, rotor speed and estimated load torque when the design parameters are set to $k = 10$, $\gamma = 4$, $c = 0.1$ and $\delta = 0.001$. The function $K = \mathrm{diag}(K_1, K_2)$ has been fixed to its lower bound given in the proof of Proposition 10.3 (see also Note 10.3). Note that for stability we require $c > 0$, however it has been observed that transient performance deteriorates for larger values of c, due to the stronger coupling between the error dynamics of the electrical and mechanical subsystems. Figure 10.7 (b) shows the time histories of the states when the speed loop gain is changed to $k = 50$. Observe that the convergence of the speed to its reference value is now faster, at the expense of the stator current which peaks at a higher value. A similar response, but with faster convergence of the load torque estimate, is obtained by increasing the adaptive gain to $\gamma = 15$, as shown in Figure 10.7 (c). Notice that the undershoot in the estimate also increases in this case, due to the "proportional" term γx_5 in (10.53).

Figure 10.8 shows the time histories of the states when we apply the simplified torque controller with $\hat{\tau} = -J_m k \tilde{x}_5$ and $k = 5$. Observe that the transient response of the rotor speed is similar to the one obtained using the adaptive controller, but there is now a noticeable steady-state error. On the other hand, due to the absence of the estimator, the transient of the stator currents is less pronounced.

Finally, Figure 10.9 shows the transient responses of the proposed controller and the simplified torque controller during a change in the load torque from 0 to 2 Nm. Notice that the adaptive controller maintains the rotor speed close to its reference value with a relatively small steady-state error.

The second experiment consists in periodically changing the speed reference from -500 rpm to 500 rpm, which corresponds to one third of the rated speed for this motor. The first and second derivatives of the speed reference, which are required to compute the control law, have been obtained via a third-order linear filter. The time histories of the motor states and load torque estimate for different values of the parameters k and γ are shown in Figure 10.10. Observe that the motor exhibits good tracking performance, most notably at low speed and around the zero crossing.

Simulation results

Simulations have been carried out using the induction motor model (10.32) in order to assess the performance of the proposed controller in tracking the rotor flux which was not measurable in the experimental setup. For consistency we have used the same parameters as in Table 10.1.

10.3 Speed/Flux Tracking for Voltage-fed Induction Motors   265

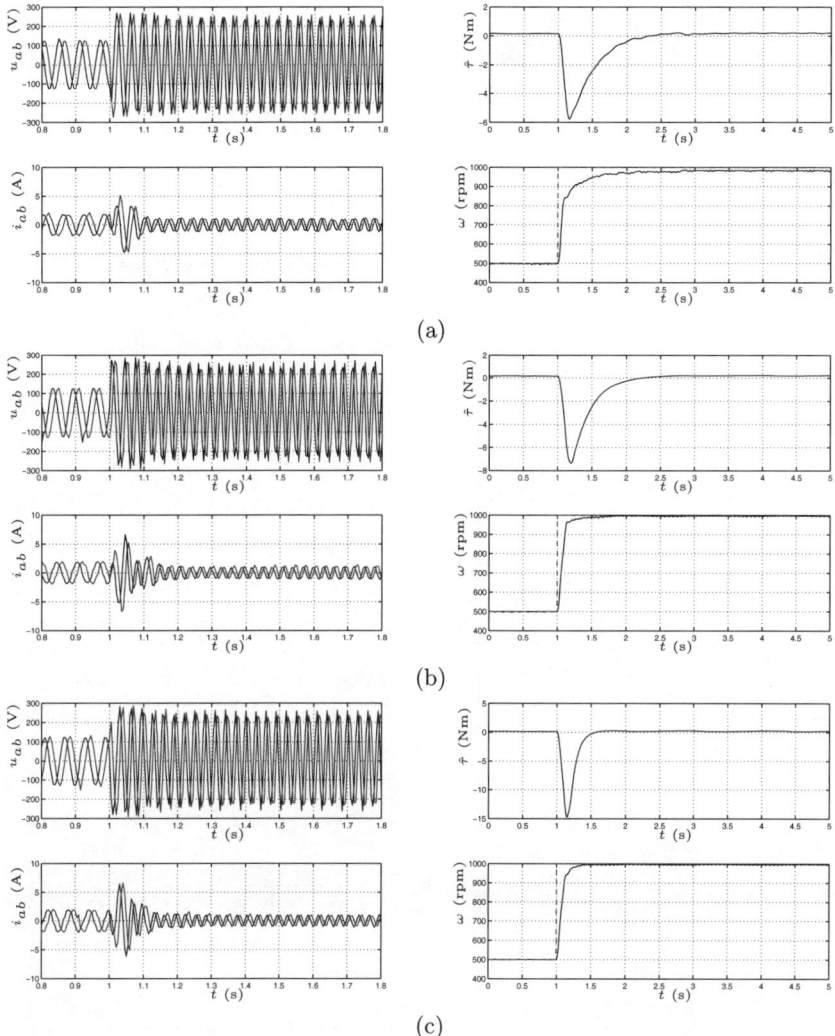

Fig. 10.7. Experimental results for a speed reference change from 500 rpm to 1000 rpm for different values of k and γ. (a) $k = 10$, $\gamma = 4$. (b) $k = 50$, $\gamma = 4$. (c) $k = 50$, $\gamma = 15$.

Figure 10.11 shows the response of the system for a periodic speed reference change from -500 rpm to 500 rpm for $k = 30$ and $\gamma = 10$. Comparing with Figure 10.10 we see that the transient responses of the stator voltages and currents have slightly larger peaks. This is due to the absence of the inverter dynamics, which have a dissipative effect. Observe that the rotor flux modulus is properly maintained while the speed accurately tracks the desired reference.

266   10 Electromechanical Systems

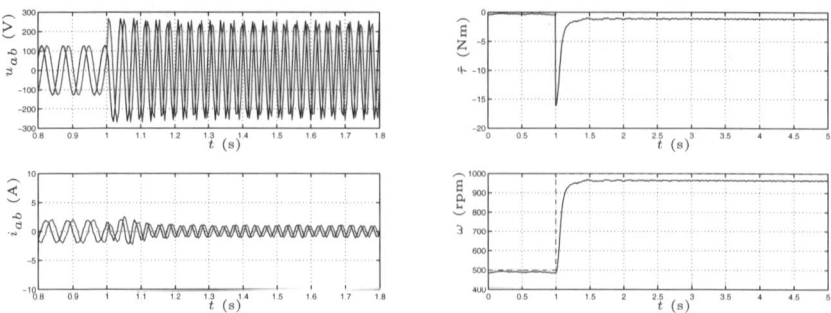

Fig. 10.8. Experimental results for a speed reference change from 500 rpm to 1000 rpm for the controller in [159].

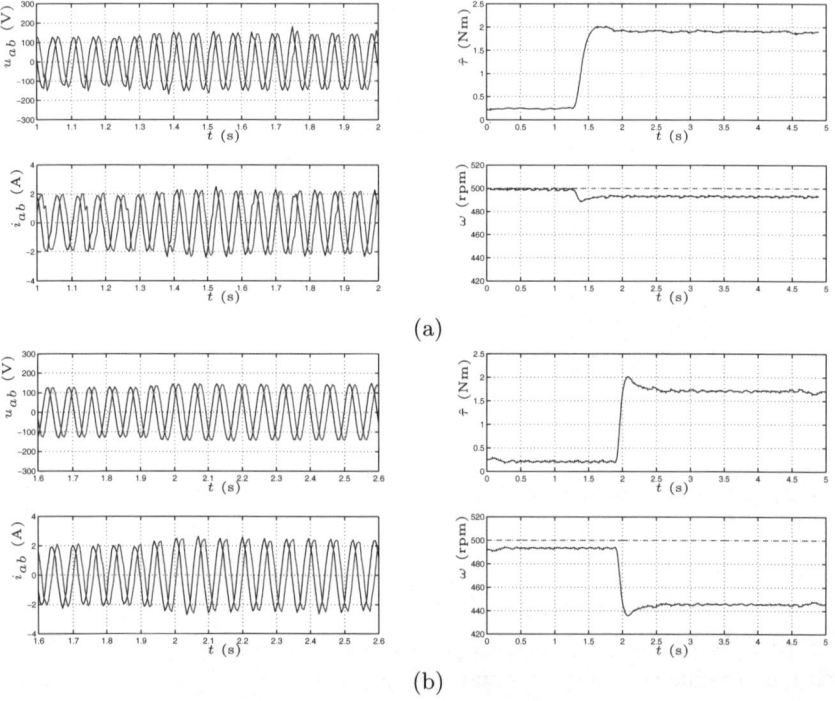

Fig. 10.9. Experimental results for a load change from 0 Nm to 2 Nm. (a) Proposed controller with $k = 50$ and $\gamma = 10$. (b) Controller in [159].

Finally, Figure 10.12 shows the response of the system to a sequence of changes in the rotor speed and flux references and in the load torque.

10.3 Speed/Flux Tracking for Voltage-fed Induction Motors   267

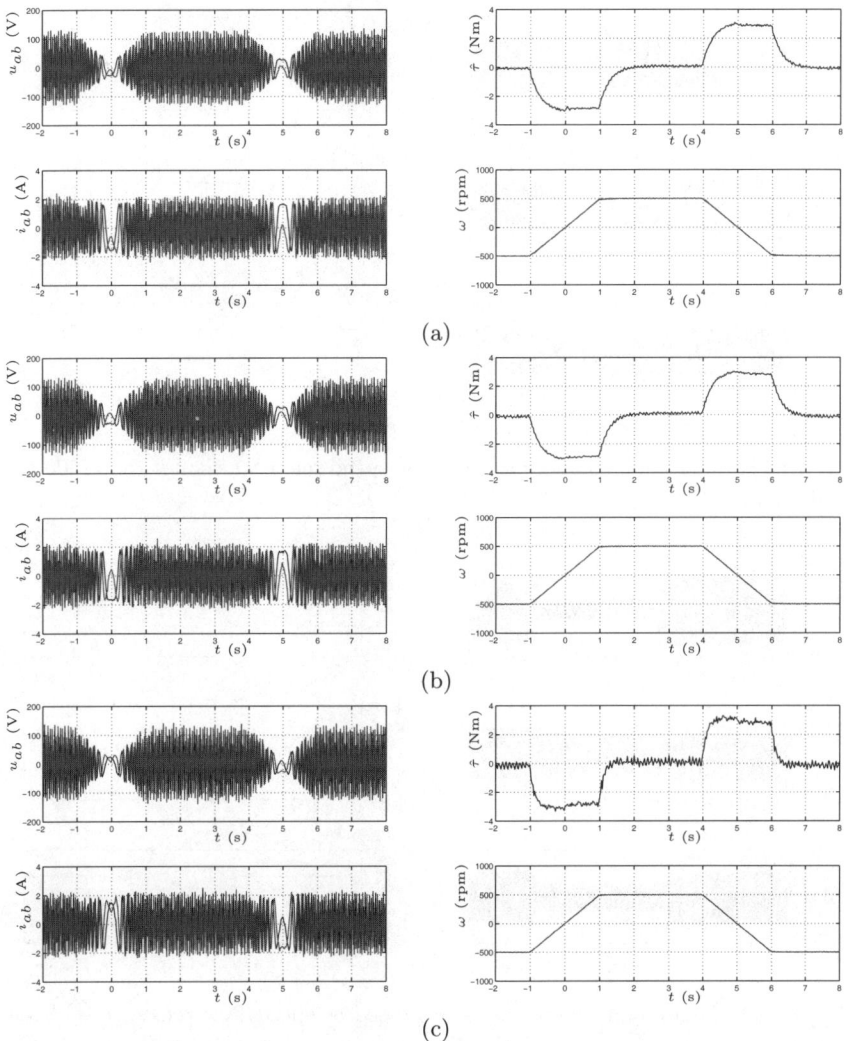

Fig. 10.10. Experimental results for a periodic speed reference change from -500 rpm to 500 rpm for different values of k and γ. (a) $k = 10, \gamma = 4$, (b) $k = 30, \gamma = 4$, (c) $k = 30, \gamma = 10$.

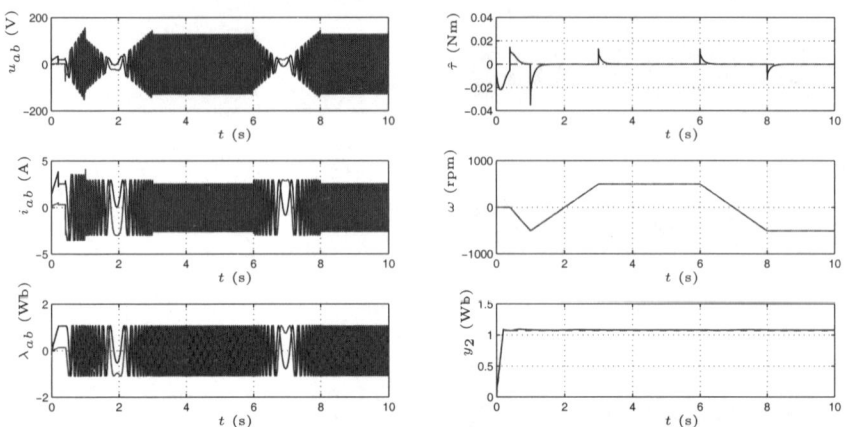

Fig. 10.11. Simulation results for a periodic speed reference change from -500 rpm to 500 rpm.

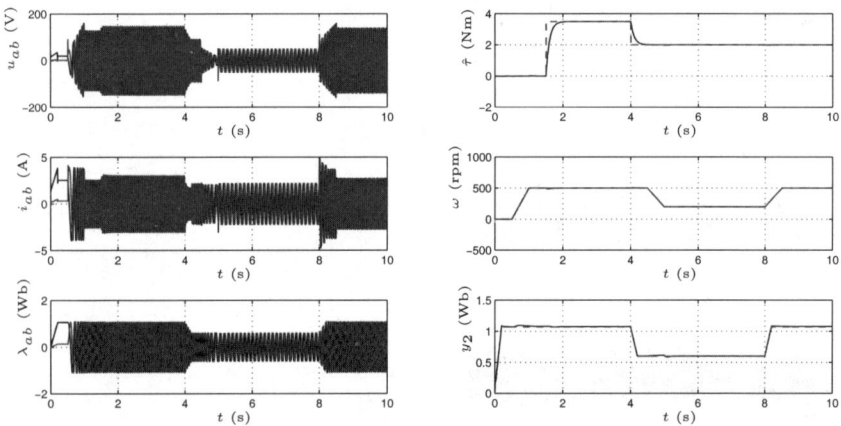

Fig. 10.12. Simulation results for a sequence of speed/flux reference and load changes.

A
Background Material

A.1 Lyapunov Stability and Convergence

For completeness, the main results of Lyapunov stability theory used in the book are briefly reviewed in this section. For a more comprehensive account the reader is referred to [112].

Definition A.1. *A continuous function $V : \mathbb{R}^n \to \mathbb{R}$ is said to be positive-definite (respectively, semidefinite) if $V(0) = 0$ and $V(x) > 0$ (respectively, $V(x) \geq 0$) for all $x \neq 0$; it is said to be negative-(semi)definite if $-V(\cdot)$ is positive-(semi)definite.*

Recall also that a (real) square matrix P is said to be positive-definite (respectively, semidefinite) if it is symmetric and the function $V(x) = x^T P x$ is positive-definite (respectively, semidefinite). Note that $P = P^\top$ is positive-definite (respectively, semidefinite) if and only if all the eigenvalues of P are positive (respectively, nonnegative).

Definition A.2. *A continuous function $\gamma : \mathbb{R}_{\geq 0} \to \mathbb{R}_{\geq 0}$ is said to be of class $\mathcal{K}$ if it is strictly increasing and $\gamma(0) = 0$; it is said to be of class $\mathcal{K}_\infty$ if in addition it is unbounded, i.e., $\lim_{s \to \infty} \gamma(s) = \infty$.*

Definition A.3. *A continuous function $\beta : \mathbb{R}_{\geq 0} \times \mathbb{R}_{\geq 0} \to \mathbb{R}_{\geq 0}$ is said to be of class $\mathcal{KL}$ if, for each fixed t, the function $\beta(s,t)$ is of class $\mathcal{K}$ and, for each fixed s, the function $\beta(s,t)$ is decreasing and $\lim_{t \to \infty} \beta(s,t) = 0$.*

Definition A.4. *A positive-definite function $V : \mathbb{R}^n \to \mathbb{R}$ is said to be radially unbounded (or proper) if there exists a class $\mathcal{K}_\infty$ function $\alpha(\cdot)$ such that $V(x) \geq \alpha(|x|)$, for all $x \in \mathbb{R}^n$.*

Lemma A.1. *Let $V : \mathbb{R}^n \to \mathbb{R}$ be a continuous positive-definite (and radially unbounded) function. Then there exist class $\mathcal{K}_{(\infty)}$ functions $\alpha_1(\cdot)$ and $\alpha_2(\cdot)$ such that*
$$\alpha_1(|x|) \leq V(x) \leq \alpha_2(|x|).$$

Using the above definitions, Lyapunov stability can be formulated as follows.

Theorem A.1. *Consider the system $\dot{x} = f(x,t)$, where $x \in \mathbb{R}^n$ and $t \geq t_0$, with an equilibrium point $x = 0$, and suppose that there exists a $\mathcal{C}^1$ function $V(x,t)$ such that*
$$\alpha_1(|x|) \leq V(x,t) \leq \alpha_2(|x|),$$
$$\frac{\partial V}{\partial x} f(x,t) + \frac{\partial V}{\partial t} \leq -\kappa(|x|),$$
where $\alpha_1(\cdot), \alpha_2(\cdot)$ are class-$\mathcal{K}_\infty$ functions and $\kappa(\cdot)$ is a positive-definite function. Then $x = 0$ is globally stable. If in addition the function $\kappa(\cdot)$ is of class $\mathcal{K}$, then $x = 0$ is globally asymptotically stable (GAS).

Definition A.5. *A function $V(x,t)$ satisfying the conditions of Theorem A.1 is called a Lyapunov function for the system $\dot{x} = f(x,t)$.*

The following theorems due to LaSalle can be used to prove asymptotic convergence to an equilibrium even when the function $\kappa(\cdot)$ is only positive-semidefinite.

Theorem A.2. *Consider the system $\dot{x} = f(x)$, where $x \in \mathbb{R}^n$, and let $\Omega \subset \mathbb{R}^n$ be a compact set that is positively invariant, i.e., $x(0) \in \Omega \Rightarrow x(t) \in \Omega, \forall t \in \mathbb{R}$. Let $V : \mathbb{R}^n \to \mathbb{R}$ be a $\mathcal{C}^1$ function such that $\dot{V}(x) \leq 0$ in Ω. Let E be the set of all points in Ω where $\dot{V}(x) = 0$. Let M be the largest invariant set in E. Then every solution starting in Ω approaches M as t goes to ∞.*

Theorem A.3. *Consider the system $\dot{x} = f(x,t)$, where $x \in \mathbb{R}^n$ and $t \geq t_0$, with an equilibrium point $x = 0$, where $f(x,t)$ is locally Lipschitz uniformly in t, and suppose that there exists a $\mathcal{C}^1$ function $V(x,t)$ such that*
$$\alpha_1(|x|) \leq V(x,t) \leq \alpha_2(|x|),$$
$$\frac{\partial V}{\partial x} f(x,t) + \frac{\partial V}{\partial t} \leq -W(x) \leq 0,$$
where $\alpha_1(\cdot), \alpha_2(\cdot)$ are class-$\mathcal{K}_\infty$ functions and $W(\cdot)$ is a $\mathcal{C}^0$ function. Then $x = 0$ is globally stable and, moreover,
$$\lim_{t \to \infty} W(x(t)) = 0.$$

The following result, known as Barbalat's lemma [174], and its corollaries are useful for proving convergence in adaptive systems.

Lemma A.2. *Let $\phi : \mathbb{R}_{\geq 0} \to \mathbb{R}$ be a uniformly continuous[1] function and suppose that*

---
[1] A function $f : \mathbb{D} \subseteq \mathbb{R} \to \mathbb{R}$ is uniformly continuous if, for any $\varepsilon > 0$, there exists $\delta > 0$ (dependent only on ε) such that $|x - y| < \delta \Rightarrow |f(x) - f(y)| < \varepsilon$, for all $x, y \in \mathbb{D}$.

$$\lim_{t\to\infty}\int_0^t \phi(\tau)\mathrm{d}\tau$$

exists and is finite. Then $\lim_{t\to\infty}\phi(t)=0$.

Corollary A.1. *Consider a function* $\phi : \mathbb{R}_{\geq 0} \to \mathbb{R}$ *and suppose that* $\phi(t) \in \mathcal{L}_2 \cap \mathcal{L}_\infty$ *and* $\dot{\phi}(t) \in \mathcal{L}_\infty$. *Then* $\lim_{t\to\infty}\phi(t)=0$.

Corollary A.2. [213] *Consider a function* $\phi : \mathbb{R}_{\geq 0} \to \mathbb{R}$ *and suppose that* $\phi(t) \in \mathcal{L}_2$ *and* $\dot{\phi}(t) \in \mathcal{L}_\infty$. *Then* $\lim_{t\to\infty}\phi(t)=0$.

Finally, the following lemma, known as Young's inequality, and its corollary are used extensively to obtain upper bounds.

Lemma A.3. *Let* $p > 1$ *and* $q > 1$ *be such that* $(p-1)(q-1)=1$. *Then, for all* $\varepsilon > 0$, $x \in \mathbb{R}$ *and* $y \in \mathbb{R}$,

$$xy \leq \frac{\varepsilon^p}{p}|x|^p + \frac{1}{q\varepsilon^q}|y|^q.$$

Corollary A.3. *For all* $\varepsilon > 0$, $x \in \mathbb{R}^n$ *and* $y \in \mathbb{R}^n$,

$$x^T y \leq \frac{\varepsilon}{2}|x|^2 + \frac{1}{2\varepsilon}|y|^2.$$

A.2 Input-to-state Stability

This section recalls the notions of input-to-state stability (ISS) and input-to-output stability (IOS) introduced in [202]; see also [203, 205].

Consider the system
$$\dot{x} = f(x,u), \qquad (A.1)$$
with state $x \in \mathbb{R}^n$ and input $u \in \mathbb{R}^m$, where $f(\cdot)$ is locally Lipschitz.

Definition A.6. *The system (A.1) is said to be input-to-state stable (ISS) if there exist a class-$\mathcal{KL}$ function* $\beta(\cdot)$ *and a class-$\mathcal{K}$ function* $\gamma(\cdot)$ *such that, for each initial condition* $x(0)$ *and each measurable, essentially bounded input* $u(t)$, *the solution* $x(t)$ *exists for all* $t \geq 0$ *and satisfies*

$$|x(t)| \leq \beta(|x(0)|,t) + \gamma(\sup_{\tau \in [0,t]}|u(\tau)|).$$

The function $\gamma(\cdot)$ is often referred to as an *ISS gain* for the system (A.1). Note that, as shown in [202], the above definition implies that an ISS system is bounded-input bounded-state (BIBS) stable and has a globally asymptotically stable equilibrium at zero when $u(t)=0$.

The ISS property can be equivalently characterised in terms of Lyapunov functions, as the following theorem shows [205].

Theorem A.4. *The system (A.1) is ISS if and only if there exists a C^1 function $V(x)$ such that*
$$\alpha_1(|x|) \leq V(x) \leq \alpha_2(|x|),$$
$$\frac{\partial V}{\partial x} f(x, u) \leq -\kappa(|x|) + \gamma(|u|),$$
where $\alpha_1(\cdot)$, $\alpha_2(\cdot)$, $\kappa(\cdot)$ and $\gamma(\cdot)$ are class-$\mathcal{K}_\infty$ functions.

Definition A.7. *A function $V(x)$ satisfying the conditions of Theorem A.4 is called an ISS-Lyapunov function for the system (A.1).*

Note that the function $\gamma(\cdot)$ in Theorem A.4 is different from the one in Definition A.6, i.e., it does not qualify as an ISS gain. However, an ISS gain for the system (A.1) is obtained from Theorem A.4 as $\alpha_1^{-1} \circ \alpha_2 \circ \kappa^{-1} \circ \gamma(\cdot)$.

Besides ISS, similar definitions have been proposed to describe the input–output properties of a system. For the purposes of the book, the notions of input-to-output stability [201] and $\mathcal{L}_2$ stability [224] are of particular interest and are recalled hereafter.

Definition A.8. *The system (A.1) with output $y = h(x)$ is said to be input-to-output stable (IOS) if there exist a class-$\mathcal{KL}$ function $\beta(\cdot)$ and a class-$\mathcal{K}$ function $\gamma(\cdot)$ such that, for each initial condition $x(0)$ and each measurable, essentially bounded input $u(t)$, the solution $x(t)$ exists for all $t \geq 0$ and satisfies*
$$|h(x(t))| \leq \beta(|x(0)|, t) + \gamma(\sup_{\tau \in [0,t]} |u(\tau)|).$$

Note that input-to-output stability implies input-to-state stability when the system is detectable via the output $y = h(x)$, see [205].

Definition A.9. *The system (A.1) with output $y = h(x)$ is said to be $\mathcal{L}_2$ stable if there exist nonnegative constants γ and β such that, for all initial conditions $x(0)$ and for any input $u(t) \in \mathcal{L}_2$, the solution $x(t)$ exists for all $t \geq 0$ and satisfies*
$$\|h(x(t))\|_2 \leq \gamma \|u(t)\|_2 + \beta.$$

The smallest constant γ that satisfies the above condition is referred to as the $\mathcal{L}_2$ *gain* of the system (A.1).

The ISS property and the notion of ISS gain have proved to be particularly useful when dealing with interconnected systems, since they allow a possible extension of the well-known small-gain theorem for linear systems to the nonlinear framework. Several such extensions have been proposed in the literature, see e.g., [87, 216]. The following is a Lyapunov formulation of the nonlinear small-gain theorem proposed in [86].

Theorem A.5. *Consider the interconnected system*
$$\dot{x}_1 = f_1(x_1, x_2, u_1), \tag{A.2}$$
$$\dot{x}_2 = f_2(x_1, x_2, u_2), \tag{A.3}$$

with states $x_i \in \mathbb{R}^{n_i}$ and inputs $u_i \in \mathbb{R}^{m_i}$, for $i = 1, 2$, where the functions $f_i(\cdot)$ are locally Lipschitz. Assume that the system (A.2) with inputs x_2 and u_1 and the system (A.3) with inputs x_1 and u_2 are ISS, i.e., there exist two $\mathcal{C}^1$ functions $V_1(x_1)$ and $V_2(x_2)$ satisfying

$$\alpha_{11}(|x_1|) \leq V_1(x_1) \leq \alpha_{12}(|x_1|), \qquad \alpha_{21}(|x_2|) \leq V_2(x_2) \leq \alpha_{22}(|x_2|),$$

and

$$\frac{\partial V_1}{\partial x_1} f_1(x_1, x_2, u_1) \leq -\kappa_1(|x_1|) + \gamma_{11}(|x_2|) + \gamma_{12}(|u_1|),$$

$$\frac{\partial V_2}{\partial x_2} f_2(x_1, x_2, u_2) \leq -\kappa_2(|x_2|) + \gamma_{21}(|x_1|) + \gamma_{22}(|u_2|),$$

where $\alpha_{ij}(\cdot)$, $\kappa_i(\cdot)$ and $\gamma_{ij}(\cdot)$, for $i, j = 1, 2$, are class-$\mathcal{K}_\infty$ functions. In addition, assume that there exist constants $0 < \varepsilon_1 < 1$ and $0 < \varepsilon_2 < 1$ such that

$$\frac{1}{1-\varepsilon_1} \kappa_1^{-1} \circ \alpha_{12} \circ \gamma_{11} \circ \alpha_{21}^{-1} \circ \left(\frac{1}{1-\varepsilon_2} \kappa_2^{-1} \circ \alpha_{22} \circ \gamma_{21} \circ \alpha_{11}^{-1}(r) \right) < r,$$

for all $r > 0$. Then the system (A.2), (A.3) with inputs u_1 and u_2 is ISS.

A.3 Invariant Manifolds and System Immersion

In this section we give the definition of invariant manifold [226] and of system immersion [34]. (Note that the latter is presented here only for completeness, since we do not explicitly use this definition in the book.)

Consider the autonomous system

$$\dot{x} = f(x), \qquad y = h(x), \tag{A.4}$$

with state $x \in \mathbb{R}^n$ and output $y \in \mathbb{R}^m$.

Definition A.10. *The manifold $\mathcal{M} = \{x \in \mathbb{R}^n \mid s(x) = 0\}$, with $s(x)$ smooth, is said to be (positively) invariant for $\dot{x} = f(x)$ if $s(x(0)) = 0$ implies $s(x(t)) = 0$, for all $t \geq 0$.*

Consider now the (target) system

$$\dot{\xi} = \alpha(\xi), \qquad \zeta = \beta(\xi), \tag{A.5}$$

with state $\xi \in \mathbb{R}^p$ ($p < n$) and output $\zeta \in \mathbb{R}^m$.

Definition A.11. *The system (A.5) is said to be immersed into the system (A.4) if there exists a smooth mapping $\pi : \mathbb{R}^p \mapsto \mathbb{R}^n$ satisfying $x(0) = \pi(\xi(0))$ and $\beta(\xi_1) \neq \beta(\xi_2) \Rightarrow h(\pi(\xi_1)) \neq h(\pi(\xi_2))$ and such that*

$$f(\pi(\xi)) = \frac{\partial \pi}{\partial \xi} \alpha(\xi)$$

and

$$h(\pi(\xi)) = \beta(\xi)$$

for all $\xi \in \mathbb{R}^p$.

Thus, roughly speaking, a system Σ_1 is said to be immersed into a system Σ_2 if the input–output mapping of Σ_2 is a restriction of the input–output mapping of Σ_1, *i.e.*, any output response generated by Σ_2 is also an output response of Σ_1 for a restricted set of initial conditions, see, *e.g.*, [34, 88].

References

1. R. Abraham, J.E. Marsden, and T. Ratiu. *Manifolds, Tensor Analysis, and Applications.* Addison-Wesley, 1983.
2. J.A. Acosta, R. Ortega, A. Astolfi, and A.D. Mahindrakar. Interconnection and damping assignment passivity-based control of mechanical systems with underactuation degree one. *IEEE Trans. Automatic Control*, 50(12):1936–1955, 2005.
3. D. Aeyels. Stabilization of a class of nonlinear systems by a smooth feedback control. *Systems & Control Letters*, 5(5):289–294, 1985.
4. N. Aghannan and P. Rouchon. An intrinsic observer for a class of Lagrangian systems. *IEEE Trans. Automatic Control*, 48(6):936–945, 2003.
5. M.R. Akella and K. Subbarao. A novel parameter projection mechanism for smooth and stable adaptive control. *Systems & Control Letters*, 54(1):43–51, 2005.
6. V. Andrieu and L. Praly. Global asymptotic stabilization by output feedback for some non minimum phase non linear systems. In *Proc. 44th IEEE Conf. Decision and Control and European Control Conference, Seville, Spain*, pages 2622–2627, 2005.
7. V. Andrieu and L. Praly. On the existence of a Kazantzis–Kravaris/Luenberger observer. *SIAM J. Control and Optimization*, 45(2):432–456, 2006.
8. V. Andrieu and L. Praly. A unifying point of view on output feedback designs. In *Proc. IFAC Symp. Nonlinear Control Systems, Pretoria, South Africa*, 2007.
9. V. Andrieu, L. Praly, and A. Astolfi. Nonlinear output feedback design via domination and generalized weighted homogeneity. In *Proc. 45th IEEE Conf. Decision and Control, San Diego, California*, pages 6391–6396, 2006.
10. V. Andrieu, L. Praly, and A. Astolfi. Homogeneous approximation, recursive observer design and output feedback. *SIAM J. Control and Optimization*, 2007. Submitted.
11. A.M. Annaswamy, F.P. Skantze, and A.-P. Loh. Adaptive control of continuous time systems with convex/concave parametrization. *Automatica*, 34(1):33–49, 1998.
12. M. Arcak and P. Kokotović. Nonlinear observers: a circle criterion design and robustness analysis. *Automatica*, 37(12):1923–1930, 2001.
13. A. Astolfi. A remark on an example by Teel–Hespanha with applications to cascaded systems. *IEEE Trans. Automatic Control*, 52(2):289–293, 2007.

14. A. Astolfi, G. Escobar, R. Ortega, and A.M. Stanković. An adaptive controller for the TCSC based on the immersion and invariance design technique. In *Proc. 14th Conf. Power Systems Computation, Seville, Spain*, 2002.
15. A. Astolfi, L. Hsu, M.S. Netto, and R. Ortega. Two solutions to the adaptive visual servoing problem. *IEEE Trans. Robotics and Automation*, 18(3):387–392, 2002.
16. A. Astolfi, D. Karagiannis, and R. Ortega. Power factor precompensator. International PCT patent no. WO2004027964, 2004.
17. A. Astolfi, D. Karagiannis, and R. Ortega. Stabilization of uncertain nonlinear systems via immersion and invariance. *European Journal of Control*, 13(2-3):204–220, 2007.
18. A. Astolfi, D. Karagiannis, and R. Ortega. Towards applied nonlinear adaptive control (plenary presentation). In *Proc. IFAC Workshop on Adaptation and Learning in Control and Signal Processing, St. Petersburg, Russia*, 2007.
19. A. Astolfi and R. Ortega. Energy-based stabilization of angular velocity of rigid body in failure configuration. *AIAA J. Guidance, Control and Dynamics*, 25(1):184–187, 2002.
20. A. Astolfi and R. Ortega. Invariant manifolds, asymptotic immersion and the (adaptive) stabilization of nonlinear systems. In A. Zinober and D. Owens, editors, *Nonlinear and Adaptive Control*, pages 1–20. Springer-Verlag, 2002.
21. A. Astolfi and R. Ortega. Immersion and invariance: a new tool for stabilization and adaptive control of nonlinear systems. *IEEE Trans. Automatic Control*, 48(4):590–606, 2003.
22. A. Astolfi, R. Ortega, and R. Sepulchre. Stabilization and disturbance attenuation of nonlinear systems using dissipativity theory. *European Journal of Control*, 8(5):408–431, 2002.
23. A. Astolfi and L. Praly. Global complete observability and output-to-state stability imply the existence of a globally convergent observer. *Math. Control Signals Systems*, 18:32–65, 2006.
24. J. Back, H. Shim, and J.H. Seo. An algorithm for system immersion into nonlinear observer form: forced system. In *Proc. 16th IFAC World Congress, Prague, Czech Rep.*, 2005.
25. S. Battilotti. Global output regulation and disturbance attenuation with global stability via measurement feedback for a class of nonlinear systems. *IEEE Trans. Automatic Control*, 41(3):315–327, 1996.
26. S. Battilotti. A note on reduced order dynamic output feedback stabilizing controllers. *Systems & Control Letters*, 30(2):71–81, 1997.
27. M. Becherif, H. Rodríguez, E. Mendes, and R. Ortega. Comparaison expérimentale de méthodes de commande d'un convertisseur DC–DC boost. In *Conf. Internationale Francophone d' Automatique, Nantes, France*, 2002.
28. G. Besançon. Global output feedback tracking control for a class of Lagrangian systems. *Automatica*, 36(12):1915–1921, 2000.
29. F. Blaschke. The principle of field orientation applied to the new transvector closed-loop control system for rotating field machines. *Siemens-Review*, 39:217–220, 1972.
30. A.M. Bloch, D.E. Chang, N.E. Leonard, and J.E. Marsden. Controlled Lagrangians and the stabilization of mechanical systems II: potential shaping. *IEEE Trans. Automatic Control*, 46(10):1556–1571, 2001.

31. A.M. Bloch, N.E. Leonard, and J.E. Marsden. Controlled Lagrangians and the stabilization of mechanical systems I: the first matching theorem. *IEEE Trans. Automatic Control*, 45(12):2253–2270, 2000.
32. R.T. Bupp, D.S. Bernstein, and V.T. Coppola. A benchmark problem for nonlinear control design: problem statement, experimental testbed and passive nonlinear compensation. In *Proc. American Control Conference, Seattle, Washington*, pages 4363–4367, 1995.
33. T. Burg and D. Dawson. Additional notes on the TORA example: a filtering approach to eliminate velocity measurements. *IEEE Trans. Control Systems Technology*, 5(5):520–523, 1997.
34. C.I. Byrnes, F. Delli Priscoli, and A. Isidori. *Output Regulation of Uncertain Nonlinear Systems*. Birkhäuser, Boston, 1997.
35. C.I. Byrnes and A. Isidori. New results and examples in nonlinear feedback stabilization. *Systems & Control Letters*, 12(5):437–442, 1989.
36. C.I. Byrnes, A. Isidori, and J.C. Willems. Passivity, feedback equivalence, and the global stabilization of minimum phase nonlinear systems. *IEEE Trans. Automatic Control*, 36(11):1228–1240, 1991.
37. A.J. Calise and R.T. Rysdyk. Nonlinear adaptive flight control using neural networks. *IEEE Control Systems Magazine*, 18(6):14–25, 1998.
38. C. Canudas de Wit and L. Praly. Adaptive eccentricity compensation. *IEEE Trans. Control Systems Technology*, 8(5):757–766, 2000.
39. D. Carnevale, D. Karagiannis, and A. Astolfi. Reduced-order observer design for systems with non-monotonic nonlinearities. In *Proc. 45th IEEE Conf. Decision and Control, San Diego, California*, pages 5269–5274, 2006.
40. D. Carnevale, D. Karagiannis, and A. Astolfi. Reduced-order observer design for nonlinear systems. In *Proc. European Control Conference, Kos, Greece*, pages 559–564, 2007.
41. J. Carr. *Applications of Centre Manifold Theory*. Springer-Verlag, 1981.
42. G.W. Chang, J.P. Hespanha, A.S. Morse, M.S. Netto, and R. Ortega. Supervisory field-oriented control of induction motors with uncertain rotor resistance. *Int. J. Adaptive Control and Signal Processing*, 15(3):353–375, 2001.
43. X. Chen and H. Kano. A new state observer for perspective systems. *IEEE Trans. Automatic Control*, 47(4):658–663, 2002.
44. J. Chiasson. *Modeling and High-Performance Control of Electric Machines*. John Wiley and Sons, 2005.
45. G. Ciccarella, M. Dalla Mora, and A. Germani. A Luenberger-like observer for nonlinear systems. *Int. J. Control*, 57(3):537–556, 1993.
46. D.M. Dawson, J. Hu, and T.C. Burg. *Nonlinear Control of Electric Machinery*. Marcel Dekker, New York, 1998.
47. A. De Luca. Decoupling and feedback linearization of robots with mixed rigid/elastic joints. *Int. J. Robust and Nonlinear Control*, 8(11):965–977, 1998.
48. P. De Wit, R. Ortega, and I. Mareels. Indirect field-oriented control of induction motors is robustly globally stable. *Automatica*, 32(10):1393–1402, 1996.
49. C. Desoer and M. Vidyasagar. *Feedback Systems: Input-Output Properties*. Academic Press, New York, 1975.
50. W.E. Dixon, Y. Fang, D.M. Dawson, and T.J. Flynn. Range identification for perspective vision systems. *IEEE Trans. Automatic Control*, 48(12):2232–2238, 2003.
51. R. Engel and G. Kreisselmeier. A continuous-time observer which converges in finite time. *IEEE Trans. Automatic Control*, 47(7):1202–1204, 2002.

52. D. Enns, D. Bugajski, R. Hendrick, and G. Stein. Dynamic inversion: an evolving methodology for flight control design. *Int. J. Control*, 59(1):71–91, 1994.
53. G. Escobar, D. Chevreau, R. Ortega, and E. Mendes. An adaptive passivity-based controller for a unity power factor rectifier. *IEEE Trans. Control Systems Technology*, 9(4):637–644, 2001.
54. G. Escobar, R. Ortega, and H. Sira-Ramírez. Output-feedback global stabilization of a nonlinear benchmark system using a saturated passivity-based controller. *IEEE Trans. Control Systems Technology*, 7(2):289–293, 1999.
55. G. Escobar, R. Ortega, H. Sira-Ramírez, J.-P. Vilain, and I. Zein. An experimental comparison of several nonlinear controllers for power converters. *IEEE Control Systems Magazine*, 19(1):66–82, 1999.
56. B. Etkin and L.D. Reid. *Dynamics of Flight: Stability and Control*. John Wiley and Sons, New York, 3rd edition, 1996.
57. M. Feemster, P. Aquino, D.M. Dawson, and A. Behal. Sensorless rotor velocity tracking control for induction motors. *IEEE Trans. Control Systems Technology*, 9(4):645–653, 2001.
58. R.A. Freeman and P.V. Kokotović. *Robust Nonlinear Control Design: State-Space and Lyapunov Techniques*. Birkhäuser, Boston, 1996.
59. R.A. Freeman and P.V. Kokotović. Tracking controllers for systems linear in the unmeasured states. *Automatica*, 32(5):735–746, 1996.
60. M. French, C. Szepesvari, and E. Rogers. Uncertainty, performance, and model dependency in approximate adaptive nonlinear control. *IEEE Trans. Automatic Control*, 45(2):353–358, 2000.
61. J.-P. Gauthier, H. Hammouri, and S. Othman. A simple observer for nonlinear systems applications to bioreactors. *IEEE Trans. Automatic Control*, 37(6):875–880, 1992.
62. J.-P. Gauthier and I. Kupka. *Deterministic Observation Theory and Applications*. Cambridge University Press, 2001.
63. J.-P. Gauthier and I.A.K. Kupka. Observability and observers for nonlinear systems. *SIAM J. Control and Optimization*, 32(4):975–994, 1994.
64. T.T. Georgiou and M.C. Smith. Robustness analysis of nonlinear feedback systems: an input-output approach. *IEEE Trans. Automatic Control*, 42(9):1200–1221, 1997.
65. G. Guglieri and F.B. Quagliotti. Analytical and experimental analysis of wing rock. *Nonlinear Dynamics*, 24:129–146, 2001.
66. O. Härkegård and S.T. Glad. Flight control design using backstepping. In *Proc. IFAC Symp. Nonlinear Control Systems, St. Petersburg, Russia*, pages 259–264, 2001.
67. J. Hauser, S. Sastry, and P. Kokotović. Nonlinear control via approximate input-output linearization: the ball and beam example. *IEEE Trans. Automatic Control*, 37(3):392–398, 1992.
68. N.G. Hingorani and L. Gyugyi. *Understanding FACTS*. IEEE Press, New York, 2000.
69. C.H. Hsu and C.E Lan. Theory of wing rock. *Journal of Aircraft*, 22(10):920–924, 1985.
70. L. Hsu and P.L.S. Aquino. Adaptive visual tracking with uncertain manipulator dynamics and uncalibrated camera. In *Proc. 38th IEEE Conf. Decision and Control, Phoenix, Arizona*, pages 1248–1253, 1999.

71. J. Hu and D.M. Dawson. Adaptive control of induction motor systems despite rotor resistance uncertainty. *Automatica*, 32(8):1127–1143, 1996.
72. J. Huang and C.F. Lin. Robust nonlinear control of the ball and beam system. In *Proc. American Control Conference, Seattle, Washington*, pages 306–310, 1995.
73. J.-T. Huang. An adaptive compensator for a class of linearly parameterized systems. *IEEE Trans. Automatic Control*, 47(3):483–486, 2002.
74. S. Hutchinson, G.D. Hager, and P.I. Corke. A tutorial on visual servo control. *IEEE Trans. Robotics and Automation*, 12(5):651–670, 1996.
75. A. Ilchmann. *Non-identifier-based High-gain Adaptive Control*. Springer-Verlag, London, 1993.
76. A. Ilchmann and E.P. Ryan. Universal λ-tracking for nonlinearly-perturbed systems in the presence of noise. *Automatica*, 30(2):337–346, 1994.
77. P.A. Ioannou and J. Sun. *Robust Adaptive Control*. Prentice-Hall, Upper Saddle River, 1996.
78. A. Isidori. *Nonlinear Control Systems*. Springer-Verlag, Berlin, 3rd edition, 1995.
79. A. Isidori. A remark on the problem of semiglobal nonlinear output regulation. *IEEE Trans. Automatic Control*, 42(12):1734–1738, 1997.
80. A. Isidori. A tool for semiglobal stabilization of uncertain non-minimum-phase nonlinear systems via output feedback. *IEEE Trans. Automatic Control*, 45(10):1817–1827, 2000.
81. A. Jain and G. Rodriguez. Diagonalized Lagrangian robot dynamics. *IEEE Trans. Robotics and Automation*, 11(4):571–584, 1995.
82. M. Jankovic and B.K. Ghosh. Visually guided ranging from observations of points, lines and curves via an identifier based nonlinear observer. *Systems & Control Letters*, 25(1):63–73, 1995.
83. Z.-P. Jiang. A combined backstepping and small-gain approach to adaptive output feedback control. *Automatica*, 35(6):1131–1139, 1999.
84. Z.-P. Jiang and I. Kanellakopoulos. Global output feedback tracking for a benchmark nonlinear system. *IEEE Trans. Automatic Control*, 45(5):1023–1027, 2000.
85. Z.-P. Jiang, E. Lefeber, and H. Nijmeijer. Saturated stabilization and tracking of a nonholonomic mobile robot. *Systems & Control Letters*, 42(5):327–332, 2001.
86. Z.-P. Jiang, I. Mareels, and Y. Wang. A Lyapunov formulation of the nonlinear small-gain theorem for interconnected ISS systems. *Automatica*, 32(8):1211–1215, 1996.
87. Z.-P. Jiang, A.R. Teel, and L. Praly. Small-gain theorem for ISS systems and applications. *Math. Control Signals Systems*, 7:95–120, 1994.
88. P. Jouan. Immersion of nonlinear systems into linear systems modulo output injection. *SIAM J. Control and Optimization*, 41(6):1756–1778, 2003.
89. G. Kaliora and A. Astolfi. Nonlinear control of feedforward systems with bounded signals. *IEEE Trans. Automatic Control*, 49(11):1975–1990, 2004.
90. G. Kaliora, A. Astolfi, and L. Praly. Norm estimators and global output feedback stabilization of nonlinear systems with ISS inverse dynamics. *IEEE Trans. Automatic Control*, 51(3):493–498, 2006.
91. I. Kanellakopoulos, P.V. Kokotović, and A.S. Morse. Systematic design of adaptive controllers for feedback linearizable systems. *IEEE Trans. Automatic Control*, 36(11):1241–1253, 1991.

92. D. Karagiannis and A. Astolfi. Nonlinear adaptive control of systems in feedback form: an alternative to adaptive backstepping. In *Proc. IFAC Symp. Large Scale Systems, Osaka, Japan*, pages 71–76, 2004.
93. D. Karagiannis and A. Astolfi. A new solution to the problem of range identification in perspective vision systems. *IEEE Trans. Automatic Control*, 50(12):2074–2077, 2005.
94. D. Karagiannis and A. Astolfi. Nonlinear observer design using invariant manifolds and applications. In *Proc. 44th IEEE Conf. Decision and Control and European Control Conference, Seville, Spain*, pages 7775–7780, 2005.
95. D. Karagiannis and A. Astolfi. Rotor resistance estimation for current-fed induction motors. In *Proc. 16th IFAC World Congress, Prague, Czech Rep.*, 2005.
96. D. Karagiannis and A. Astolfi. A robustly stabilising adaptive controller for systems in feedback form. In *Proc. American Control Conference, Minneapolis, Minnesota*, pages 3557–3562, 2006.
97. D. Karagiannis and A. Astolfi. Adaptive state feedback design via immersion and invariance. In *Proc. European Control Conference, Kos, Greece*, pages 553–558, 2007.
98. D. Karagiannis and A. Astolfi. Nonlinear adaptive flight control of autonomous aircraft. In *Proc. European Control Conference, Kos, Greece*, pages 571–576, 2007.
99. D. Karagiannis, A. Astolfi, and R. Ortega. Two results for adaptive output feedback stabilization of nonlinear systems. *Automatica*, 39(5):857–866, 2003.
100. D. Karagiannis, A. Astolfi, and R. Ortega. Nonlinear stabilization via system immersion and manifold invariance: survey and new results. *SIAM J. Multiscale Modeling and Simulation*, 3(4):801–817, 2005.
101. D. Karagiannis, A. Astolfi, and R. Ortega. Output feedback stabilization of a class of uncertain systems. In A. Astolfi, editor, *Nonlinear and Adaptive Control: Tools and Algorithms for the User*, pages 55–77. Imperial College Press, 2006.
102. D. Karagiannis, A. Astolfi, R. Ortega, and M. Hilairet. A nonlinear tracking controller for voltage-fed induction motors with uncertain load torque. *IEEE Trans. Control Systems Technology*, 2007. Submitted.
103. D. Karagiannis, Z.-P. Jiang, R. Ortega, and A. Astolfi. Output feedback stabilization of a class of uncertain non-minimum-phase nonlinear systems. *Automatica*, 41(9):1609–1615, 2005.
104. D. Karagiannis, Z.-P. Jiang, R. Ortega, and A. Astolfi. Output feedback stabilisation of uncertain nonminimum-phase systems. In *Proc. European Control Conference, Kos, Greece*, pages 565–570, 2007.
105. D. Karagiannis, E. Mendes, A. Astolfi, and R. Ortega. An experimental comparison of several PWM controllers for a single-phase AC–DC converter. *IEEE Trans. Control Systems Technology*, 11(6):940–947, 2003.
106. D. Karagiannis, R. Ortega, and A. Astolfi. Nonlinear adaptive stabilization via system immersion: control design and applications. In F. Lamnabhi-Lagarrigue, A. Loría, and E. Panteley, editors, *Lecture Notes in Control and Information Sciences*, volume 311, pages 1–21. Springer-Verlag, London, 2005.
107. J. Kassakian, M. Schlecht, and G. Verghese. *Principles of Power Electronics*. Addison-Wesley, Reading, 1991.
108. N. Kazantzis and C. Kravaris. Nonlinear observer design using Lyapunov's auxiliary theorem. *Systems & Control Letters*, 34(5):241–247, 1998.

109. N. Kazantzis and C. Kravaris. Singular PDEs and the single-step formulation of feedback linearization with pole placement. *Systems & Control Letters*, 39(2):115–122, 2000.
110. R. Kelly. Robust asymptotically stable visual servoing of planar robots. *IEEE Trans. Robotics and Automation*, 12(5):759–766, 1996.
111. H.K. Khalil. High-gain observers in nonlinear feedback control. In H. Nijmeijer and T.I. Fossen, editors, *New Directions in Nonlinear Observer Design*, pages 249–268. Springer-Verlag, London, 1999.
112. H.K. Khalil. *Nonlinear Systems*. Prentice-Hall, Upper Saddle River, 3rd edition, 2002.
113. H.K. Khalil and A. Saberi. Adaptive stabilization of a class of nonlinear systems using high-gain feedback. *IEEE Trans. Automatic Control*, 32(11):1031–1035, 1987.
114. P.V. Kokotović. Recent trends in feedback design: an overview. *Automatica*, 21(3):225–236, 1985.
115. P.V. Kokotović. The joy of feedback: nonlinear and adaptive. *IEEE Control Systems Magazine*, 12(3):7–17, 1992.
116. G. Kreisselmeier and R. Engel. Nonlinear observers for autonomous Lipschitz continuous systems. *IEEE Trans. Automatic Control*, 48(3):451–464, 2003.
117. A.J. Krener and A. Isidori. Linearization by output injection and nonlinear observers. *Systems & Control Letters*, 3(1):47–52, 1983.
118. A.J. Krener and W. Respondek. Nonlinear observers with linearizable error dynamics. *SIAM J. Control and Optimization*, 23(2):197–216, 1985.
119. A.J. Krener and M.Q. Xiao. Nonlinear observer design in the Siegel domain. *SIAM J. Control and Optimization*, 41(3):932–953, 2002.
120. A.J. Krener and M.Q. Xiao. Observers for linearly unobservable nonlinear systems. *Systems & Control Letters*, 46(4):281–288, 2002.
121. P. Krishnamurthy and F. Khorrami. Dynamic high-gain scaling: state and output feedback with application to systems with ISS appended dynamics driven by all states. *IEEE Trans. Automatic Control*, 49(12):2219–2239, 2004.
122. P. Krishnamurthy, F. Khorrami, and Z.-P. Jiang. Global output feedback tracking for nonlinear systems in generalized output-feedback canonical form. *IEEE Trans. Automatic Control*, 47(5):814–819, 2002.
123. M. Krstić, I. Kanellakopoulos, and P. Kokotović. *Nonlinear and Adaptive Control Design*. John Wiley and Sons, New York, 1995.
124. A. Kugi. *Non-linear Control Based on Physical Models*. Springer-Verlag, London, 2000.
125. P. Kundur. *Power System Stability and Control*. McGraw-Hill, 1994.
126. I.D. Landau. *Adaptive Control: the Model Reference Approach*. Marcel Dekker, New York, 1979.
127. H. Lei, J. Wei, and W. Lin. A global observer for observable autonomous systems with bounded solution trajectories. In *Proc. 44th IEEE Conf. Decision and Control and European Control Conference, Seville, Spain*, pages 1911–1916, 2005.
128. W. Leonhard. *Control of Electrical Drives*. Springer-Verlag, Berlin, 1985.
129. D.G. Luenberger. Observing the state of a linear system. *IEEE Trans. Military Electronics*, 8:74–80, 1964.
130. D.G. Luenberger. An introduction to observers. *IEEE Trans. Automatic Control*, 16(6):596–602, 1971.

131. I. Mareels and J.W. Polderman. *Adaptive Systems: An Introduction.* Birkhäuser, Berlin, 1996.
132. R. Marino, S. Peresada, and P. Tomei. Exponentially convergent rotor resistance estimation for induction motors. *IEEE Trans. Industrial Electronics*, 42(5):508–515, 1995.
133. R. Marino, S. Peresada, and P. Tomei. Adaptive output feedback control of current-fed induction motors with uncertain rotor resistance and load torque. *Automatica*, 34(5):617–624, 1998.
134. R. Marino, S. Peresada, and P. Tomei. Global adaptive output feedback control of induction motors with uncertain rotor resistance. *IEEE Trans. Automatic Control*, 44(5):967–983, 1999.
135. R. Marino and P. Tomei. Global adaptive output-feedback control of nonlinear systems, part I: linear parameterization. *IEEE Trans. Automatic Control*, 38(1):17–32, 1993.
136. R. Marino and P. Tomei. Global adaptive output-feedback control of nonlinear systems, part II: nonlinear parameterization. *IEEE Trans. Automatic Control*, 38(1):33–48, 1993.
137. R. Marino and P. Tomei. *Nonlinear Control Design: Geometric, Adaptive and Robust.* Prentice-Hall, London, 1995.
138. R. Marino and P. Tomei. A class of globally output feedback stabilizable nonlinear nonminimum phase systems. *IEEE Trans. Automatic Control*, 50(12):2097–2101, 2005.
139. R. Marino, P. Tomei, and C.M. Verrelli. A nonlinear tracking control for sensorless induction motors. *Automatica*, 41(6):1071–1077, 2005.
140. B. Martensson. *Adaptive stabilization.* PhD thesis, Dept. Automatic Control, Lund Institut of Technology, Sweden, 1986.
141. P. Mattavelli, A.M. Stanković, and G.C. Verghese. SSR analysis with dynamic phasor model of thyristor-controlled series capacitor. *IEEE Trans. Power Systems*, 14(1):200–208, 1999.
142. F. Mazenc. Stabilization of feedforward systems approximated by a non-linear chain of integrators. *Systems & Control Letters*, 32(4):223–229, 1997.
143. F. Mazenc and L. Praly. Adding integrations, saturated controls, and stabilization for feedforward systems. *IEEE Trans. Automatic Control*, 41(11):1559–1578, 1996.
144. A.N. Michel, K. Wang, and B. Hu. *Qualitative Theory of Dynamical Systems: The Role of Stability Preserving Mappings.* Marcel Dekker, New York, 2nd edition, 2001.
145. N. Mohan, T.A. Undeland, and W.P. Robbins. *Power Electronics: Converters, Applications and Design.* John Wiley and Sons, New York, 2nd edition, 1995.
146. M.M. Monahemi and M. Krstić. Control of wing rock motion using adaptive feedback linearization. *AIAA J. Guidance, Control and Dynamics*, 19(4):905–912, 1996.
147. R. Morici, C. Rossi, and A. Tonielli. Variable structure controller for AC/DC boost converter. In *Proc. 20th Int. Conf. Industrial Electronics, Control and Instrumentation*, pages 1449–1454, 1994.
148. A.S. Morse. Towards a unified theory of parameter adaptive control. *IEEE Trans. Automatic Control*, 37(1):15–29, 1992.
149. A.S. Morse. Supervisory control of families of linear set-point controllers—part 1: exact matching. *IEEE Trans. Automatic Control*, 41(10):1413–1431, 1996.

150. A.S. Morse. Supervisory control of families of linear set-point controllers—part 2: robustness. *IEEE Trans. Automatic Control*, 42(11):1500–1515, 1997.
151. B. Morton, D. Enns, and B.-Y. Zhang. Stability of dynamic inversion control laws applied to nonlinear aircraft pitch-axis models. *Int. J. Control*, 63(1):1–25, 1996.
152. K.S. Narendra and A.M. Annaswamy. *Stable Adaptive Systems*. Prentice-Hall, Englewood Cliffs, 1989.
153. M.S. Netto, A.M. Annaswamy, R. Ortega, and P. Moya. Adaptive control of a class of non-linearly parametrized systems using convexification. *Int. J. Control*, 73(14):1312–1321, 2000.
154. P.J. Nicklasson, R. Ortega, and G. Espinosa-Pérez. Passivity-based control of a class of Blondel–Park transformable electric machines. *IEEE Trans. Automatic Control*, 42(5):629–647, 1997.
155. H. Nijmeijer and A. van der Schaft. *Nonlinear Dynamical Control Systems*. Springer-Verlag, 1990.
156. H. Olsson, K.J. Åström, C. Canudas de Wit, M. Gafvert, and P. Lischinsky. Friction models and friction compensation. *European Journal of Control*, 4(3):176–195, 1998.
157. R. Ortega, A. Astolfi, and N.E. Barabanov. Nonlinear PI control of uncertain systems: an alternative to parameter adaptation. *Systems & Control Letters*, 47(3):259–278, 2002.
158. R. Ortega and G. Espinosa. Torque regulation of induction motors. *Automatica*, 29(3):621–633, 1993.
159. R. Ortega and G. Espinosa-Pérez. Passivity-based control with simultaneous energy-shaping and damping injection: the induction motor case study. In *Proc. 16th IFAC World Congress, Prague, Czech Rep.*, 2005.
160. R. Ortega, M. Galaz, A. Astolfi, Y. Sun, and T. Shen. Transient stabilization of multimachine power systems with nontrivial transfer conductances. *IEEE Trans. Automatic Control*, 50(1):60–75, 2005.
161. R. Ortega and A. Herrera. A solution to the continuous-time adaptive decoupling problem. *IEEE Trans. Automatic Control*, 39(7):1639–1643, 1994.
162. R. Ortega, L. Hsu, and A. Astolfi. Immersion and invariance adaptive control of linear multivariable systems. *Systems & Control Letters*, 49(1):37–47, 2003.
163. R. Ortega, A. Loría, P.J. Nicklasson, and H. Sira-Ramírez. *Passivity-based Control of Euler-Lagrange Systems*. Springer-Verlag, London, 1998.
164. R. Ortega, P.J. Nicklasson, and G. Espinosa-Pérez. On speed control of induction motors. *Automatica*, 32(3):455–460, 1996.
165. R. Ortega and Y. Tang. Robustness of adaptive controllers–A survey. *Automatica*, 25(5):651–677, 1989.
166. R. Ortega, A. van der Schaft, I. Mareels, and B. Maschke. Putting energy back in control. *IEEE Control Systems Magazine*, 21(2):18–33, 2001.
167. R. Ortega, A. van der Schaft, B. Maschke, and G. Escobar. Interconnection and damping assignment passivity-based control of port-controlled Hamiltonian systems. *Automatica*, 38(4):585–596, 2002.
168. R. Outbib and G. Sallet. Stabilizability of the angular velocity of a rigid body revisited. *Systems & Control Letters*, 18(2):93–98, 1992.
169. E. Panteley, A. Loria, and A. Teel. Relaxed persistency of excitation for uniform asymptotic stability. *IEEE Trans. Automatic Control*, 46(12):1874–1886, 2001.

170. E. Panteley, R. Ortega, and P. Moya. Overcoming the detectability obstacle in certainty equivalence adaptive control. *Automatica*, 38(7):1125–1132, 2002.
171. A.V. Pavlov and A.T. Zaremba. Real-time rotor and stator resistances estimation of an induction motor. In *Proc. IFAC Symp. Nonlinear Control Systems, St. Petersburg, Russia*, pages 1252–1257, 2001.
172. S. Peresada and A. Tonielli. High-performance robust speed-flux tracking controller for induction motor. *Int. J. Adaptive Control and Signal Processing*, 14(2):177–200, 2000.
173. S. Peresada, A. Tonielli, and R. Morici. High-performance indirect field-oriented output-feedback control of induction motors. *Automatica*, 35(6):1033–1047, 1999.
174. V.M. Popov. *Hyperstability of Control Systems*. Springer-Verlag, New York, 1973.
175. L. Praly. Adaptive regulation: Lyapunov design with a growth condition. *Int. J. Adaptive Control and Signal Processing*, 6(4):329–352, 1992.
176. L. Praly. Asymptotic stabilization via output feedback for lower triangular systems with output dependent incremental rate. *IEEE Trans. Automatic Control*, 48(6):1103–1108, 2003.
177. L. Praly and M. Arcak. A relaxed condition for stability of nonlinear observer-based controllers. *Systems & Control Letters*, 53(3-4):311–320, 2004.
178. L. Praly and A. Astolfi. Global asymptotic stabilization by output feedback under a state norm detectability assumption. In *Proc. 44th IEEE Conf. Decision and Control and European Control Conference, Seville, Spain*, pages 2634–2639, 2005.
179. L. Praly and Z.-P. Jiang. Linear output feedback with dynamic high gain for nonlinear systems. *Systems & Control Letters*, 53(2):107–116, 2004.
180. L. Praly and I. Kanellakopoulos. Output feedback asymptotic stabilization for triangular systems linear in the unmeasured state components. In *Proc. 39th IEEE Conf. Decision and Control, Sydney, Australia*, pages 2466–2471, 2000.
181. L. Praly, R. Ortega, and G. Kaliora. Stabilization of nonlinear systems via forwarding mod$\{L_g V\}$. *IEEE Trans. Automatic Control*, 46(9):1461–1466, 2001.
182. C. Qian and W. Lin. Output feedback control of a class of nonlinear systems: a nonseparation principle paradigm. *IEEE Trans. Automatic Control*, 47(10):1710–1715, 2002.
183. W. Ren and R.W. Beard. Trajectory tracking for unmanned air vehicles with velocity and heading rate constraints. *IEEE Trans. Control Systems Technology*, 12(5):706–716, 2004.
184. B.D. Riedle and P.V Kokotović. Integral manifolds of slow adaptation. *IEEE Trans. Automatic Control*, 31(4):316–324, 1986.
185. H. Rodríguez, A. Astolfi, and R. Ortega. On the construction of static stabilizers and static output trackers for dynamically linearizable systems, related results and applications. *Int. J. Control*, 79(12):1523–1537, 2006.
186. H. Rodríguez, R. Ortega, and A. Astolfi. Adaptive partial state feedback control of the DC-to-DC Ćuk converter. In *Proc. American Control Conference, Portland, Oregon*, pages 5121–5126, 2005.
187. H. Rodríguez, R. Ortega, G. Escobar, and N. Barabanov. A robustly stable output feedback saturated controller for the boost DC-to-DC converter. *Systems & Control Letters*, 40(1):1–8, 2000.

188. S.R. Sanders and G.C. Verghese. Synthesis of averaged circuit models for switched power converters. *IEEE Trans. Circuits and Systems*, 38(8):905–915, 1991.
189. S.R. Sanders and G.C. Verghese. Lyapunov-based control for switched power converters. *IEEE Trans. Power Electronics*, 7(1):17–24, 1992.
190. S.R. Sanders, G.C Verghese, and D.F. Cameron. Nonlinear control laws for switching power converters. In *Proc. 25th IEEE Conf. Decision and Control, Athens, Greece*, pages 46–53, 1986.
191. S. Sastry and M. Bodson. *Adaptive Control: Stability, Convergence and Robustness*. Prentice-Hall, London, 1989.
192. S.S. Sastry and A. Isidori. Adaptive control of linearizable systems. *IEEE Trans. Automatic Control*, 34(11):1123–1131, 1989.
193. J.M.A. Scherpen, D. Jeltsema, and J.B. Klassens. Lagrangian modeling and control of switching networks with integrated coupled magnetics. In *Proc. 39th IEEE Conf. Decision and Control, Sydney, Australia*, pages 4054–4059, 2000.
194. R. Sepulchre, M. Janković, and P. Kokotović. *Constructive Nonlinear Control*. Springer-Verlag, Berlin, 1996.
195. H. Shim and A.R. Teel. Asymptotic controllability and observability imply semiglobal practical asymptotic stabilizability by sampled-data output feedback. *Automatica*, 39(3):441–454, 2003.
196. B. Siciliano and L. Villani. *Robot Force Control*. Kluwer Academic Publishers, 2000.
197. S.N. Singh and M. Steinberg. Adaptive control of feedback linearizable nonlinear systems with application to flight control. *AIAA J. Guidance, Control and Dynamics*, 19(4):871–877, 1996.
198. H. Sira-Ramirez. Sliding motions in bilinear switched networks. *IEEE Trans. Circuits and Systems*, 34(8):919–933, 1987.
199. E. Skafidas, A. Fradkov, R.J. Evans, and I.M.Y. Mareels. Trajectory-approximation-based adaptive control for nonlinear systems under matching conditions. *Automatica*, 34(3):287–299, 1998.
200. J.-J. Slotine and W. Li. Adaptive manipulator control: A case study. *IEEE Trans. Automatic Control*, 33(11):995–1003, 1988.
201. E. Sontag and Y. Wang. Lyapunov characterizations of input to output stability. *SIAM J. Control and Optimization*, 39(1):226–249, 2000.
202. E.D. Sontag. Smooth stabilization implies coprime factorization. *IEEE Trans. Automatic Control*, 34(4):435–443, 1989.
203. E.D. Sontag. On the input-to-state stability property. *European Journal of Control*, 1(1):24–36, 1995.
204. E.D. Sontag and M. Krichman. An example of a GAS system which can be destabilized by an integrable perturbation. *IEEE Trans. Automatic Control*, 48(6):1046–1049, 2003.
205. E.D. Sontag and Y. Wang. On characterizations of the input-to-state stability property. *Systems & Control Letters*, 24(5):351–359, 1995.
206. M.W. Spong, S. Hutchinson, and M. Vidyasagar. *Robot Modeling and Control*. John Wiley and Sons, 2006.
207. M.W. Spong, K. Khorasani, and P.V. Kokotović. An integral manifold approach to the feedback control of flexible joint robots. *IEEE Journal of Robotics and Automation*, 3(4):291–300, 1987.
208. R.F. Stengel. *Flight Dynamics*. Princeton University Press, 2004.

209. B.L. Stevens and F.L. Lewis. *Aircraft Control and Simulation*. John Wiley and Sons, 2nd edition, 2003.
210. Y.Z. Sun, Q.J. Liu, Y.H. Song, and T.L. Shen. Hamiltonian modelling and nonlinear disturbance attenuation control of TCSC for improving power system stability. *IEE Proc. Control Theory and Applications*, 149(4):278–284, 2002.
211. R. Talj. Comparaison expérimentale des lois de commande nonlinéaire sur une machine asynchrone. Master's thesis, LSS–LGEP, Supélec, July 2006.
212. Y. Tan, I. Kanellakopoulos, and Z.-P. Jiang. Nonlinear observer/controller design for a class of nonlinear systems. In *Proc. 37th IEEE Conf. Decision and Control, Tampa, Florida*, pages 2503–2508, 1998.
213. G. Tao. A simple alternative to the Barbălat lemma. *IEEE Trans. Automatic Control*, 42(5):698–698, 1997.
214. D.G. Taylor. Nonlinear control of electric machines: an overview. *IEEE Control Systems Magazine*, 14(6):41–51, 1994.
215. A. Teel and L. Praly. Tools for semiglobal stabilization by partial state and output feedback. *SIAM J. Control and Optimization*, 33(5):1443–1488, 1995.
216. A.R. Teel. A nonlinear small gain theorem for the analysis of control systems with saturation. *IEEE Trans. Automatic Control*, 41(9):1256–1270, 1996.
217. A.R. Teel and J. Hespanha. Examples of GES systems that can be driven to infinity by arbitrarily small additive decaying exponentials. *IEEE Trans. Automatic Control*, 49(8):1407–1410, 2004.
218. A.R. Teel and L. Praly. Global stabilizability and observability imply semiglobal stabilizability by output feedback. *Systems & Control Letters*, 22(5):313–325, 1994.
219. P. Tomei. A simple PD controller for robots with elastic joints. *IEEE Trans. Automatic Control*, 36(10):1208–1213, 1991.
220. M. Torres and R. Ortega. Feedback linearization, integrator backstepping and passivity-based controller design: a comparison example. In D. Normand-Cyrot, editor, *Perspectives in Control: Theory and Applications*. Springer-Verlag, London, 1998.
221. S. Townley. An example of a globally stabilizing adaptive controller with a generically destabilizing parameter estimate. *IEEE Trans. Automatic Control*, 44(11):2238–2241, 1999.
222. J. Tsinias. Further results on the observer design problem. *Systems & Control Letters*, 14(5):411–418, 1990.
223. V.I. Utkin. *Sliding Modes in Control and Optimization*. Springer-Verlag, Berlin, 1992.
224. A. van der Schaft. L_2-*Gain and Passivity Techniques in Nonlinear Control*. Springer-Verlag, London, 2nd edition, 2000.
225. H.F. Wang and F.J. Swift. A unified model for the analysis of FACTS devices in damping power system oscillations part I: single-machine infinite-bus power systems. *IEEE Trans. Power Delivery*, 12(2):941–946, 1997.
226. S. Wiggins. *Introduction to Applied Nonlinear Dynamical Systems and Chaos*. Springer-Verlag, New York, 1990.
227. J. Willems and C. Byrnes. Global adaptive stabilization in the absence of information on the sign of the high frequency gain. In A. Benssousan and J.L. Lions, editors, *Proc. 6th Int. Conf. Analysis and Optimization of Systems*, pages 49–57. Springer-Verlag, Berlin, 1984.
228. W.M. Wonham. *Linear Multivariable Control: A Geometric Approach*. Springer-Verlag, 3rd edition, 1985.

229. A.R.L. Zachi, L. Hsu, R. Ortega, and F. Lizarralde. Dynamic control of uncertain manipulators through immersion and invariance adaptive visual servoing. *Int. J. Robotics Research*, 25(11):1149–1159, 2006.

Index

$\mathcal{L}_2$ stability, 67, 70, 270

Adaptive backstepping, 39, 56, 67, 72
Aircraft
   airspeed control, 231
   attitude control, 224
   trajectory tracking, 227
   wing rock, 72

Backstepping, 22, 26, 64, 123, 138
Ball and beam, 112
Barbalat's lemma, 268

Cart and pendulum, 9, 28
Composite control, 20, 26

Detectability, 39
Dynamic scaling, 74

Eccentricity compensation, 152, 160

Feedback form, 10, 21, 57
Feedforward form, 10, 23
Flexible joints robot, 8, 27, 209
Friction compensation, 153, 161

Hamiltonian systems, 225, 237, 252
High-frequency gain, 78, 81, 161
High-gain observers, 92

I&I
   adaptive stabilisation, 34
   stabilisation, 19
Immersion, 4

Indirect field-oriented control, 243
Input-to-output stability, 270
Input-to-state stability, 269
Invariance, 4
ISS-Lyapunov function, 270

LaSalle invariance principle, 58, 268
Linear systems, 78
Linearly parameterised systems, 121
Lyapunov
   function, 268
   stability, 267

Magnetic levitation, 6, 25, 99
Model-reference adaptive control, 81
Monotonic nonlinearities, 100, 124

Neural networks, 153, 161
Nonlinear observers, 93, 103, 135
Nonlinear PI control, 13

Observers, 12
Output feedback form, 129

Passivity, 5, 87, 100, 102, 167
Passivity-based control, 148, 237, 252
Pendulum, 48
Power converter
   Ćuk, 179
   AC–DC boost, 200
   DC–DC boost, 188
Power factor correction, 200

Range estimation, 95

Robust output feedack stabilisation, 133

Separation principle, 121
Singularly perturbed systems, 19
Small-gain theorem, 136, 270

Thyristor-controlled series capacitor, 41, 46, 173

Translational oscillator/rotational actuator, 145
Two-DOF mechanical systems, 107, 216
Two-link manipulator, 216

Unknown control gain, 61, 81, 153, 161, 165

Visual servoing, 49, 167

Printing: Krips bv, Meppel, The Netherlands
Binding: Stürtz, Würzburg, Germany